DRAW IN ORDER TO SEE

A Cognitive History of Architectural Design

Voglio vedere le cose, non mi fido che di questo.
Le metto qui davanti a me sulla carta, per poterle vedere.
Voglio vedere, e per questo disegno.
Posso verdere una immagine solo se la disegno.

—Carlo Scarpa, lecture to students in Venice

Every artist has to know and construct a schema
before he can adjust it to the needs of portrayal.

—Ernst Gombrich, *Art and Illusion*

Ainsi les gestes multipliaient le savoir, avec une
variété de touche et de dessin dont une habitude
millénaire nous cache la puissance inventive.

—Henri Focillon, *L' Eloge de la main*

I want to imagine things, but I don't trust my imagination. I put them in front of myself on a piece of paper in order to visualize them. I want to see, and so I draw. I can only see an image if I draw it.

—Scarpa (translation of above quote)

Thus (an artist's) hand gestures multiplied his understanding, with a variety of touch and draftsmanship, during a thousand years of habituation, in which was hidden his power of invention.

—Focillon (translation of above quote)

ORO Editions
Publishers of Architecture, Art, and Design
Gordon Goff: Publisher

www.oroeditions.com
info@oroeditions.com

Published by ORO Editions

Text by Mark Alan Hewitt
Book Design by Pablo Mandel
Managing Editor: Jake Anderson

10 9 8 7 6 5 4 3 2 1 First Edition

ISBN: 978-1-943532-83-4

Color Separations and Printing: ORO Group Ltd.
Printed in China.

ORO Editions makes a continuous effort to minimize the overall carbon footprint of its publications. As part of this goal, ORO Editions, in association with Global ReLeaf, arranges to plant trees to replace those used in the manufacturing of the paper produced for its books. Global ReLeaf is an international campaign run by American Forests, one of the world's oldest nonprofit conservation organizations. Global ReLeaf is American Forests' education and action program that helps individuals, organizations, agencies, and corporations improve the local and global environment by planting and caring for trees.

Cover illustration: Karl Friedrich Schinkel, "View from the window of my lodging in Rome, St. Peter's in the distance," 1803. Berlin, Nationalgalerie, SMB (SM 4.56); pen and brown ink, 295 x 488 cm. Linecut from published engraving, in the public domain.

MARK ALAN HEWITT

DRAW IN ORDER TO SEE

A Cognitive History of Architectural Design

ORO EDITIONS

Contents

Preface and Acknowledgments

When I began this book I intended to write a conventional history of architectural representation by summarizing the work done by contemporary scholars on a subject I had studied throughout my career. I began thinking about it as a graduate student during the 1970s, after I had read the work of James J. Gibson concerning perception and the environment, and continued to develop my ideas in a series of essays, initially published in the mid-1980s and early 1990s. I did not know then that I would expand my knowledge of psychology and cognitive science by following the work of Gibson to its logical conclusions in the twenty-first century.

The first premise for my thinking came via the idea of "affordances" in the physical environment that could be assessed by the brain-body in maintaining homeostasis. When I began reading the work of Antonio Damasio, Daniel Dennett, and Steven Pinker, among others, I recognized the power of this new "embodied" schema in studying the minds of architects as they pursued their designs. My book began to take on a new character, one that used cognitive science to unveil the habits of builders from the earliest days of human sedentary settlements to the present day. After initially planning to look at perception, conception, and representation in architecture I recognized that all artistic processes are organized in the brain to understand and manifest the relationship between humankind and nature, much as John Dewey theorized in his groundbreaking work in the early twentieth century.

In his great synthesizing book of 1925, *Experience and Nature*, Dewey criticized what we would now call logical positivism and scientific "naturalism" by exploring the "psycho-physical" nature of organisms, specifically *Homo sapiens*. Like Damasio, he recognized the limitations and epistemological falsehoods in Cartesian dualism and instead followed the Greek philosophers in emphasizing the pursuit of knowledge through practical, artistic, and empirical endeavors that did not emphasize instrumental means over open experience—in other words a repudiation of the modern view of technology, science, and "fine art" as manifestations of "progress" or "invention." Dewey revived an Aristotelian view of art, craft, science, and technics as related aspects of humanity's need to understand the natural world—"experience [as] equivalent to art." Many of today's philosophers and neuroscientists have embraced similar views, influenced by new discoveries about the brain and its relationship to the body it supports.

As he wrote, "modern thought also combines exaltation of science with eulogistic appreciation of art, especially fine or creative art. At the same time it retains the substance of the classic disparagement of the practical in contrast with the theoretical, although formulating it in somewhat different language: to the effect that knowledge deals with objective reality as it is in itself, while in what is 'practical,' objective reality is altered and cognitively distorted by subjective factors of want, emotion and striving."[1] He did not fathom, nor did he predict, the degree to which these views would come to strangle architecture, the most practical of the arts, during the remaining years of the century.

In the final chapters of this book I address the crisis that the profession faces as a result of our obsession with technology and "conceptual" art. Today, in many schools of architecture the subject is divided into "theory" and "pragmatics," with the former occupying the high ground and commanding the attention of most of the faculty, and the latter reduced to a sideshow. Aristotle, the first philosopher to distinguish these two aspects of knowledge, did not assume that they were exclusive, but rather that a virtuous person should possess both. Cognitive neuroscience demonstrates that the brain acquires knowledge along the lines that Aristotle predicted, not as Locke, Hume, Kant, and Hegel surmised centuries later.

Within these pages the reader will find a history of architectural design that does not accept many common understandings about material progress and artistic genius, but instead focuses on the cognitive requirements needed by designers of buildings as they pursued their goals of making good shelter, splendid monuments, and well-crafted artifacts throughout recorded history. In particular it notes the relationship between external memory aids, such as drawings and models, and internal conceptual schemas that arose among communities of designer/builders at particular times, in particular places. Broadly, the historical narrative follows humans in their early attempts to craft buildings from the materials at hand, proceeds to the Renaissance discovery of depiction through new drawing types, and through the industrial revolution, where machine assembly was emphasized as a cognitive model. The narrative includes as much relevant neuroscience and social science as is appropriate to the historical argument. There are, however, some limitations to the use of this research.

It is essential to recognize the broad significance of new discoveries in cognitive neuroscience as well as the rapid pace at which the science is advancing. It is premature for any author who cites experimental data or theories to claim that they are "settled." In this book I attempt to use as much settled science as possible, recognizing that even relatively well-tested hypotheses may be overturned by new discoveries during the next decades. Embodied cognition is one such contemporary discovery, and there are several competing hypotheses about its reach and significance. I have cited the positions in the text below, but do not presume to know which will prevail in the discourse to come.

My belief is that the preponderance of evidence in this research will prove that humans, including architects, think with their bodies and the surrounding environment when being creative, and that embodiment offers a new paradigm for studying cognitive activity. Moreover, the integration of perception, conception and representation in the brain makes it imperative that architects and those that study design thinking revise their notions of how we create buildings.

Our thinking involves drawing, model making, and crafting—all mimetic forms of representation. To limit the use of these external memory resources is to curtail the creative capacity that humans, as organisms, have always possessed. Computer aided design has begun to constrain the creative process, and should be more critically evaluated among teachers of architecture and design. Only by acknowledging the mimetic nature of design can we regain the humanistic, and embodied, nature of architecture as a discipline.

I could not have reached my conclusions, nor could I have even acquired the knowledge to pursue them, without the help of friends, colleagues, and mentors whom I met during the long gestation period of this project. I will try my best to acknowledge all of them, though I cannot thank everyone who may have helped in small ways over more than ten years.

I must first thank my late wife, Mia Kissil Hewitt (1966–2018), who maintained our house and guided our two daughters to maturity while I was at my computer struggling with a difficult subject for many years. My family has stood beside me in every endeavor and I deeply appreciate their love and support.

When I began to write I sought the opinions of two architect colleagues, Philip Kennedy-Grant and Brian Oschwald, who live nearby, and bothered them with many inane questions about form and content. They were the first to encourage my efforts to expand the book and discard old material in favor of a new approach to my subject. I also poked two friends and architectural historians, Jeff Cohen and Keith Morgan, about the advisability of writing such a book—they too were encouraging. Stefano Gulizia, a scholar of Alberti and Renaissance science, also gave valuable advice on sources. When I despaired and wanted to junk the project they prodded me to keep going.

When I had drafts of the book I did not hesitate to circulate them to both trusted colleagues and to new acquaintances, especially in the field of cognitive neuroscience, who could offer salient advice and correct errors in my knowledge of a broad subject that I have never formally studied. Juhani Pallasmaa, Harry Mallgrave, and Michael Arbib filled this role graciously and I cannot thank them enough. I was fortunate to meet two younger scientists, Sergei Gepshtein (Salk Institute) and Jennifer Groh (Duke University), through the Place-Science research group in 2016. Ann Sussman, a Boston architect, also became collaborator. They have given me advice and encouragement and deserve a hearty thank you.

Two of the earliest scholars to see the manuscript were Marc Treib and Dana Cuff, and I appreciate their candid thoughts about how I might improve it. John Lang, Branko Mitrovic, Robert A. M. Stern, John Onians, David Dunster, William McGill Thompson, and Jeff Cohen reviewed early drafts and provided valuable insights. I gave a short paper at Duncan Stroick's symposium, "The Art of Architecture," at the University of Notre Dame in 2016, and there met several people who were helpful in refining ideas. The most important was Vinod Goel of York University, who graciously reviewed a draft and met with me to discuss its merits and weaknesses.

Though I did not publish with their presses, several editors were kind in reviewing my manuscript and suggesting options for publishing: Michelle Komie, Fran Ford, and Boyd Zenner. The editors at Mimesis International in Milan were gracious in considering the book as well.

Last and certainly not least, I want to thank my old friend and fellow drawing enthusiast George Dodds of the University of Tennessee for looking at several versions of the book and steering me in the right direction. Without him I should certainly be wandering in the wilderness. He is a deeply knowledgeable scholar and a trusted resource on many aspects of architectural design.

For assistance with the illustrations, I must particularly thank Brian Oschwald, who helped me devise charts and diagrams to explain difficult concepts. For particular permissions I thank the Galleria degli Uffizzi in Florence, the Alvar Aalto Museo in Finland, the British Museum in London, the Avery Architectural Archives in New York, the Museo Civico and Bibioteca Nationale in Turin, the Morgan Library in New York, Sir John Soane's Museum in London, and ETH in Zurich. Anthony Panzera, David Esterly, and Harley Jessup were gracious in granting permission to use their wonderful drawings and sculptures. Last and most importantly I thank Gordon Goff and his staff at ORO Editions in San Francisco for their faith in this project as publishers.

1 John Dewey, *Experience and Nature* (New York, Dover Editions: 1958): 355.

[1]

NEW THINKING ABOUT DESIGN

What does it mean when someone says: "I see now what I have to do?" It certainly does not pertain to "seeing" with one's eyes, though whatever is contemplated may as well be as clear as a picture in the mind's eye. Humans use the verb "to see" when talking about all sorts of understandings about the world. Our visual senses are the most highly developed among the five modes of apprehending the environment outside our bodies, so it makes perfect sense for us to explain many aspects of thought using visual terms or metaphors.[1]

In all sorts of design endeavors, from creating furniture to envisioning the next spacecraft to explore distant planets, thinking involves visualization. Creative sessions are often done at a table, with markers, pencils, and any sort of stylus available to record a pictorial idea of the thing we want to make. What gets drawn or sketched is a record of a cognitive process.[2] Visual thinking has been studied by artists, psychologists, and even philosophers, and much has been learned by simply analyzing what artists and designers leave on paper to record their efforts to solve a problem or make a new work.[3] However, very little was known about visualization inside the brain until the last half century.[4] Scientists simply didn't have the means, in technological and theoretical terms, to unlock the mystery of cognition.

Things changed when various kinds of scanning and metering devices were aimed at the cortex and subcortical regions of the brain during the second half of the twentieth century. Some have called the 1990s the "decade of the brain," because so much revolutionary work was published about many aspects of neuroscience. Today, scientists have an astounding array of new methods and theories with which to run experiments on the most important problems in many biological fields, but particularly in cognitive science and neurology.[5]

During the past ten years alone, fascinating work has been done to bring humanistic and scientific insights to bear on some of the mysteries of creativity and artistic invention.[6] There have been anthropological and neuro-scientific theories of how humans evolved to produce art, music, and literature.[7] Some scholars believe that an "art instinct" evolved out of our ancestors' experiences with the natural world, and their need to understand themselves as a part of the cosmos.[8] Others are probing the brain's internal workings to find aesthetic preferences that may be universal, not culturally biased.[9] It is a propitious time to look at how designers, and architects in particular, use their brains when doing creative work.

Ten years ago the American Institute of Architects established a small research program that morphed into the Academy of Neuroscience for Architecture, located in San Diego, California. The academy is small but growing, and a few leading neuroscientists studying perception and cognition have turned their attention to architecture.[10] However, given the size of the cognitive science community worldwide, these researchers are insignificant and hardly noticed. More importantly, there are few articles and fewer books devoted to the relationship between architectural design and cognitive neuroscience.[11] This book addresses that significant gap in the literature by outlining both the science and the historical knowledge available on that subject at the present moment.

Since I am both a practicing architect and an architectural historian, I have chosen to look at the subject of architectural design historically rather than from any prescribed theoretical position, as is so often the case with writing about the built environment today. Such a narrative approach avoids some of the philosophical entanglements that have discouraged architects from using all kinds of empirical science in their endeavors. I have subtitled the book, a "cognitive history" of architectural design, because I employ well-established tools and concepts from cognitive science in analyzing and interpreting themes from the history of architectural representation. I have spent my career teaching design, writing about architectural practice, studying individual architects, and researching historic buildings. I have taught courses on the history of drawing, on the classical language of architecture, and on the development of architectural theory since Vitruvius.

Studies of the brain show that the individual experiences of creative people build their cognitive "architectures" into unique working habits and conceptions of the world. I believe that my own life experience has shaped my views of architecture and architectural history, and I share some of that experience with the reader in the narrative below. In particular, I want to offer the reader the perspective of someone who has both taught and practiced architecture for four decades. I have been fortunate to be able to write and publish my observations throughout that period.

My scholarly work has consistently focused on how architects and designers work, particularly on the social history of architectural practice.[12] In several books and a handful of articles, I delved deeply into the cognitive styles of individual designers and groups of architects who shared similar education and professional experiences. I began to recognize that cognitive habits and drawing styles followed some recognizable patterns that could be studied psychologically as well as historically. Emerging breakthroughs in neuroscience were providing a means of unlocking these patterns and schemata.

In 2006 I was invited to present a paper at a conference organized by Professor Marc Treib at the University of California at Berkeley, which later appeared in the anthology, *Drawing/Thinking*. My interaction at the conference with artists, architects, delineators, teachers, landscape architects, and animators piqued my curiosity. All were lamenting the absence of hand drawing in the training of young artists and designers, and emphasizing the importance of hand skills in their own work.

Shortly afterward I began considering how I might write a book about the experience of designing three-dimensional objects, such as buildings, from a cognitive point of view. As with many journeys in life, that curiosity and need to understand my own design habits led to years of additional inquiry, and changed my entire attitude toward what I do.

Drawing As Seeing

Apple's industrial designers, Pixar Studios, the film composer John Williams, and the artist Jeff Koons don't need paper and pencils to bring their astounding creations to life. They have animation software, 3-D printers, Sibelius music software and synthesizers, digital tablets running Adobe Illustrator, and a host of other "smart" media platforms on which to play with ideas and concepts. For these artists, digital media make it possible to freely invent without interference from tediously slow analog media like pianos and music paper or pencils with broken lead.

More significantly, young people don't have time to learn such disciplines as making a charcoal line thin or thick, mixing gesso, or drafting with a triangle and T-square. They are invested in bigger things, things made possible by technology and information in seemingly unlimited supply. Digital graphics can be generated in minutes, saving time that used to be required for polishing a design or composition before issuing the final product. Virtually everything we see on screens has some sort of sophisticated "animation" created on such software. It can be dazzling.

1. Carlo Scarpa, Banca Popolare di Verona, initial sketch of façade, 1973, 297x760 mm. (MAXXI Museum, Rome, TB39726 recto, courtesy, Centro Carlo Scarpa, Treviso)

Yet many scientists who study the brain and human behavior are finding that the furtive quest for bigger and faster media and wider bandwidth has not produced healthier or smarter humans. Information technology may not be adding to our collective wisdom or furthering the cause of art.[13] Children's attention spans and social skills are shrinking, leading to alienation and depression. The social cost of digital communication via texts and Instagram messaging appears to be high as well. In the built environment, the designs emerging from computer screens have not produced better cities, more beautiful parks, or housing that inspires praise from non-designers.[14] In fact, much of what has been built during the past three decades has been bombastic, monotonous, or obscenely expensive. Many who study digital technology and design are concerned that these tools have inhibited, rather than enhanced, visual judgment.

The latest research on design thinking suggests that architecture students should use soft pencils and sketch paper if they want to learn to think visually. In order to feed their brains a healthy diet, humans need to use their eyes, hands, and spatial awareness simultaneously, especially when learning an art or craft. To learn to design, architects must first learn to draw. As the Italian architect Carlo Scarpa put it: "I want to see, and that's why I draw. I can see an image only if I draw it." In fact, what we are learning about the brain suggests that Scarpa was onto something: visual artists and architects can't really create without producing "pictures" of their thoughts on paper or in some other external form. Though experienced architects, like other artists, may internalize visual ideas before recording them in some external medium, most admit that "sketching realizes things, it does make real what your thoughts are, it makes your thoughts concrete so you can go and test them."[15] The act of designing requires an external adjunct to internal representations in the brain, activating an action-perception loop, as we shall see below.

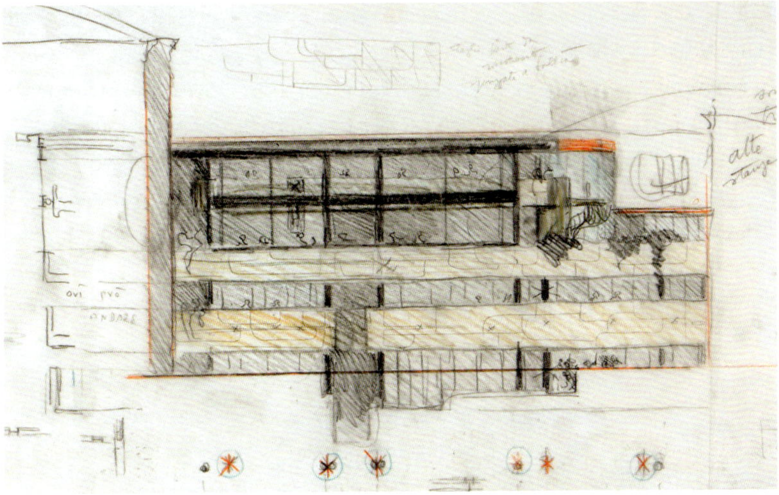

2. Carlo Scarpa, Banca Popolare di Verona, 1973, section sketch of courtyard façade, 300x610 mm. (MAXXI Museum, Rome, TB397339 recto, courtesy, Centro Carlo Scarpa, Treviso)

Scarpa often said that he learned architecture by studying painting and fine art, and that this freed him from the kind of ideological posturing common to other Italian architects during most of his career around Venice. He referred to all of his work in terms of "visual logic," avoiding complex jargon and theoretical conceits. He admired and socialized with painters and fine artists, and designed many museums for their work. He was a compulsive draftsman, studying everything around him and weaving the light, materials, and artifacts of the Veneto into all of his work. It's clear that nothing Scarpa designed would be as rich and complex without his many sketches. [1, 2] One of his last great works, many years in the making, was the Banca Popolare di Verona in the heart of the historic city.[16] [3]

Though relatively small at three stories and less than 20,000 square feet, the bank headquarters features nearly all of Scarpa's trademark motifs and design conceits. In the case of the bank the architect proceeded slowly after his invitation in 1973 to join three disparate buildings into one facility in the center of Verona's historic district. Eventually he focused primarily on the façades of a linking building, using ideas from earlier works such as the Castelvecchio and Brion Vega tombs.

Scarpa was acutely aware of both his visual acuity and of the physical qualities he wished to impart to his designs. But he was no idle dreamer. Every day he retired to his drafting table in Vicenza to make sketches of the latest projects under design or construction. [2] Rafael Moneo of Madrid and Harvard pointed out that Scarpa's thoughts seemed to rise to the surface as he drew, so as to "become the immediate and sincere reflection of his inner thoughts." All of his admirers, students, and fellow designers have remarked on the uncanny felicity between his sketches and his finished buildings.[17]

This was especially true in the bank project. An early (summer 1973) elevation sketch shows the architect recording his intention to create a dynamic interplay between geometric openings, a stepped stone revetment, and a glazed inner plane. [1] Drawn to scale, the sketch has a jittery, unfinished quality that Scarpa liked in all of his preliminary studies. Most telling is the typical group of doodles at the bottom of the vellum sheet. Most are stepped lines that appear to be plans or sections of the stone outer wall, indicating Scarpa's wish to see as many changes in plane as possible, despite a thickness of less than a foot. Obviously, these are mental "notes" about future details that can't be resolved in a sketch elevation. Also only schematic are the shapes, placement, and projections of the window openings. One, near the top, seems to use Scarpa's trademark echelon system that many see as an influence from Wright's tile block houses of the 1920s. Like Borromini, Scarpa liked to repeat formal motifs in as many areas as possible, so we see the stepped pyramids in window plinths, the cornice, the walls, etc. [3, 4]

It is easy to misread Scarpa's compulsive drawing of details as an obsession with the craft of making his buildings, or with materials, or with tiny connections rather than with the larger composition. This is wrong. He was ultimately trying to understand the experience of the building in as many sensory aspects as possible, weaving this cognizance into his abstract concepts about the work. This is what Moneo meant about the "availability" (disponibilita) provided in his drawings for a study of what he wanted to "see" in the finished work. Scarpa "built" the artifact in his brain by producing sketches of both part and whole. With each loop between hand, eye, and brain, he could see more clearly what the object would become. Eventually he would visit the site to work with the stuccatore rubbing the marmorino until its color, depth, and polish were exactly as he conceived them.

As most of us who design buildings know, these loop sequences are not linear like engineering studies—they can't be done on a computer. Scarpa had to jump ahead to see some things before going back to resolve the whole composition. It is frustrating to see design theorists, often computer science experts, diagram the process as if it were a line, a matrix, or some other "rational" schema. One feature of a cognitive loop is invariant: the give and take between the image on the page and the concept in the brain. Let me be clear: they are not the same, even though we retain much of the drawing in our working memory. That is why architects continue to refine their drawings until some kind of resolution is reached. Further, only when the building is built does the designer truly understand what she has conceived.

Though no comprehensive study has yet been published on this building, even a cursory perusal of the drawings shows how intricate the conceptual process can become when a master architect is working on a major commission.[18]

It makes an interesting case study because we know a good deal about the architect's working methods and preferred sequence of drawings and models. Throughout the book I will present similar case studies, using both drawings/models and finished buildings, to explore the brain's inner workings. However, rather than simply accepting the architect's representations as direct records of "concepts," I will attempt to unravel the complex process of cognitive wrestling that precedes concept formation, and then proceeds as the design becomes better defined in the mind of its creator.

3. Banca Popolare di Verona. Courtyard view (copyright, Yukio Futagawa, used by permission of GA Monographs, Tokyo).

The Man Who Drew Too Much

Leonardo da Vinci has been credited with so many wondrous achievements that it is hard to see how his genius might be under appreciated by contemporary historians. Yet Walter Isaacson, the biographer of Steve Jobs, views the greatest artist of his age as a scientist, philosopher, writer, architect, and scene designer in addition to praising his remarkable graphic art. Leonardo will appear as the protagonist of this book for reasons that will become obvious as the narrative progresses. Several leading scholars have called him the first neuroscientist because he understood so much about his own cognitive processes. I will argue that he was the paragon of all designers from the fifteenth century onward, and remains so today. And yet, many would contend that he was not an architect because he produced no built work that we can visit today. He was more like a shaman, conjuring images of things not yet experienced but powerful enough to change the world.

4. Banca Popolare di Verona. Entry and plaza elevation. Photo by Tad Jusczyk, from his blog, Longroadtovenice.com.

Two aspects of Leonardo's approach to design merit attention with respect to cognitive science. One was his experience as a left-handed writer, perhaps even a dyslexic, resulting in relative illiteracy. Moreover, as an illegitimate child he did not have recourse to extensive tutelage in Latin, and therefore did not read extensively, sometimes making fun of his own lack of book learning. He was therefore compelled to observe the world around him and record his impressions in drawings and brief notes. As a right-brained, left-handed draftsman he engaged the visual cortex more directly than a left-brained, right-handed student. His mirror writing was an outgrowth of this unusual mental handicap. He also had to draw cross hatching from right to left in order to avoid smearing ink as his hand passed over the paper.

Leonardo was self-conscious about his verbal literacy, but consistently professed the superiority of the painter over other artists, because of his direct observation of the world. He was an advocate for sensory experience of the environment, and consistently pressed himself to use all of his senses to record that experience. He was a singer, harp player, orator, sculptor, scene designer, and mechanic as well as a visual artist, and used all of these skills to advance his understanding of the world in which he lived. Leonardo is said to have been one of the first artists to draw over his sketches until he reached a resolution of his mental ideas. With more abundant paper he took advantage of the opportunity to produce several drafts of a composition, something he likened to a poet's writing and crossing out words in order to find the right solution to a metric or rhyme scheme. Michelangelo would later employ a similar routine in creating his concept sketches, and was himself an accomplished poet. He often wrote sonnets in the margins of his drawings.

Once he left his native Florence to work for the Sforzas in Milan, Leonardo acted as a court entertainer as well as engineer and artist. He was charged with designing and producing numerous celebrations, theatrical events, and lectures during his tenure in Milan, and continued to do so throughout the rest of his career. He relished this work, as it allowed him to create art that was immediate and ephemeral, unburdened by the tedium of fabrication associated with sculpture or painting. Moreover, he could invent machines for hoisting and flying light scenery or even performers, venting his desire to defy gravity. His sketchbooks are full of theatrical inventions, costumes, and scene designs, though they are not often featured in books on his art. As we shall learn in Chapter 6, mimetic forms such as theater and dance merged with visual art during the Renaissance to expand the vision of architects as well as artisans inventing interior decoration. For this man, the whole world was a stage on which to create magic, drama, and illusions.

5. Leonardo da Vinci, wooden model of one of the central plan church designs, Hayward Gallery, London, 1980. (Author collection)

6. Leonardo da Vinci. Two central plan church sketches, 1487–90. (Paris, Institute de France, Codex B, 17v, Dover reprint)

Leonardo's preferred modality in making his thousands of drawings was a kind of hybrid perspective, often with cut away views of the object to be depicted, revealing its inner organs or workings. Trained in the workshop of Andrea del Verrochio in Florence, he was one of the first painters to become adept at both linear perspective construction and the wide variety of new media available following the advent of paper making in Italy. Prior to the mid-fifteenth century artists were still limited by the availability of parchment, difficult to make and expensive to procure. Students of Verocchio and Ghirlandaio were exposed to various media such as silverpoint, black and red chalk, graphite, ink, and ink wash during their studies. Moreover, these new media gave illustrators more latitude in depicting aspects of the natural world such as animals, plants, insects, and landscapes. Leonardo began producing designs for buildings during his first years in Milan, after 1482. Scholars know that he interacted with the architect Donato Bramante during this period. Though he never had the opportunity to erect a church during his lifetime, he nevertheless made important contributions to the practice of architectural design via his interactions with painter/designers who would build more extensively than he. Only recently have scholars been able to trace the interrelationship between this wide-ranging genius and his contemporaries, including his work on anatomy, geology, mathematics, optics, and engineering. Leonardo's notebooks are also

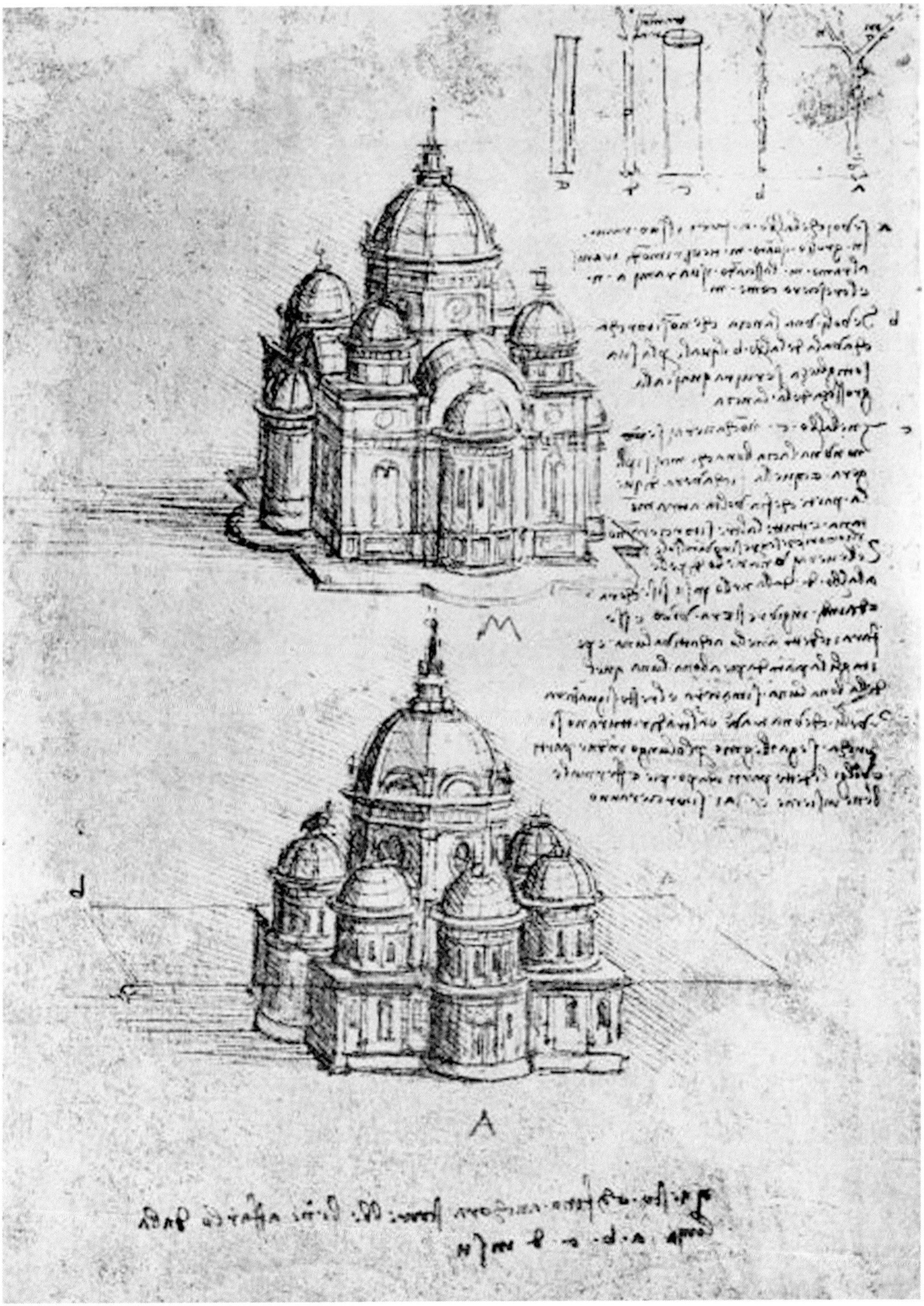

filled with drawings of unrealized ideas for buildings, particularly religious buildings, and one form caught the eye of historians because of its relationship to the subsequent designs produced for the Vatican. Leonardo made fascinating, astounding sketches of central plan churches, some no larger than a thumbnail, throughout his career. Like his designs for unrealized inventions, these images are convincing enough to have been used as ideas for many, many executed works, though it is difficult to trace their direct influence. Indeed, the organizers of a major retrospective on the artist in 1979 were so intrigued as to construct a model of one of these designs, proving that its form was completely understood by Leonardo [5]

About a dozen of Leonardo sketches depict fully developed designs for central plan churches. In such a church the altar is placed at the very center of a regular geometric figure such as a circle, square, or octagon. In a martyrium such as Bramante's Tempietto at San Pietro di Montorio (c. 1510) above the Trastevere in Rome, the center point may be above a tomb or sacred location, in this case the place of Peter's crucifixion. One of his first sketches shows the exterior of a church that is square in plan, with a dome and four ancillary domes, much like a typical Russian Orthodox church (1487–90). This relatively conservative scheme gave way to another idea, drawn below it, for a similar plan with eight domes that more directly expressed drum and dome volumes in the exterior massing [Paris, Institute de France, Codex B, 17v]. [6] He may also have been referring to the multi-domed design of mosques in Constantinople.

Arguing with other scholars who found such speculative drawings unimportant, James Ackerman pointed to the extraordinary power of these sketches to render space and mass, leading to a "major evolutionary leap in architecture during Leonardo's lifetime." The same album, now in Paris, contains at least two other sheets exploring similar design themes. One attaches the central plan form to a longitudinal nave [Codex B, 24r]. Another, described by Ackerman as a preaching theater, combines three sketch plans with a small perspective [Codex B, 52r]. The most sophisticated [Codex B, 22r] juxtaposes a shaded perspective with a shaded plan. Leonardo appears to be stretching drawing conventions in order to imagine both the interior and the exterior at once, but the plan cannot give him a fully volumetric understanding of the designs.

Several apparently later sketches in other notebooks indicate that he found an answer to his representational conundrum. The most often reproduced of these drawings is also one of the smallest: two sketches on a larger sheet in Milan's Biblioteca Ambrosiana [Codex Atlanticus 547v/205v]. [7] The right hand sketch shows a central plan church with a free-standing drum and four almost free-standing chapels attached on the cardinal axes of the circle. Next to it, at approximately the same scale, is a sectional perspective, from the same high angle, showing a cut-away across the long axis of the building. It is unlikely that any

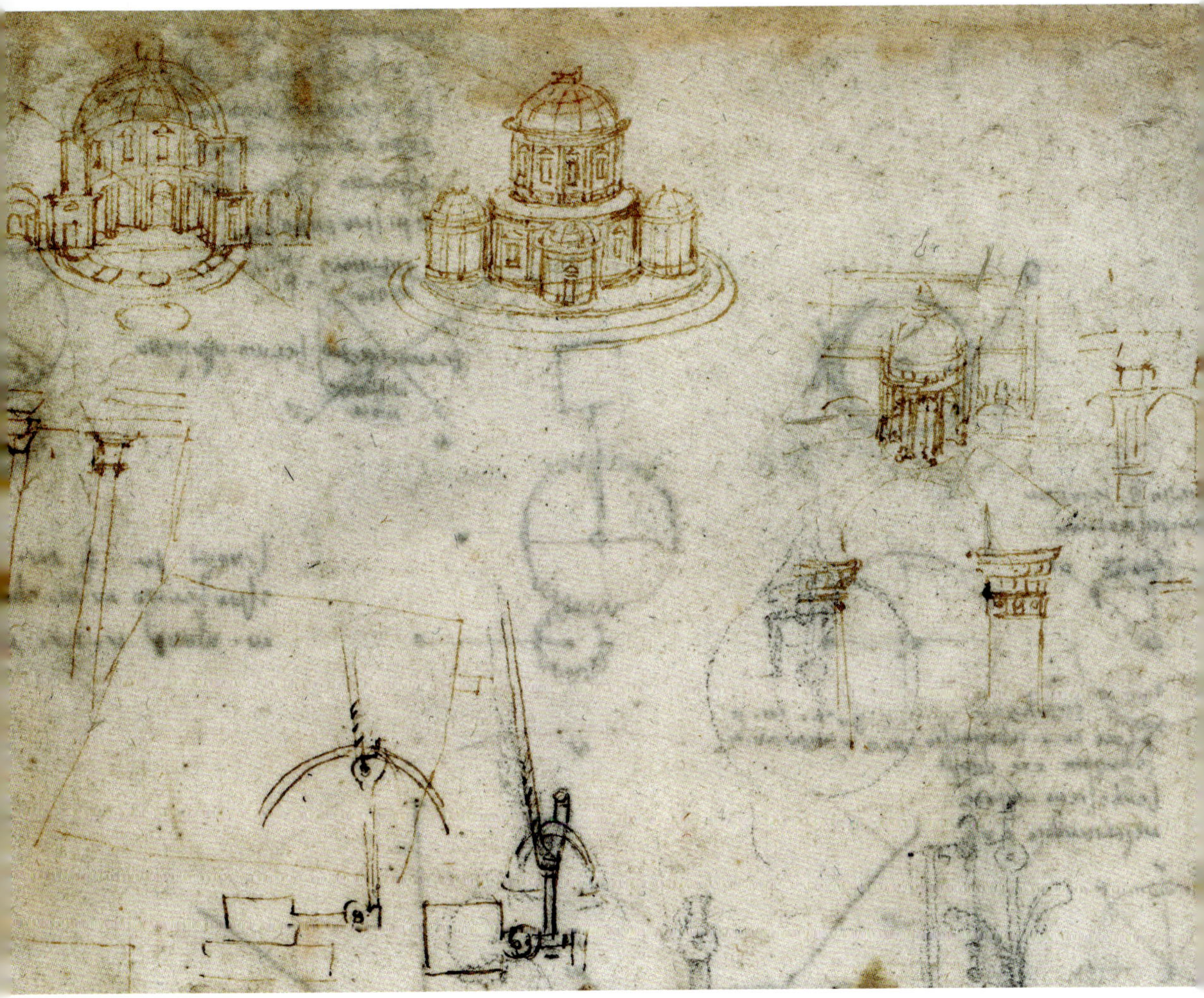

7. Leonardo, cut away section and perspective of a central plan church. Milan, Codex Atlanticus. (Veneranda Biblioteca Ambrosiana, Milan, Italy).

drawing of this kind had been made before, though other perspective-sections were common among drawings of existing Roman buildings in the mid-1400s. Astounding in its accuracy, this drawing is not a constructed perspective, contrary to what Ackerman believes, but rather a mental synthesis of plan, section, and perspective that uniquely captures the design problem at hand. Leonardo was unable to envision the walls of the dome to determine their structural role; therefore he constructed the volume in a sketch form. The resulting drawing is unprecedented in the history of architecture.

This artistic genius was more than a gifted painter or sculptor. He was the first artist in Europe to use small conceptual sketches to study the natural world, to think about its phenomena, and to invent physical artifacts for human use. None of his contemporaries attempted anything so ambitious with their preparatory drawings. None kept voluminous notes in books that were later bound for posterity. And none were so emphatic about the necessity for drawing the world in order to comprehend it.

Revising the Story of Design

In this book I present two interwoven narratives. The first explains, in broad strokes, the recent discoveries in neuroscience that have direct bearing on how architects do the mental work of developing their designs. I concentrate particularly on the theory of embodied cognition, a revolutionary scientific and philosophical revision of long held beliefs about the brain. The second story recasts the history of architectural representation (how architects use drawings, models, and other tools in their studios) in light of embodiment, and suggests that as soon as designers had a full range of drawing types, after the Renaissance, a powerful and consistent practice evolved, maturing during the late nineteenth century. That practice can be analyzed now from the point of view of cognitive neuroscience, noting how closely traditional and contemporary methodologies track with what we know about how the brain works.

How would such a new rendering of the history of design influence the way architects practice today? One take-away would be that "modern" architecture was indeed developed during the fifteenth century by such masters as Leonardo da Vinci, Leone Battista Alberti, and Francesco di Giorgio Martini. Their innovations, long seen as essential but no longer applicable to current practice, might prove more persuasive as we consider the role of visual judgment in making buildings that are beautiful and commodious. More important, these revolutionary designers may have been the first humans to use cognitive loops between images in the brain and drawings on paper to study multiple aspects of three-dimensional objects.

Though architectural historians have consistently pointed to the decisive break between pre-Renaissance and post-Renaissance practices, until recently there has been no frame of reference to distinguish these methods on a psychological level from the kinds of drawings and other forms of representation used by say, Roman architects in 150 CE, Gothic master masons in 1300, or Mughal builders in 950. Neuroscience provides such a lens for looking at the cognitive modes followed by different designer/builders throughout recorded history. As long as both narrative descriptions and physical artifacts are available for study, scholars will be able to draw inferences about cognitive modalities. I will argue here that the crucial cognitive division between fully depicted building designs and artisan crafted buildings came during the mid-1400s, and that there has been no similar break until today's digital revolution.

Both of the narratives in the text are provisional, as science has only touched upon the interaction between perception, memory, and concept formation during aesthetic and artistic apprehension of the environment. I believe, however, that enough is known to make the necessary connections between how architects see their designs in relationship to the environment, and how the history of design has been molded by the structures and workings of the brain. I am not alone in this conviction, as many architects and researchers are teaming with neuroscientists to probe different aspects of environmental awareness, perception, and even levels of well being imparted by beautiful places and buildings.

As we will see in the chapters that follow, the theory of embodiment has supplanted the Enlightenment belief that humans have a unique rational capacity for abstract thought that separates them from other mammals. This theory brings us closer to our environment, and thereby compels us to reconsider all artistic and cultural views that privilege the "naked brain" (as the philosopher Andy Clark calls logical positivism) over the whole organism.[19]

How might this affect the history of architecture, and of design in general? First, the centrality of "texts" as a means of studying history, art, and literature must be re-evaluated. If humans convey and assess meaning holistically, through signs that are not arbitrary, but are connected to bodily activities, texts and language cannot be the sole basis for meaning. Further, twentieth-century ideas about the "textuality" of things must be critically assessed according to new theories of the mind.[20]

Second, all cultural objects that humans encounter on a daily basis must be studied in relationship to not only their symbolic aesthetic value, but also in relationship to physical qualities that touch humans in their environment. Material culture, as studied by ethnologists and anthropologists, plays a more direct role in the "fine" arts than previously thought. In some respects structuralist ideas about the primitive mind have been substantiated by current research, while post-structuralist theories have not.

Lastly, the physical practice of arts and crafts, long denigrated by high culture theorists, must be given more prominent consideration as generators of "intellectual" ideas among historians and critics. During the Renaissance, artists were not entirely cut off from the medieval guild system, nor did they jettison kinetic habits of design and production in favor of "abstract" concept formation. This fact is an unacknowledged myopia in nearly all post-Renaissance art and architecture studies. As recent scholars have shown, the "science" of art was based upon direct experience with the physical world before the time of Bacon and Descartes. Only gradually did we come to the technocratic system commonly used today.

Our revised history of architectural design will begin long before the fifteenth century, with ancient civilizations such as Greece and Rome, and proceed forward rapidly to consider the "visual" revolution of the Quattrocento in terms of embodiment and studio practices. That revolution did not simply offer a new perspective-based system of depicting the world; it also made the world more "rounded" in other ways. Artists such as Leonardo da Vinci and Filippo Brunelleschi expanded the capacity of designers by encouraging the complete depiction and physical de-construction of buildings as objects in the environment. Exploded and dissected views of plants, animals, and inanimate objects filled the notebooks of artist-architects from the late 1400s onward, making it possible to design more and more complex machines and useful objects. These depictions were not only helpful in understanding the physical world and constructing larger buildings; they also expanded the brain's image-making capacity exponentially. In this way the cultural "modes of conception" used by designers became more and more sophisticated following the professionalization of architecture during the sixteenth and seventeenth centuries. By the end of the nineteenth century a complete system of design education had integrated embodied modalities of drawing and aesthetic judgment into Beaux Arts curricula.

The Action-Perception Cycle

When artists or architects begin a design, they initiate a continuous feedback loop of actions (drawing, sculpting, painting) and perceptions (seeing and judging the artifacts in front of them). Psychologists, neuroscientists, and architects have recognized these cycles of cognition, but until recently there has been limited experimental research documenting their existence. Michael Arbib discusses the most recent findings in his book, *How the Brain Got Language*, noting that the *Homo* genus evinced action-perception patterns thousands of years before tool-making, artistic culture, or language emerged in our nearest ancestors. His simple example of an action-perception cycle involves a frog seeing a flying insect, determining its location, and catching the insect with its long tongue. Without both perception (seeing, sensing) and action (motor neurons and muscles controlling the tongue) the frog could not get food. It would seem that drawing what is in one's imagination is a good deal more complex, and perhaps different, from catching an insect.[21]

Almost everything humans do is contingent on the development of skills that involve action-perception loops. Scientists studying perception now recognize that apprehension of the physical world is "enacted" rather than passively sensed; some would say it is "mediated" by the environment.[22] What this means is that when a human or other creature surveys and moves about its immediate habitat, it is continuously using perceptions and motor actions to get an adequate reading of what objects are available for manipulation, and which are too far away to be useful. When an object is recognized, a single position or view will not provide a complete understanding of the shape, size, color, or position of the object in the field of sensation (what James Gibson calls the "ambient optic array"). To get a better "view" of its prey or predator, most animals will move around the organism. Humans tend to do the same thing, but sometimes make assumptions about objects without a full three-dimensional apprehension of their character.

A simple example of this is our general understanding of the roundness of fruits and vegetables. Rounded objects are hard to understand in perspective, as their surfaces appear the same from virtually any angle. Visual perception is sometimes akin to touch in that we "feel" the size, shape, and orientation of objects by comparing views from a limited number of positions, then make judgment about its probable characteristics. Memories of fruit may help to assess the character of say, a plum, when one has only seen apples and pears previously. Likewise, seeing a building section drawing, when one has previously seen only elevations, may help an architect understand both the drawing and the building it depicts.

Writing and research on drawing and model making among architects has generally followed the procedures that designers use in doing studio work, but the cycles of perception and action have not been noted until relatively recently. This book makes this aspect of cognition an explicit subject. Sketches and study drawings are the subjects of some interesting previous research that will also figure into the discussion below.[23] It is important to recognize that when a designer makes a preliminary drawing to begin a design, he or she begins a loop-like cycle of drawing, seeing (the image in mind and the drawing before him), memory retention, discernment, re-drawing, and new perceptions that follow his or her first attempt.[24] This cycle allows the architect to "survey" the problem before him by physically rendering part of its perceptual content in some external form, such as a drawing or model. In this way he is able to see the problem more clearly, thereby understanding what steps to take in order to advance that understanding to the next level.

Notes

1 The metaphoric nature of thought has been explored by George Lakoff and Mark Johnson since the 1990s. See George Lakoff and Mark Johnson, *Metaphors We Live By* (Chicago, Univ. of Chicago Press: 2003). The simple metaphor, Knowledge is Vision, is one of many explored in this influential book.

2 See Vinod Goel, *Sketches of Thought*, (Cambridge, MIT Press: 1995) for the best treatment of this subject to date.

3 Rudolf Arnheim, *Visual Thinking* (Berkeley, Univ. of California Press: 1969) is one of the earliest studies.

4 David Hubel and Torsten Wiesel were awarded the Nobel Prize in 1981 for their work on the fundamentals of vision in the eye and brain. David Marr also did pioneering research during the 1970s. See David Hubel, *Eye, Brain and Vision*, Scientific American Library Vol. 22 (New York, W.H. Freeman: 1988) and David Marr, *Vision* (San Francisco, W.H. Freeman: 1982).

5 See C. Frith and K. Friston, "Studying brain function with neuroimaging," in Michael D. Rugg, Ed., *Cognitive Neuroscience* (London, Taylor & Francis: 1997): 169–92.

6 Some of this is documented in O. Vartanian, A. Bristol and J.C. Kaufman, Eds., *Neuroscience of Creativity* (Cambridge, MIT Press: 2013).

7 See for instance Merlin Donald, *A Mind So Rare: the evolution of human consciousness* (New York, Norton: 2002).

8 Denis Dutton, *The Art Instinct: Beauty, pleasure and human evolution* (London, Bloomsbury: 2009).

9 Anjan Chatterjee, *The Aesthetic Brain: How we evolved to desire beauty and enjoy art* (New York, Oxford Univ. Press: 2014).

10 For the views of the founder, see J.P. Eberhard, *Brain Landscape: the coexistence of neuroscience and architecture.* (New York, Oxford Univ. Press: 2008). This book does not present a coherent or compelling view of the new alliance.

11 The best to date is Juhani Pallasmaa and Sarah Robinson, Eds. *Mind In Architecture: neuroscience, embodiment and the future of design* (Cambridge, MIT Press: 2015).

12 Mark Alan Hewitt, *The Architect and the American Country House, 1890–1940* (New Haven, Yale University Press: 1990) examined patronage, practice, and house design during the Progressive Era.

13 For an incisive critique, see Nicholas Carr, *The Shallows: What the Internet is doing to our brains* (New York, W.W. Norton: 2011).

14 Charlene Spretnak makes this clear in her book, *Relational Reality: New discoveries of interrelatedness that are transforming the modern world* (Topsham, Maine, Green Horizons Books: 2011): 113–46. Her book focuses on the benefits of community, pointing to the numbing alienation in much contemporary architecture.

15 This quote is from a remarkable study conducted on experienced architects in Australia. Zafer Bilda, John S. Gero, and Terry Purcell, "To sketch or not to sketch? That is the question," *Design Studies* 27 (2006): 587–613. The authors concluded that although sketching was not necessary as a means of generating ideas among expert designers, the subjects considered them essential to their working process. They also confirmed earlier studies on working memory limitations in architects.

16 For photographs and plans see, Carlo Scarpa, *Banca populare di Verona head offices, Verona, Italy 1973–81* (Tokyo, A.D.A. Edita Tokyo: 1983).

17 Francesco dal Co et. al., Eds. *Carlo Scarpa: Opera completa* (Milan, Electa Editrice: 1984).

18 The drawings are preserved in both the Castelvechhio Museum in Verona and in the Centro Carlo Scarpa, in Treviso.

19 See for instance, Andy Clark, *Bring There: Putting Brain, Body and World Together Again* (Cambridge: MIT Press: 2001).

20 One book that purports to discuss the interaction between the brain and the architect, but continues the text-based critical approach, is Alberto Perez Gomez, *Attunement: Architectural Meaning After the Crisis of Modern Science* (Cambridge: MIT Press: 2016). The author consistently cites theoretical and philosophical works, insisting that obscure texts have had a determinative impact on modern architectural practice, with little evidence to support his assertions.

21 Michael A. Arbib, *How the Brain Got Language: The Mirror System Hypothesis* (New York, Oxford Univ. Press: 2012): 4–16.

22 I find one of the clearest explanations in Alva Noe, *Action In Perception* (Cambridge, MIT Press: 2004). As a philosopher Noe does not lay out the most recent scientific theories, but summarizes them.

23 An excellent compendium of articles is in Werner Oechslin, Ed. *Daidalos 5: The First Sketch*, 15 September 1982. Berlin Architectural Journal, Bertelsmann F. GambH.

24 Daniel M. Herbert, in his book *Architectural Study Drawings* (New York, Van Nostrand Reinhold: 1993), described the cycle as: Perception, Memory, Schematization, Deliberate Action, Automatic Action, New Mental Information, and New Graphic Information Forwarded to Next Cycle, 78–79. His analysis is a bit more complex than that of subsequent researchers, but is informed by a number of case studies of individual architects at work. It is the only study of its kind known to me at this time.

[2]

BRAINS, BODIES, AND IMAGES

Seeing is so vital to our daily experience that we often take it for granted, failing to notice the role that our body plays in all kinds of perceptions of the world around us. The human visual system is a complex component in a larger, even more complex, central nervous system. In the process of perceiving and making sense of the physical world, humans engage stimuli from five senses, including the eyes and optic channel, and miraculously manage to find their way around a bewildering environment, often fraught with threats and problems.

The lenses of the eyes focus millions of photons from the environment onto a sensitive membrane, that in turn sends it on a path that organizes the light data into dark and light patches, shapes, and colors, finally sending this data to the middle of the brain for more sophisticated processing into an "image" of the outside world. Moreover, the entire brain-body organism participates in assessing the environment, integrating many kinds of sensory inputs into a coherent "umwelt" of the world, based upon its own shape and orientation.[1] As the philosopher Alva Noe puts it, "Perception is not something that happens to us. It is something we do."[2] In this chapter we will explore the process of perception as it relates both to the brain and to the body that it supports. We will then discuss "images" in terms of both cognition and design.

Cartesian Movies

Until quite recently it was common to think of the eye as a camera, with a lens or cornea refracting light on a curved retina just as a Kodak box camera did on roll film. The raw signals absorbed by rods and cones could be immediately developed in the brain's darkroom, resulting in a lifelike image of the visual field somewhere in the cortex or outer brain. This image might then be recorded in short- or long-term memory, much as computers and smart phones store digital images in their data banks. It was the eye that did the seeing, sending a fully comprehensible image to a passive brain.[3] What Daniel Dennett called "the Cartesian theater" was located somewhere in the brain, and it was there that humans could view the phenomenological movie that was projected by the senses.

In this model the basic optics were correct, but just about everything else was a mystery. Just how the brain might attach meaning, associate previous memories, or otherwise process an image was not often discussed by artists and visual theorists, even as late as the 1970s.[4] It wasn't until the last quarter of the twentieth century, and particularly the 1990s, that scientists began to understand the entire visual system, including the eyes, the thalamus, the optic channel, and the visual cortex in the brain's occipital lobes. Functional magnetic resonance imaging, direct electrochemical monitoring of mouse neurons, PET scans, EEG, MEG, and other forms of "neuro-imaging" revealed that the visual system was much more complex than previous theories had postulated. By more precisely mapping the functional areas of the brain, and studying neural networks in more detail, scientists made astounding discoveries about the brain and its connections to both sensory systems and the body itself.[5]

The brain is not merely a circuit board that processes zeros and ones like a computer chip; it is also a supporting actor in a complex network of organs, nerves, chemicals, and electrical signals that we know as a human organism.[6] The human brain, as Baruch Spinoza surmised more than four centuries ago, evolved to maintain the organism in a complex and hostile environment, not only by controlling motor functions, appetite, and other internal organs, but also by recognizing threats from other humans, building social awareness, and generally allowing *Homo sapiens* to ascend the food chain.[7] As the neuroscientist Antonio Damasio has explained it: "brain activity is aimed primarily at assisting with the regulation of the organism's life processes both by coordinating internal body proper operations, and by coordinating the interactions between the organism as a whole and the physical and social aspects of the environment." He further emphasizes, "that in complex organisms such as ours, the brain's regulatory operations depend on the creation of mental images (ideas or thoughts) in the process we call mind." To neuroscientists, images are not merely visual.

The brain gathers these images not in one conscious center, but via multiple "micro-conscious" processes that run virtually parallel to each other. The three branches of the central nervous system—sensory, motor, and limbic—each process information in this multi-modal way. Often several areas of the brain, including the outer cortex and inner sub-cortical sections, act together in a

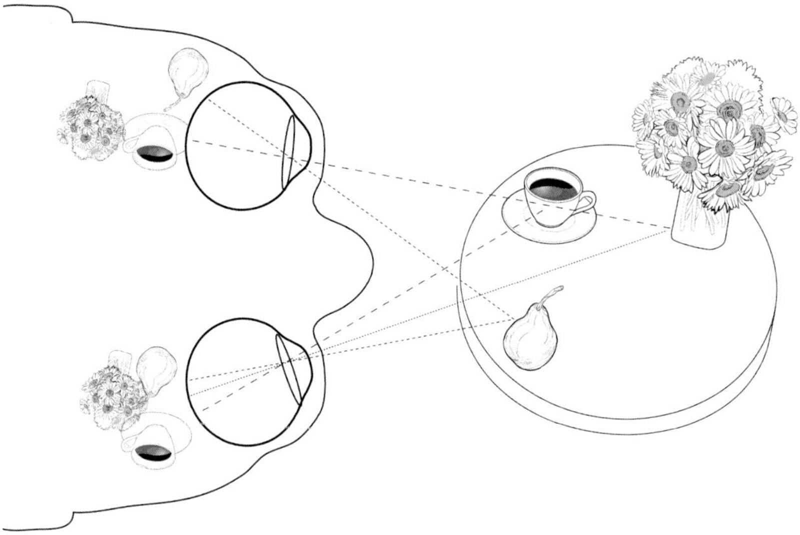

8. Diagram showing stereopsis, from Jennifer Groh, *Making Space* (Courtesy, Jennifer Groh). This system is one of several that allow the brain to assess distance and depth.

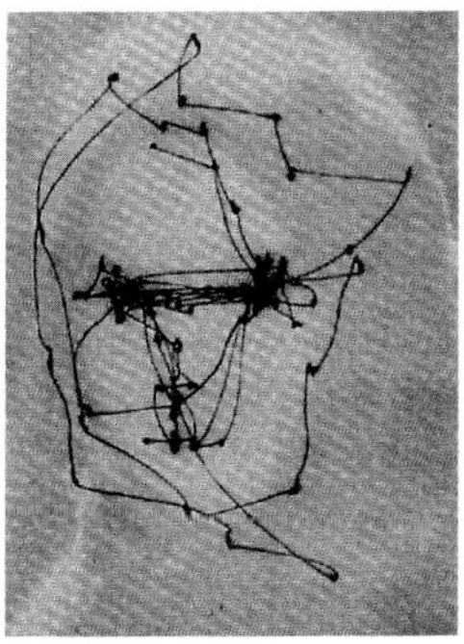

9. Photo and eye tracking data showing saccades across a typical human face. After Kandel, et. al, *Principles of Neural Science.*

complex interchange of signals. Signals from the five senses—sight, touch, hearing, smell, and taste—first arrive in specialized functional areas, then spread to other locations for higher level processing, all in a split second. Moreover, all of the sections and functional areas of the brain consist only of neurons and supporting tissue, and all of its information processing occurs at the neuronal level through electrochemical signals between these simple cells. There is no "central processing unit" in the brain that synthesizes this information.[8]

The visual system, which shares data with other functional areas, is among the best understood portions of the brain, but its operations are quite complex. To understand its workings one must first appreciate the considerable limitations imposed by our two eyes on the things we can perceive. [8] Each eye has an effective viewing angle of between 8 and 15 degrees. However, the central foveal area has only a 5-degree spread, and it is this area that provides our detailed images of the environment. Scientists now believe that there are essentially two visual regions that are mapped in the cortex: one for the fovia and another for the peripheral areas of the retina.[9]

The retina, with millions of receptors, has a concentration of color sensitive cones in only its center. The effective field of vision is rather narrow, forcing the brain to compensate by making "assumptions" about the signals it receives. Without these cognitive interpolations, and the use of memory, humans cannot adequately perceive the world through vision. The same is true of the other senses, though scientists believe that visual data trumps all other perceptual signals in sighted individuals.[10]

Moreover, each eye must move from one side to the other in rapid *saccades* to understand both the wider scene and the objects within it. [9] Recognition

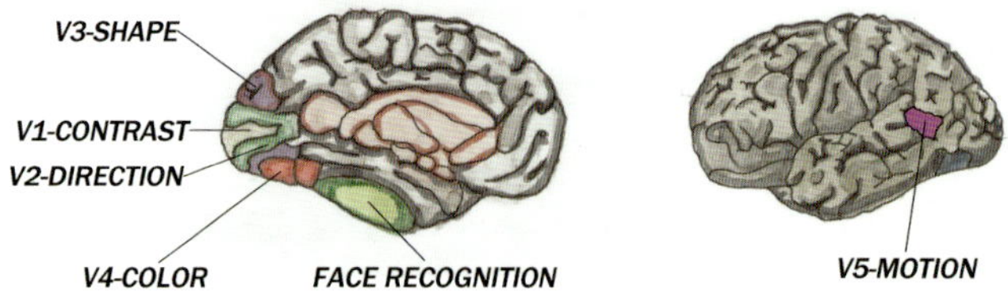

10. Functional areas of visual cortex, with rough descriptions of the processing done in each. In reality, each area shares certain functions during low level and middle level processing, before signals are sent to the midbrain for high-level recognition. (Author)

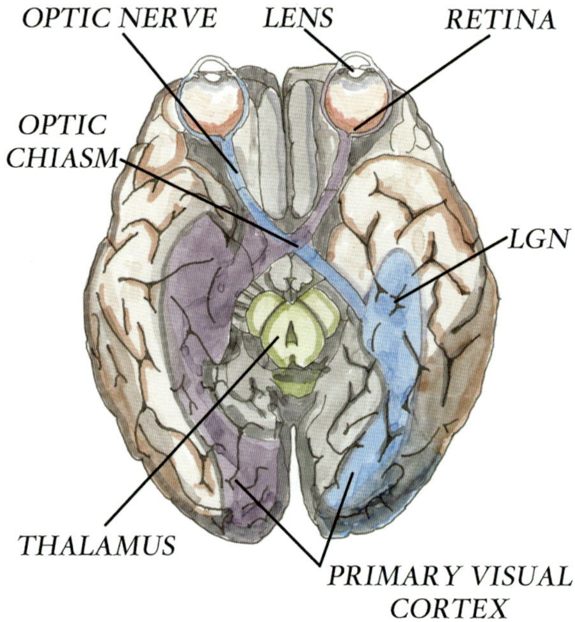

11. Section through the midbrain showing the optical system connecting the eyes to the visual cortex. (Author)

of even the simplest objects requires repeated micro-saccades scanning the object (about 2 degrees in width), and these intense scans are activated only after a general scan of the overall scene. Objects are first recognized and sorted according to their significance to the individual viewer on a semantic level. Experiments indicate that the eye then moves repeatedly over the chosen object in a burst of saccades to burn its characteristics into short-term memory. A good deal of information from the initial scan is not retained, allowing illusionists to fool viewers by placing incongruous objects in the scene for microseconds. When quizzed about their presence, few have any recollection of the non-significant objects. This phenomenon is known as change blindness. Like many aspects of visual perception, scientists have only studied its effects in detail in recent decades.[11]

12. Two systems for following moving objects, after R. Gregory, *Eye and Brain.* In A the eye remains stationery, with the object moving across the retina. In B the head moves, leaving the object in the same place on the retina. (Author)

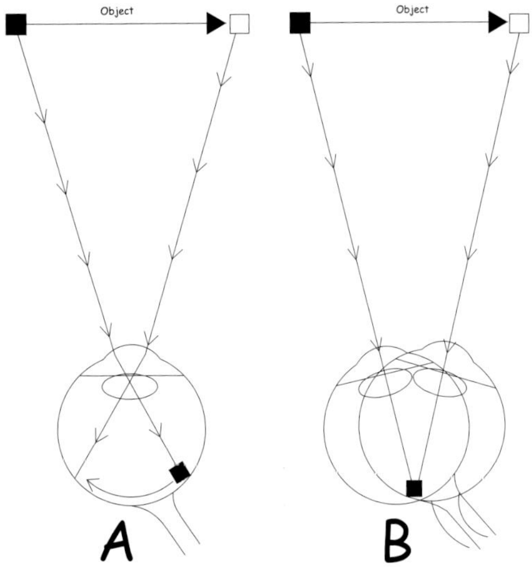

When a significant object is recognized, the eye may fixate on it. Fixation density can be measured by the number of rapid saccades across the object in the foveal area– many times per second. Diagrams of eye movement over a scene contain dense scribbly lines over the important objects in the scene, such as faces. It takes a flurry of such passes in order to attain what is called "high level" scene recognition—roughly the kind of resolution achieved in a fine-grained photograph. This is the kind of intense perception required during any artistic activity.[12]

The entire visual system consists of two eyes, a bundle of neurons that link them to the brain (the LGN or lateral geniculate nucleus), and millions of cells that process the signals they receive from those neural networks. [10] When we focus on a particular part of our environment, the brain goes to work in microseconds to make sense of what arrives in the visual cortex located at the back of the brain. Signals arrive in area V1, the primary visual cortex, which sorts the information according to which specific cortical area must process it—a kind of central mailroom.

Area V2 seems to be reserved for assessing direction or orientation of objects in the visual field. V3 is concerned with shapes. V4 is mainly occupied with color recognition. V5 cells differentiate kinds of motion. V6 seems to process facial features, a key to understanding whether individuals are hostile or friendly. Other areas appear to process even more specialized visual cues. [11] Evolutionary biologists have noted that the hierarchy of processing areas and their temporal sequences track closely with the information that *Homo sapiens* needed to ward off dangerous predators and find food. All of this occurs quickly enough to trigger fight or flight reactions and other means of avoiding danger.[13]

Like other areas of "evolutionary psychology" these observations about functions in the brain point to adaptations made in the course of thousands of years.

What this means to those who want to understand design is that the brain has evolved to assist our two eyes, with their binocular parallax, in making sense of the signals they receive—not only information about colors, shading and depth, but even more confusing things involving speed, parallel lines, distance, and even the nature of animate beings. [12] Designers don't think like cameras, and the visual field does not contain pictures. Even the design of colored things like rugs and dishes engages the brain in very particular ways.[14]

All of this information was vital to our survival during the long evolution of our species. One of the most extraordinary faculties in humans is what is called "approximate color constancy." Except in near darkness, or highly unusual short or long wavelength illuminations, virtually anyone can accurately identify primary and secondary colors. Edwin Land's revolutionary experiments with "Mondrian" arrays were key in identifying how the brain recognizes colors despite varying conditions of luminance, reflectance, and wavelength.[15] His work with light shining on colored cards proved that the brain actually compensates for low light, contrasting colored illumination, and other conditions that could easily "fool" a computer, photometer, or a camera. This kind of basic recognition was helpful to early humans when tracking or being approached by animals of various hues under different lighting conditions. According to neuroscientists, it is an innate neuronal property.[16] [13] The eye also has three types of cones that respond to overlapping wavelengths, allowing it to quickly sort and "mix" red, blue, and green hues. Artists sometimes use these characteristics to produce color illusions, or to emphasize emotional qualities of one color in contrast to others.[17]

13. Edwin Land's "Mondrian experiment." Projectors below shine light on the colored array. Even in relatively low luminance conditions the visual system can identify colors (Author collection).

Architects should pay particular attention to the ways in which this sorting is done, and which areas of the visual cortex handle key information that designers must use in their work. Stereo vision would not be possible without corrective processing of the crossing stimuli that each eye receives when looking at objects at varying distances. Occlusion, the covering of one object by another, is critical to establishing depth and distance. The outline, or contour, of the object in the foreground is also an important signal that must be separately apprehended if the brain is to make sense of what is happening in three-dimensional space.[18] One scientist has even cautioned his colleagues not to use the common Cartesian description of space, preferring to describe vision as a 2 ½ dimensional system.[19] The brain is engaged in a constant process of compensation for the imperfections of the eyes but also the ears, nose, and the haptic system as a whole.

To simplify somewhat, we know that our eyes provide information about the environment around us in a rather imperfect mix of visual signals that must be interpreted by the brain before any meaningful perception can occur.[20] The brain

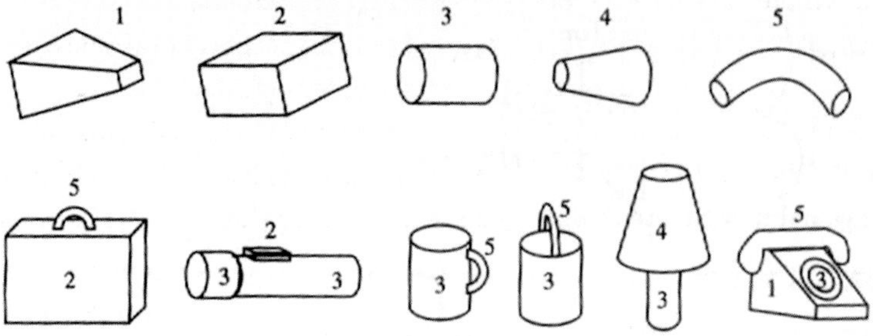

14. Irv Biederman: "geons" or basic geometric forms. The brain may store these building blocks in memory to allow it to distinguish different shapes.

sorts these signals according to which information is most vital for understanding risks in the surrounding environment. There are two main "paths" that sort the information: the "what" path and the "how" path. The path for identifying objects (what) proceeds through ventral areas at the back of the brain; whereas the path locating objects in space and perceiving their movement (how) proceeds through dorsal areas, higher up in the cortex.[21] More perplexing is the fact that a true spatial understanding of the scene that has been captured by the visual system cannot be reconstructed merely from the information projected on the retina. What is called "the inverse optics problem" has not been explained by brain science (though hypotheses have been around for years).[22]

The eyes and brain work together, along with other sensory systems, to maintain our physical and psychological wellbeing. The basic information received is not detailed, as we think initially in abstractions. Neuroscientist Semir Zeki believes that abstraction is a necessary aspect of neural activity, since the brain must schematize objects and environments before dealing with fine-grain information.[23] Erich Kandel, a Nobel laureate, has theorized that a "reductionist" approach to visual processing ties perception to various types of abstract art.[24] Likewise, Rudolph Arnheim and Ernst Gombrich both wrote extensively about the importance of abstraction in perception.[25] Essentially, the brain constructs an image of the visual field from a wide spectrum of sensory data.

Constructing a Mental Image

Once a basic level of processing is complete, experiments have shown that the brain stores information in memory files that are used to determine the characteristics of visual signals on more specific criteria than just motion, color, shape, or direction. Where this activity is most intense has not been fully identified, but it is clear that memory plays a large role in processing signals from all the sensory cortices. In language, ideas stored in memory are known as *memes*. In the visual cortex, there appears to be an analogous dictionary of shapes and forms. According to one scientist, shape recognition may be made possible by a storehouse of *geons*, or basic geometric building blocks.[26] [14]

A full image (or idea) about the content of perceptual stimuli is not formed merely from immediate information recognized in certain key areas of the brain. Rather, the brain must compare this fresh information with memories of previous visual images and other sensory information about balance, smells, sounds, and the like. This comparison is similar to *mimesis*, or imitation, an artistic concept recognized by the ancient Greeks. As the neuroscientist Martha Farah has argued, "[mental] images appear able to recruit attention to perceptual representations. This suggests that visual imagery is not simply 'like' visual perception at a functional level, but that the two systems actually share representational structures." She and her colleagues have published persuasive research about mental "imagery" that links perception directly to cognitive activity in the brain.[27] This will be important as we look at how architects use representations of images to design buildings.

15. Author, after Charles Bridgeman's figure studies showing geometric "building blocks." All traditional figure-drawing courses use the brain's natural mechanisms to assess the shapes and form of the human body.

Many drawing and design teachers fail to understand a key facet of perception: nothing that the brain can reproduce as an image is processed without recourse to an encyclopedic database of similar forms, colors, scenes, and cues about direction, motion, etc. What neuroscientists call "top down processing" influences the way we understand our environment. An external scene or object triggers neurons in specific ways, providing "affordances" for the organism to act upon or store in memory. Such external images can be better internalized in memory if captured in a drawing, providing a mimetic (embodied) adjunct for use in the visual system when another similar image is perceived or imagined.

There are dozens of illustrated books describing a more imitative or mimetic process of learning to draw, but most were denigrated by Modernists in the mid-twentieth century. Traditional texts on drawing admonish students to construct their first figure drawings from known models or basic shapes. Virtually all art supply stores sell wooden models of human figures that are constructed from them. In this way the memory is engaged in learning to represent complex forms from smaller components. As the brain learns to compare basic shapes to those in the human figure, it no longer requires schematic models, but can work with more specific data when representing parts of the body. The pedagogy precisely reflects mental processes. [15]

The most famous example of this mimetic analysis of the human form comes from the notebooks of Leonardo da Vinci. Throughout his life the artist drew the human figure, noting measures, attributes, and proportions of each limb, the torso, and the head. He devised a complete system of proportions for the human figure, recorded not only in his famous drawing of a "Vitruvian" man in a circle, but also in many other drawings.[28] Other artists used his system to refine their own patterns and archetypes for the figure. [16] They were using memory and drawing to record their images for future use.[29] Recently Anthony Panzera, a painter teaching at Hunter College in New York,

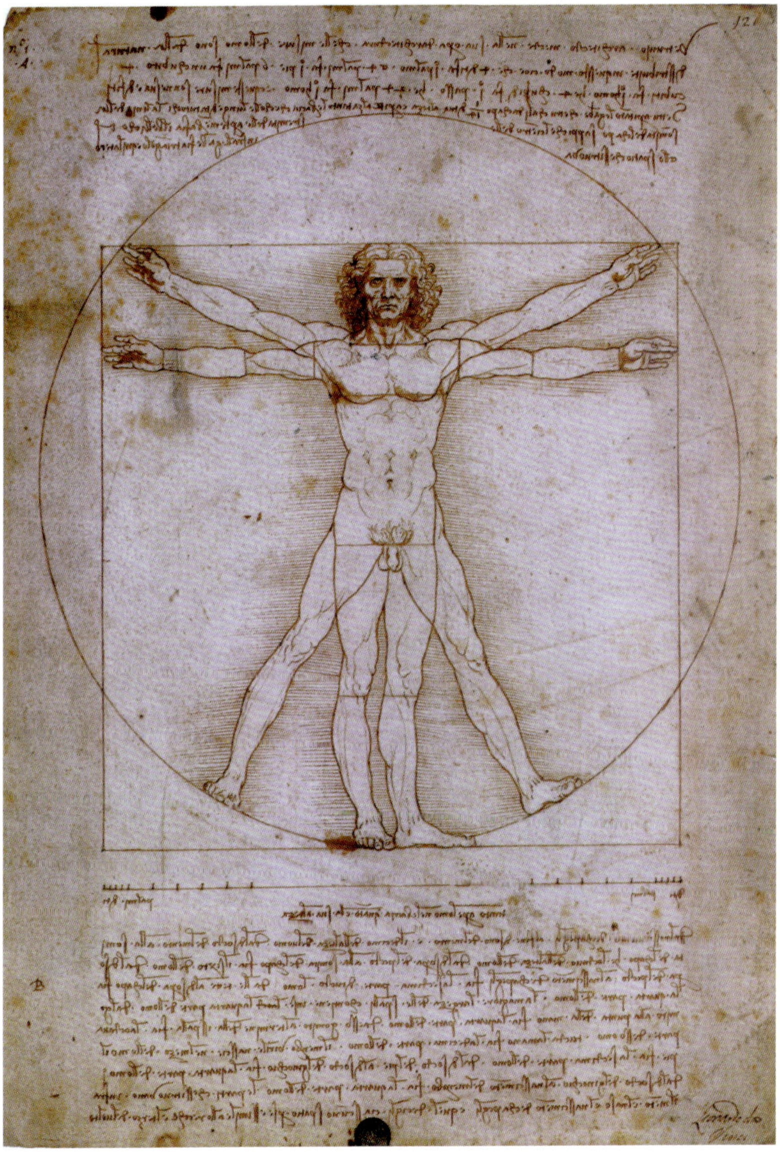

16. Leonardo da Vinci, *Proportions of the human figure after Vitruvius*, metalpoint, brown ink, with touches of watercolor, on brown paper, 344 x 245 mm., c. 1490. (Archivio fotografico G.A.VE, su concessione del Ministero dei beni e delle attività culturali - Gallerie dell'Accademia di Venezia)

published a complete graphic explication of Leonardo's system, which is as comprehensive as any anatomical compendium published since.[30] Panzera used Leonardo's system to teach figure drawing for decades, despite its "outdated" information, earning scorn from some of his colleagues. [17, 18]

We now know more about how the interaction between basic perception and memory works. Though many artists imagine that they conceive complete "images" or "pictures" in their heads before starting to draw, they are actually fooled by their memories of such images. No brain is capable of constructing a

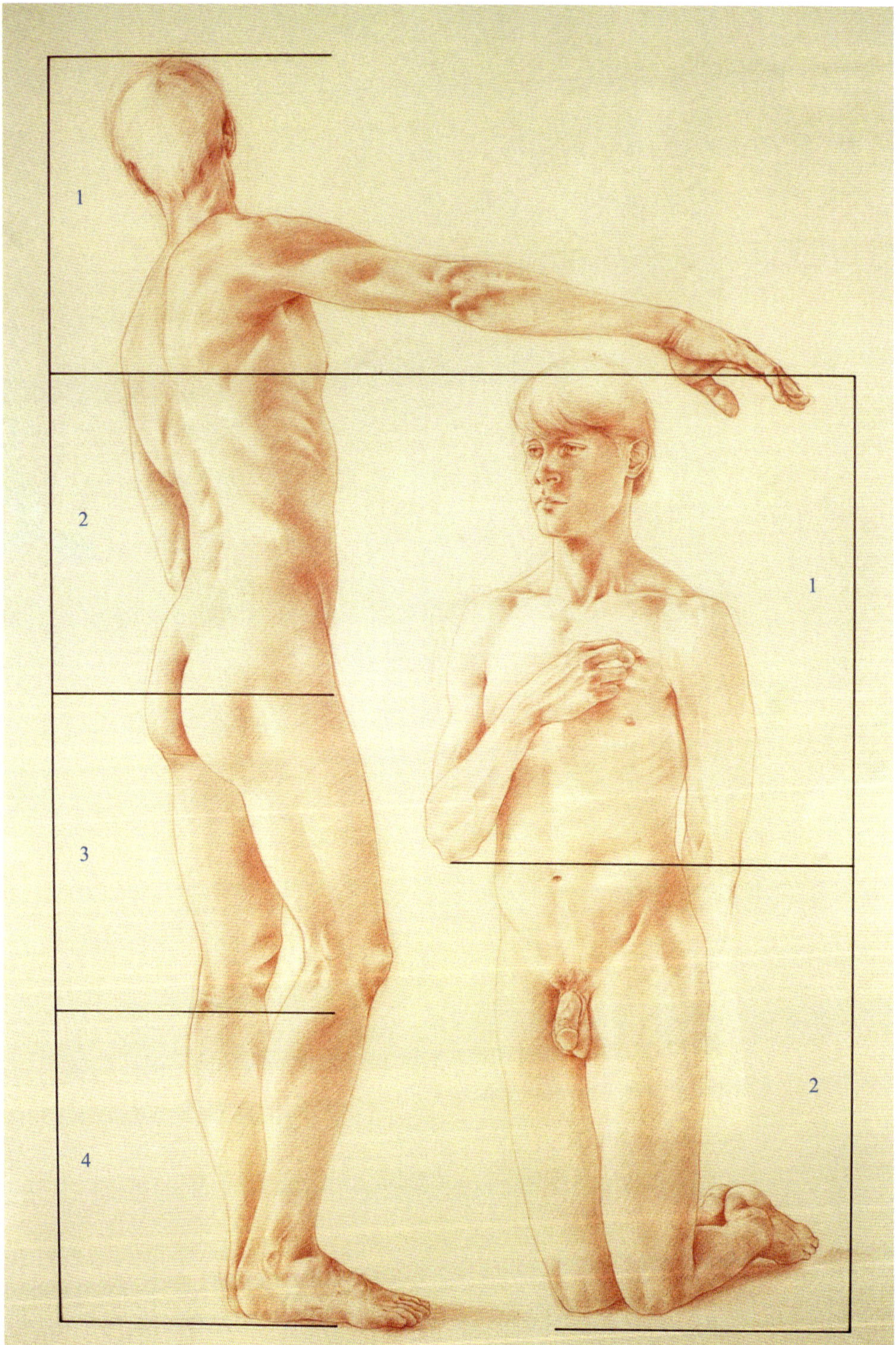
1
2
3
4
1
2

17. Anthony Panzera, adult male and adolescent male figures, A 143, from *The Leonardo Series*. (Courtesy of the artist)

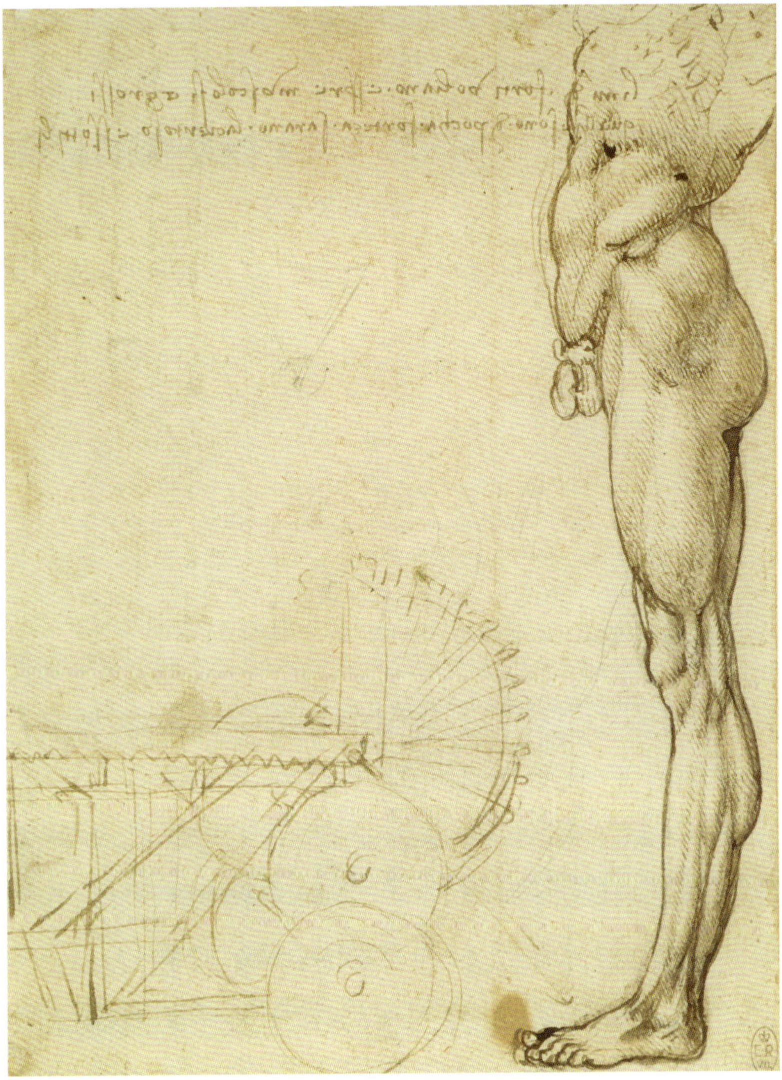

18. Leonardo, lower torso and wooden machine, Panzera Plate 17. The preceding drawing illustrates the proportions shown in this one. (Courtesy Royal Collections, Windsor Castle)

complete image of anything in its visual cortex. What we call "images" are really rather fuzzy concepts or ideas about what we want to draw, compose (as music), or sculpt. Such images (and I will use this term in the neurological sense from here forward) are actually incomplete bundles of neural signals from various parts of our sensory memory banks (smells, tastes, sounds, locations, feelings, pictures, maps).[31] These signals are organized via control circuits into data that can be used to negotiate the environment, interact with other beings, or create artifacts—to do whatever humans do.

As the Harvard neuroscientist, Steven Pinker, explains it: "Images are fragmentary. We recall glimpses of parts, arrange them in a mental tableau, and then

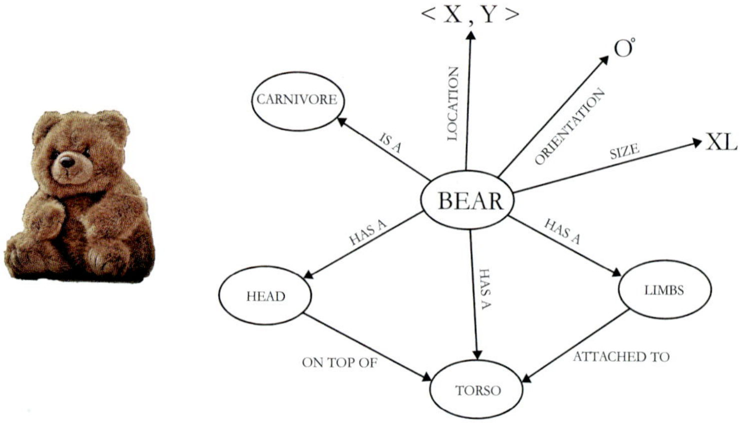

19. Steven Pinker's diagram of visualization (Author).

do a juggling act to refresh each image as it fades. Worse, each glimpse records only the surfaces visible from one vantage point, distorted by perspective. … To remember an object, we turn it over or walk around it, and that means our memory of it is an album of separate views. An image of a whole object is a slide show or pastiche." [19] Alva Noe goes further, contending that it is the object itself that provides the brain-body with triggers for memory of similar objects.

One of the things that has led us to assume a greater imaginative capacity than we actually possess is the fact that there are so many examples of artistic geniuses who have composed works of extraordinary complexity for us to marvel over. Did Mozart "write" his symphonies in his head before committing them to paper? Probably not, though he had an extraordinary memory. Did Frank Lloyd Wright "design" Fallingwater the night before Edgar Kaufman came to his office to see the completed drawings? His concepts were fresh, but he probably refined them in his memory while putting pencil to paper. Such imaginative leaps were only possible because these artists had vast warehouses of material in their memory banks prior to undertaking their masterpieces. They were able to draw on previous models while working on their creations. And the act of committing ideas to paper was a necessary step in the creative loop.[32]

What is most fascinating about the ability of the brain to form these increasingly complex images (not merely visual ones) is that every increase in sensory, motor, and emotional activity builds stronger and larger neural networks that correspond to the activities in question. Even more intriguing is the fact that individuals who have been deprived of sensory experiences during their formative developmental years have difficulty making sense of their perceptions later in life.[33] The environment is critical, as is our individual experience of the world around us, and the people who we admire as teachers and colleagues. As the Nobel laureate Eric Kandel put it, "the reason the great violinist and child prodigy Jascha Heifetz became Heifetz was not simply because he had the right genes, but also because he began practicing his musical skills very early in life, when his brain was most sensitive to modification by experience." The brain changes continuously with new experiences, including aesthetic ones.[34] We shall come back to this factor in Chapter 3.

It isn't hard to understand why this is the case. Musical geniuses have far larger capacity in their auditory networks than normal people, and visual artists likewise build an outsized capacity to process data in the visual cortex relative to most others. Harry Francis Mallgrave has demonstrated that many architects trained their brains to think more profoundly about space, form, proportions, ornament, etc. than non-architects.[35] Michael Baxandall believes that schools of artists formed neural networks that helped them produce their particular kinds of art. We will learn about how architects exercised their visual cortex in the chapters that follow.

What this means to artists and designers is that the division between the perceptual sphere and the creative one is far less distinct than any theorist or historian has yet surmised. *Concepts*, the building blocks of all creative individuals, and *percepts*, the necessary stuff of everyday life, are in many respects the same thing.[36] In fact the theory of "grounded cognition," suggests that areas of the brain process semantic information (concepts) and sensory-motor information (percepts) together, almost instantaneously.[37] Artists carry concepts in their subconscious for use in their everyday practices, much as all humans carry beliefs and behavioral norms. These concepts influence how art is created, especially among individuals who have shared formative experiences, as we shall see in Chapter 5.

Two Kinds of Images

As an architect, I have a clear idea of how images function when I design buildings. They are the raw material that I use to imagine everything I create. I get one in my head and then draw it on a piece of paper, or perhaps on a computer screen. I used to believe that there was a strong correspondence between the image in my brain and whatever I managed to represent in my drawings. When I was young I was frustrated that I couldn't just "snap a photo" of the wonderful pictures that came into my conscious mind, or occupied my dreams at night. Learning to draw was an inconvenient and tedious chore that, when mastered, could only produce a fraction of what I had in my head at any given moment.

As we will see in the following chapters, my naive understanding of my own cognitive machinery no longer makes sense in light of recent discoveries about visual perception, imagination, creativity, and design thinking. And, although most architects are vaguely aware of new discoveries about the visual brain, few are cognizant of the revolutionary nature of those breakthroughs. Hence the need for this book, and others.[38]

I have come to believe that one fundamental understanding lies at the heart of solving the problem of how humans use their imaginations to manipulate the physical world around them. The "images" that our brains process, create, and manipulate are very different, and infinitely more varied, than even the best-informed designers recognize when they work. Moreover, all conceptualization involves embodied representations of the external environment. From the very emergence of our species we have modeled the physical world in relation to our senses, our bodies, and the things around us.

When we design artifacts two kinds of image schemas are active in our brains. Each serves to give us a broad sense of the relationship between the environment, the object, and our own bodies.

One kind of image is permanently stored in both procedural memory and various functional areas of the brain. We might call these "memes" or inherited concepts, terms used by Daniel Dennett and Semir Zeki,[39] respectively. Neurons develop into interconnected webs via the process of synaptic electro-chemical actions over our entire lifetime. Emotions create stronger neuronal connections and more vivid memories. Learning contributes to the development of webs and networks. Habits and repeated physical activities also build up larger networks. Artists must practice their art or craft in order to improve, create new things, and develop personal or collective styles. The personal *modes of thinking* that every architect uses when beginning a design form such a conceptual schema: when we enter a studio we quickly understand how an artist or architect works.

The other kind of image comes about only when we are being creative or solving problems of some kind. Cognitive scientists call this process "ideation" or "concept formation," though I like the word "conception" because ideas take some time to be born, some gestation. When an image emerges in the brain it may be inchoate or vague, just a feeling or fleeting perception that can't be located consciously. Most architects are aware that they did not have a full "concept" of particular buildings in their oeuvre until some external representation process was underway—they got the "parti" after some indeterminate period of "designing," using whatever habits or methods they preferred for the project at hand. They were engaged in "modes of conception" (my term) that involved drawings, models, or external representations that could be placeholders for images in their minds. Psychologists call these images "exograms," and they will be further defined in Chapter 3.

By activating these two kinds of cognitive imaging in our brains we are able to invent or depict the artifacts we make before they are realized in final, physical form. Most previous studies of the cognitive process of designing or "creating" art or architecture do not distinguish between these types of thought. Many books have been written about drawings, models, and photographs, and about how these representations can be used in imagining things to be produced.[40] Most have been useful in getting us to our current understanding of design thinking, but we need to go further with the new tools available now.

Likewise, lots of research has gone into how the brain takes in visual data, how humans solve problems, how the mind creates meaning and symbols, how cognitive processes resemble computing or logic, and how language has developed to help us think about abstract things. There are studies of creativity in living individuals, and among groups of like-minded or culturally similar people.[41] Neuroscientists and psychologists publish studies in journals devoted solely to these topics. This too is useful, but of limited value if unconnected to the history of creative endeavors in different social and cultural contexts.

Of course, design thinking can't be described as merely drawing pictures. Choices must be made based upon quantitative information, lists of words, abstract concepts, social exigencies, and even personal biases among the team of designers. Yet the art of design has followed historical patterns that made it possible for humans to incorporate all these forms of knowledge into mental images that could be used to make useful things. Ergo, a study of the cognitive processes used by designers should assess how such images are generated, manipulated, and externally conveyed. As the late Michael Graves reminds us, "Representation is seeing something again, it is an idea, an artifact, a reoccurrence, a replica of something else. It is important to remember that representations do not act so much as substitutions for, as interpretations of the thing represented."[42]

Memory, Images, and Creativity

The cognitive process of creating objects of any type, whether familiar or entirely new, involves both the rather mechanical mental activity of perception (less straightforward than we previously thought), and the more "imaginative" activity of conception—the birth of new ideas, whether visual, verbal, or auditory. As more neuroscience research reaches acceptance in the academic community, it is becoming clear that the brain combines the semantic or symbolic understanding of the visual world with purely sensory understanding of its objects, flora, fauna, and environment. Moreover, in contrast to problem solving and scientific reasoning, the brain uses more of its functional activity centers during architectural design than in most other kinds of thinking. This fact, based on neuro-imaging, scientists have come to believe that the sensory motor cortex is working along with pre-frontal and parietal cortices during visual design sessions.[43] During creative pursuits these regions of the cortex are bursting with synaptic activity.[44] As we will see such internal schemas interact with external triggers known as exograms, which include drawings and models.

How do architects construct, or conceive, new works from material in their brains? How much is already present in their memories, and how much emerges as new during the conceptual process? And, ultimately, does creative thinking involve only a cognitive process or do the sensory systems (and memories) contribute to conceptual and abstract thought?[45] According to the theory of embodiment, creative thinking is actively connected to the objects, processes, and materials of everyday experience, or to the ultimate goal of making something useful.[46] Architects were once the masters of embodied thinking, designing shelters that used things in the everyday environment for their construction.

Previous theorists about visual thinking have generally discounted the role of memory, and the use of exemplary precedents, in the creative process. One of the key tenets of Modernism, first promulgated in the Bauhaus *Vorkurs* (introductory course), was the idea that artists begin with a "blank slate" before undertaking detailed studies in their medium of expression. Such a state of primitive or naïve awareness was thought to increase inventiveness. In the search for the novel or new, the artist was encouraged to throw out memories of specific objects related to his or her field of study. Old chair designs, such as those of Chippendale (actually among the most comfortable ever made), were discarded by Marcel Breuer and his fellow designers as they searched for new forms. Breuer's chairs were designed to improve on both the aesthetics and function of previous models, but in fact were largely inferior in nearly every way to wooden chairs made by even the least competent traditional craftsmen.[47] Discarding proven archetypes proved to be foolish, partially because the brain uses such archetypes in order to process information that stimulates creativity. [20]

Another attribute of remembered ideas in the brain has also broken with old theories about perception and cognition. Our strongest and most tenacious memories are those that correspond most closely to strong emotions, pleasurable or traumatic experiences, and even instances of high aesthetic awareness. The brain seems to encode these fleeting images into its tissue with particularly strong signals. In evolutionary terms, such strong memories serve to protect the organism from pain and harm, but also encourage more pleasurable impulses like love, the quest for beauty, and the ineffable creative (sexual) drive. Since emotions, feelings, and other limbic activity cannot easily be separated from cognitive states, designers ought to be more aware of them when undertaking their work.[48] As the primatologist Frans de Waal argues, "Emotions are an essential part of the intellect. The idea that they are separate remains so ingrained, however, that it survives in full force in many circles," (which include architects). "We celebrate the cerebral, believe in such a thing as 'pure reason,' and have a low opinion of the emotions, the body, and any species not our own."[49]

We might start paying attention to what Alain de Botton has called "the architecture of happiness."[50] Roger Scruton, in a once controversial book, described architectural aesthetics in terms of common pleasurable experiences

20. Above, Wassily chair by Marcel Beuer. Below, Chippendale side chair. The chair below is more comfortable than the one above, which pinches the leg under the knee and puts it to sleep. (Wikimedia Commons)

that could be appreciated by everyone, not just by cognoscenti, but also stressed the development of taste through repeated, memorable, encounters with great buildings.[51]

One of the most important implications behind the encoding of memories into neural networks is that repetition creates greater density among neurons sharing synaptic activity, and therefore stronger, more accessible memories and motor facilities. In other words, an artist may persist in using the same types of imagery from memory, while also developing increasing facility in his or her medium of expression (motor skills or thinking). The necessary sensory-motor linkages develop along with emotional, formal, and image "archives." When firing in parallel during a creative session, these brain areas eventually coordinate to produce competent artistic works, later leading talented artists to produce masterworks. In addition, recent research has proven that even non-designers remember images and ideas more effectively after drawing them than when using other mnemonic aids.[52]

If stored images, feelings, and experiences contribute to our aesthetic pleasure, it seems probable that these same things will also be present when we form concepts in advance of a creative act. Most observers of artists know that the most accomplished painters have a highly developed "eye" or sensitivity to things like light, shapes, proportions, and characteristics of various animate and inanimate objects. Though innate talent contributes to this sensitivity, it is likely that the best painters have also spent long hours charging their neural networks with memories, motor reflexes, recognition exercises, and the like, as a prelude to their mature work.

Studies have shown that every accomplished musician must put in approximately the same number of hours of practice on his or her instrument before attaining mastery (about 10,000).[53] When musicians create a particularly powerful composition or perform a piece with extraordinary virtuosity, all recognize that emotion, and emotional memory, we're in some measure responsible for their success. While teaching architecture students I became convinced that the standard three-semester foundation studio sequence was exactly the minimum needed to teach young designers to use and understand plan, section, and elevation projections. As we learn more about the brain, we will no doubt connect more of these faculties to specific parts of our vast cortical encyclopedia.

Sensation, Mapping, and the Body in Space

While research on the visual cortex has revealed extraordinary things about mental images and concepts, neuroscientists have not neglected research on the auditory, somatosensory, and limbic systems. It is clear that the brain's images are not merely formed out of stimuli from the eyes but from the entire body as it moves through the environment.[54] James J. Gibson was one of the first psychologists to present a case for looking at the interaction between sensory "modes" during the 1960s.[55] His theories, then controversial, are gaining credibility with neuroscientific discoveries like the ones we've discussed here: there is no intermediate step between the acquisition of sensory data and our perception of the world around us.[56] Most important, for further research, he presented a case for perception grounded in the entire organism as it apprehends its own position in space. Among philosophers, pragmatists such as John Dewey and William James were also early proponents of an ontology grounded in direct sensory experience.[57]

One of the truly extraordinary recent discoveries concerns the communication between neurons that process stimuli as metered (strong and weak) electrochemical signals versus neurons that process sensory data as maps or physically oriented signals such as right, left, up and down.[58] If the brain processes sensory data in both ways simultaneously, many scientists tend to favor an integrated or body-based theory of space perception, sometimes called "grounded cognition." Much like the body-image theory of the 1960s, this view of consciousness places the "center" of the mind along the gravitational axis of the body in its vertical orientation.[59]

Experiments of various kinds have indicated a common theme in the brain's architecture: the simultaneous coordination of signals from multiple senses that allow the body to assess its position in space via "maps" that abstractly describe that position. The concept of mental mapping was a common theme in research in urban design during the 1960s, when Kevin Lynch published his groundbreaking book, *The Image of the City*.[60] Anticipating the discoveries occurring in the twenty-first century, Lynch tracked subjects in three cities and noted their techniques for way finding in both familiar and unfamiliar urban environments. Lynch's five "elements," nodes, paths, edges, districts, and landmarks, captured the mental constructs that subjects used to place themselves in environments in order to move toward both familiar and unfamiliar goals. Such body-oriented mapping is re-emerging as a key construct in neuroscience today.[61]

It is also becoming clear that the brain processes information about space and location through multi-sensory information processing that does not occur only at the conscious level. Just as bees can indicate routes to pollen sources to their comrades by simply moving their bodies, so humans use somatosensory cues and signals to make accurate maps of their positions in space. Some of these maps are quite precise, as the eyes, ears, and brain are capable of fine calibration to determine the angle of light on the retina and the direction of sounds entering the cochlea beyond the ears. There is exciting research identifying specialized neurons in the hippocampus that respond to specific places or locations, both in animals and in humans.[62] In addition, there appear to be "grid cells" that help animals navigate through their environments in a kind of coordinate system of equilateral triangles.[63] The science writer Nicholas Carr has noted that computer assisted navigation can circumvent or even hamper the brain's spatial representation system.[64]

Lawrence Barsalou, a neuroscientist at Glasgow University, posited a radical re-definition of cognition as "grounded" in sensory perception.[65] He argued that cognition (thinking) is more than the mere "computation" in the cerebral cortex utilizing innate or amodal symbols that do not in any way correspond to the world as perceived. Rather, he posits the radical position that symbol systems in the cortex are not amodal—that is, they are not divorced from the body's sensory systems—touch, vision, hearing, feeling, and taste. Like Damasio, who presents his findings at a finer level of neuronal interaction, Barsalou rejects the "homunculus" and the phenomenal view that symbolic reasoning occurs without inputs from the sensory experience of the environment. And like Gibson, who offers "affordances" in that physical environment as direct prompts for behavior, Barsalou surveys the research that documents the give and take between the senses and the environment, leading to awareness and eventually cognitive activity. Moreover, Barsalou points to issues examined in this book concerning "representation" of perceptual data which he calls "simulations" of modal activity—what we will encounter below as a feedback loop in conceptualizing architectural form.[66]

More importantly, scientists are testing hypotheses regarding grounded cognition on two tracks—linguistic and sensory-modal. A number of successful experiments have proven the thesis that the brain processes both pure sensory information and abstract/conceptual knowledge simultaneously. One study regarding colors used a now familiar coupling of words describing colors with images of the same and different hues. As the researchers explain it:

Lime **Lime**

> Unlike images depicting objects, understanding written words requires access to knowledge from an arbitrary symbol (i.e., the word form). The physical properties of these symbols bear no relation to the physical properties of their referents (the monochromatic printed word "lime" has no relation to a lime's greenness). Whereas color-selective fMRI activity triggered by a word (e.g., an object name) is silent as to whether the voxel was activated during or subsequent to access of color knowledge, a demonstration that the timing of access to color knowledge is modulated by a purely visual variable (visual contrast) would constitute evidence uniquely for the grounded view.[67]

Researchers in San Diego did indeed prove this hypothesis in 2014. The so-called amodal or purely cortical thesis on cognitive knowledge acquisition leaves no room for a concurrent recognition of these two types of information. As more and more experimental corroboration enters the literature, proponents of grounded cognition and artisanal knowledge acquisition can pursue their historical research with more confidence. Mimetic forms of art and material culture are thus validated by neuroscience.

Though not often consulted today, one book from the 1970s, written by two of my teachers at Yale, relates some of the concepts in modern neuroscience to architecture. *Body, Memory and Architecture* appeared in 1977 and was quickly tossed aside when Charles Moore became a target for angry Modernists attacking "post-modernism." Yet Moore and his co-teachers at Yale, Kent Bloomer and Buzz Yudell, boldly asserted that if architects considered basic human perceptual parameters before undertaking their designs, none of the dominant functionalist or machine-aesthetic precepts could maintain their sway over the profession. Indeed, the authors were years ahead of their time in stressing that the body was the center of all spatial experience, and that all five sensory systems analyzed by Gibson should be considered when looking at behavior in the environment. They even presented a clear exposition of what would later be called brain mapping of the body's position in space.[68]

"We propose here," they wrote, "to look again at the architectural building blocks in the existential space that surrounds us, to pursue them from the boundary of the individual body to the first shared boundary (the house) and beyond to boundaries shared by larger and larger communities, seeing how they can be a means of extending inner order outward, of making a world that is an extension of our sense of ourselves." Like many of Moore's writings, he concluded the book with examples from both architectural history and his own work, noting the paucity of recent architecture that was attendant to these matters.[69]

Like Gibson and Barsalou, these Yale architects were fully aware of the relationship between the brain/body and the surrounding environment, stressing that "we make places that are an expression of our haptic experiences even as these experiences are generated by the places we have already created. Whether we are conscious or innocent of this process, our bodies and our movements are in constant dialogue with our buildings."[70] In this book, we will examine examples of this process of give and take between individual creative minds, groups of designers, and the places in which they work and live, noting the "affordances" and influences that architecture imprints on architects.

Notes

1 For a recent review of the science, see Joost Maier and Jennifer Groh, "Multisensory guidance of orienting behavior," *Hearing Research* 258 (2009), 106–112. See also Wandell, B.A., Dumoulin, S.O., Brewer, A.A., 2007. Visual field maps in human cortex. *Neuron* 56, 366–83.

2 Alva Noe, *Action In Perception* (Cambridge, MIT Press: 2004): 1.

3 James J. Gibson was an early skeptic of these theories. See "The Information Available in Pictures," in *Leonardo*, v. 4 (Pergamon Press): 27–55. Rather than confining himself to the laboratory, Gibson ventured far afield in his brain research.

4 David H. Hubel and Torsten Wiesel were the first to map the visual cortex using electrodes in the brains of cats and monkeys, during the late 1950s. Hubel describes the eye as more of a movie camera. "The eye has often been compared to a camera. It would be more appropriate to compare it to a TV camera attached to an automatically tracking tripod—a machine that is self-focusing, adjusts automatically for light intensity, has a self-cleaning lens and feeds into a computer with parallel-processing capabilities so advanced that engineers are only just starting to consider similar strategies for the hardware they design." D. Hubel, *Eye, Brain and Vision*, Chapter 3, page 1, online edition. (1988)

5 There is continuous debate about the efficacy and precision of fMRI and other techniques for seeing brain activity. See for instance, Karl J. Friston, "Modalities, Modes and Models in Functional Neuroimaging," *SCIENCE* New Series Vol. 326, No. 5951 (Oct. 16, 2009): 399–403.

6 Damasio, 194–95.

7 Damasio, 268–70.

8 The so-called mind-body problem, or substance dualism that has guided post enlightenment science and philosophy "is no longer mainstream in science or philosophy, although it is the view that most human beings today would regard as their own" according to Antonio Damasio and other brain scientists. Antonio Damasio, *Looking for Spinoza: Joy, Sorrow and the Feeling Brain* (New York, Harcourt, 2003): 187.

9 See Matthew Schneps, L. Todd Rose, and Kurt W. Fischer, Visual Learning and the Brain: Implications for Dislexia, *Mind Brain and Education* (V. 1, no.3, Harvard School of Education): 128–39. The authors argue that in dislexics the perceptual attention and acuity that others have in the fovial area is often shifted to the peripheral region.

10 For a concise analysis of how the visual system constructs various "images" from retinal stimulus see Donald D. Hoffman, *Visual Intelligence: How We Create What We See* (New York: W.W. Norton: 1998).

11 For a good summary of its importance see Noe, *Action In Perception*, 49–55.

12 See John M. Henderson and Andrew Hollingsworth, "High-Level Scene Perception," *Annual Review of Psychology* 50 (1999): 243–71.

13 Samir Zeki, *A Vision of the Brain* (London, Blackwell: 1993): 94–114. I find Zeki's explanation of the visual cortex to be clearer than others I have read, though by now it is somewhat dated.

14 As Donald D. Hoffman puts it, "What happens when you see is not a mindless process of stimulus and response, as behaviorists thought for most of the twentieth century, but a sophisticated process of construction whose intricacies we are just now beginning to understand." 1.

15 Steven Pinker, *How the Mind Works* (New York: Norton: 1999): 246–48.

16 Zeki, *Vision*, 230–40.

17 A particularly good explanation for this is in Kandel, *Reductionism*, 143–54.

18 See James J. Gibson, *The Ecological Approach to Visual Perception* (New York, Taylor & Francis Group: 1986): 147–70. Gibson was the first psychologist to posit a persuasive theory of location and movement in visual perception, thus establishing spatial orientation in neurological terms.

19 Pinker, 256–58.

20 For an excellent description of this, see Donald D. Hoffman, *Visual Intelligence: How We Create What We See* (New York, Norton: 1998): 1–16.

21 See Shifferl, 74–76.

22 Dale Purves et al, op cit.

23 Samir Zeki, *Splendors and Miseries of the Brain* (London, Wiley-Blackwell: 2009): 13–25.

24 Eric R. Kandel, *Reductionism in Art and Brain Science: Bridging the Two Cultures* (New York, Columbia Univ. Press: 2016).

25 See for instance Arnheim, *Visual Thinking*, 153–88, on the concept of abstraction.

26 See Irv Biederman, "Recognition by Components: A Theory of Human Image Understanding," *Psychological Review* 94:2 (1987): 115–47. Biederman's pathbreaking study has not been surpassed or superseded.

27 Martha J. Farah, "Psychological evidence for a shared representational medium for mental images and percepts" *Journal of Experimental Psychology: General* 114 (1985) 91–103. Quote from page 96. See also Farah, *The Cognitive Neuroscience of Vision* (Malden, MA, Blackwell: 2000).

28 Martin Kemp, Ed. *Leonardo da Vinci* (New Haven and London, Yale Univ. Press, 1989): 185–209 on measuring.

29 In many respects Leonardo was the first "neuroscientist" to understand the visual system as a whole. In his writings on the "science" of painting, he explored many of the critical issues on light, the eye, illusions, perception of space, perspective, and geometry. See, for instance, Martin Kemp, Ed., *Leonardo on Painting* (New Haven, Yale Univ. Pres: 1987).

30 Anthony Panzera, *The Leonardo Series: Drawings by Anthony Panzera Based on Leonardo's da Vinci's Work on Human Proportion* (Albany, SUNY Press: 2015).

31 James J. Gibson recognized this in his research. See "The Ecological Approach to the Visual Perception of Pictures," *Leonardo*, Vol. 11, No. 3 (Summer, 1978), 227–35.

32 Pinker, 361.

33 Zeki, *Vision*: 209–25. Gregory, 153–58 on the case of "S.B" a man congenitally blind from infancy to early adulthood, after which surgery gave him sight. He died three years later of acute depression, having never assimilated the ability to understand the world through eyesight. Only be closing his eyes and touching objects could he "see" them.

34 Kandel, *Reductionism*, 55.

35 Harry Francis Mallgrave, *The Architect's Brain* (New York, Wiley-Blackwell: 2007).

36 The *Shorter Oxford English Dictionary* defines "concept" as mental representation, from the Old French *concevoir* and the Latin *conspiere* (to take in or understand). Conversely, concept is listed as a synonym for image, which comes from the verb, to imagine, or form a mental concept.

37 See Carrie Esopenko, *A Cognitive Neuroscience Examination of Embodied Cognition*, Ph.D. Dissertation, University of Saskatchewan, 2011, accessed, 2015.

38 A recent study introduces some of this material, but does not explain it adequately. See Sarah Williams Goldhagen, *Welcome to Your World: How the built environment shapes our lives* (New York, Harper Collins: 2017).

39 On this definition, see Samir Zeki, "Neural Concept Formation in Art: Dante, Michelangelo, Wagner," *Journal of Consciousness Studies* 9:3 (2002): 53–76.

40 C.f. Omer Akin and Eleanor Weinel, Eds., *Representation in Architecture* (Silver Spring, MD, Information Dynamics: 1982).

41 Mihaly Csikszentmihalyi, *Creativity: the psychology of discovery and invention* (New York: Harper Perennial: 1996) is one of the best books on the subject.

42 Akin & Weinel, op. cit., 28.

43 See Katerina Alexiou, T. Zamenopoulos and J. H. Johnson, "Exploring the neurological basis of design cognition using brain imaging: some preliminary results," *Design Studies* 30 (2009): 623–47. "On the whole, the findings suggest that there is a more extensive neural network involved in the activity of understanding and resolving design tasks than the network involved in problem-solving tasks." 641.

44 Semir Zeki, *Splendors and Miseries*: 44–45.

45 Kandel, *Reductionism in Art and Brain Science*, explores some of these questions. See 109 ff.

46 A fascinating and provocative view of art along these lines is in Alva Noe, *Strange Tools: Art and Human Nature* (New York, Hill and Wang: 2015).

47 Allan Greenberg, "Design Paradigms in the Eighteenth and Twentieth Centuries," *Via 3: Ornament* (University of Pennsylvania, 1977): 62-81.

48 Cognitive scientists refer to this as the somatic marker hypothesis. It was developed by Antonio Damasio. See *The Feeling of What Happens* (New York, Harcourt: 1999), his excellent book on this subject.

49 Frans De Waal, *Mama's Last Hug: Animal Emotions and What They Tell Us About Ourselves* (New York, W.W. Norton: 2019): 205, 206.

50 Alain de Botton, *The Architecture of Happiness* (London: 2006).

51 See Roger Scruton, *The Aesthetics of Architecture* (Princeton, Princeton Univ. Press: 2013).

52 See Myrna A. Fernandez, Jeffrey D. Wammes, Melissa E. Meade, "The Surprisingly Powerful Influence of Drawing on Memory" *Current Directions in Psychological Science* (Volume: 27 issue: 5): 302–08. Also Park, J., Jin, Y., Ahn, S., & Lee, S. "The Impact of Design Representation on Visual Perception: Comparing Eye-Tracking Data of Architectural Scenes Between Photography and Line Drawing." *Archives of Design Research*, 32 (1), (2019): 5-29.

53 The neuroscientist Gary Marcus, in his entertaining book on music, *Guitar Zero* (New York and London, Penguin: 2012) uses this statistic (page 12). It was established in the early 1990s by K. Anders Ericsson. See K.A. Ericsson, et al., "The role of deliberate practice in the acquisition of expert performance," *Psychological Review* 100 (3): 363–406.

54 See, for instance, Juhani Pallasmaa, "Body, Mind and Imagination: The Mental Essence of Architecture," in *Mind in Architecture*, 51–74.

55 James J. Gibson, *The Senses Considered as Perceptual Systems* (1966).

56 For a contemporary architect's view of the Gibsonian position, see Robinson, "Nested Bodies," in *Mind in Architecture*, 137–59.

57 John Dewey, *Art As Experience* (New York, G.P. Putnam: 1980 (1934)); William James, *Principles of Psychology* (Cambridge, Harvard Univ. Press: 1981).

58 See Jennifer Groh, *Making Space: how the brain knows where things are* (Cambridge, Harvard Univ. Press: 2014). Cited below.

59 See the work of Seymour Fisher, especially *Body and Experience in Fantasy and Behavior* (New York, Appleton-Century-Crofts: 1970), and Seymour Fisher and Sidney E. Cleveland, *Body Image and Personality*, 2nd Ed. (New York, Dover Publications: 1968).

60 Kevin Lynch, *The Image of the City* (Cambridge, MIT Press: 1960). His studies were based on work in Los Angeles, Jersey City, and Boston.

61 See, for instance, Emanuele Coluccia, Andrea Bosco, Maria A. Brandimonte, "The role of visuo-spatial working memory in map learning: new findings from a map drawing paradigm" (*Psychological Research* (2007) 71: 359–72.

62 John O'Keefe, "A Review of the Hyppocampal Place Cells," *Progress in Neurobiology* 13, no. 4, (2009): 419–39.

63 Edvard I. Moser, et al., "Place Cells, Grid Cells, and the Brain's Spatial Representation System," *Annual Review of Neuroscience* 31 (2008): 69–89.

64 Nicholas Carr, *The Glass Cage: How our computers are changing us* (New York, W.W. Norton: 2014): 125–52.

65 Lawrence Barsalou, "Grounded Cognition," *An. Rev. Psychol.* (2008) 59: 617–45.

66 For a measured review of he literature on embodied cognition versus the more accepted views, see Paul E. Tibbets, "Where Does Cognition Occur? In One's Head or in One's Embodied/Extended Environment," *Quarterly Review of Biology*, Vol. 89, No. 4 (December 2014), 359–68.

67 Ben D. Amsel, Thomas P. Urbach, Marta Kutas, "Empirically Grounding Grounded Cognition: The Case of Color," *NeuroImage* 99 (2014): 149–56, quote on 150.

68 See Kent Bloomer and Charles W. Moore, *Body, Memory and Architecture* (New Haven, Yale Univ. Press: 1977): 39–44. The authors describe all of the brain's basic systems of orientation: front-back, right-left, up-down, and here-in-the-center, as psychophysical. They posit a number of emotional adjuncts to body image and suggest some universal laws of design that should relate to perception. Their book, though still in print, has had little impact on architectural theory or education since its publication.

69 Ibid, 77.

70 Ibid. Robert J. Yudell, Chapter 7, "Body Movement," 57. Yudell notes that dancers use objects in imaginary space as a means to experience boundaries outside their own "centers." He goes on to give numerous examples of how architecture offers stimulus for movement by providing visual and haptic cues.

[3]

MIMESIS, MEMORY, AND ENACTMENT

"Monkey see, monkey do." This common adage is the basis for all sorts of negative stereotypes about mimicry, copying, imitation, and a general lack of creativity in contemporary society. It is also the basis for one of the most important discoveries in neuroscience: canonical and mirror neurons. Primates, including *Homo sapiens*, cannot make sense of themselves in their environments without them.

In the early 1990s a team of scientists at the University of Parma were metering neural responses in macaque monkeys as their brains processed various kinds of visual scenes and stimuli. They noticed some peculiar spikes in groups of neurons that were in the motor and pre-motor cortices of these primates. When a researcher grabbed a peanut in order to give it to a monkey, the identical types of neurons in their motor cortices spiked as when the same monkey grabbed the peanut himself. Apparently something in his brain was "mirroring" the actions of the "monkey" opposite him. It wasn't long before scientists performed similar experiments on humans and found these "mirror" neurons. In addition certain canonical neurons were primed for motor activity even when looking at an object that could be grasped or used. The arm and hand activity did not even have to be visualized. For a decade or so neuroscientists scoffed at the notion that seeing and doing were connected in the brain, but recently researchers proved that individual neurons are coded for this kind of response.[1]

More fascinating still, and awaiting additional research, was the connection between human emotions and visualizations of other living beings. Empathy, the basic human capacity for sensing the emotions of others (called intersubjectivity in neuroscience literature), was connected to mirroring in the same way that sensory-motor activity could be perceived in other primates. Vittorio Gallese of the University of Parma pioneered these areas of research and an entire branch of neuroscience followed his discoveries since the 1990s. He has also been a contributor to studies on embodied cognition, believing that this work substantiates this theory of mind and its philosophical ramifications. The most recent research in this area suggests that when humans perceive animals and other objects in the environment, neurons in many areas of the brain light up to create a mimetic schema of the things around them. In this way they can assimilate the environment into their bodies, comparing the objects they perceive with their own shape, motor actions, orientation, spatial awareness, etc.[2]

We can see why this would be the case when looking at primates' relationships with other living things in the prehistoric era. Early hominids and modern primates required mirror neurons not only to learn about their fellow beings, but also to explore the world around them. Merlin Donald and other evolutionary scientists have proved that the defining characteristics of modern humans emerged through mimetic activity in hominids long prior to *Homo erectus*. This includes aspects of language, society, and symbolic behavior.

Donald emphasizes the non-linguistic structure of mimesis, which is common not only to *Homo sapiens* but also to earlier hominids. "Mimetic skill or mimesis rests on the ability to produce conscious, self-initiated, representational acts that are intentional but not linguistic. These mimetic acts are defined primarily in terms of their representational function." He further distinguishes mimesis from mimicry or imitation of actions or behaviors. "Thus, mimesis is fundamentally different from imitation and mimicry in that it involves the *invention* of intentional representations." He cites many kinds of mimetic representations, including childhood games, charades, ritual dances, pantomime, and visual tableaux. On a cognitive level, such actions do not presuppose a linguistic capacity in the actor.[3]

The Italian concept of *disegno* presents a strong analogy to mimesis. If, as Semir Zeki has suggested, visual art is an extension of the brain's activity as it acquires knowledge of the environment, then *mimesis* is a part of human creativity. He believes that art itself, as a mimetic activity, is necessary to humans as a means of recording constant and universal aspects of experience. Likewise, Donald refers to mimesis as the "center of artistic activity." The difference between primitive mimetic culture and that of modern artists is one of degree rather than essence. We know that Aristotle, in his *Nichomachean Ethics*, distinguished between *pronesis*, or practical knowledge, and *theoria*, or "scientific" knowledge. The former is similar to mimetic forms of knowing the world.[4]

21. Lionesses, from Chauvet cave, France (Wikimedia Commons).

22. Rhinos, from Chauvet cave, France (Wikimedia Commons).

We now know, because of experiments like those above, that humans are not the only animals to possess a capacity for symbolic representation, or for basic communication using symbols. Mimicry, imitation of gestures and activities, and other forms of mimetic cognition are all around us in the biosphere. We cannot locate ourselves, our bodies and minds, our place in the world, without mimetic cognition. Nor have we ceased using this part of our brains as we have developed language, logical reasoning, and other "higher" forms of thought, such as the development of metaphorical concepts.[5] In this chapter we will explore the relationship between, mimesis, artistic creativity, and habitual thinking in architectural design.

Mimesis and Imitation

When the brain compares a new set of visual signals to those recorded in memory, such "matching" (Gombich's term) of like things builds a taxonomy of stored ideas that can be used in the conception of new things.[6] Moreover, we have seen that the brain of *Homo sapiens* developed a capacity for using external symbolic memories in understanding the world and performing higher order thinking. About 32,000 years ago, our ancestors produced very convincing paintings of animals on the walls of the Chauvet Caves in France, indicating a strong capacity for creative thinking.[7] [21] One of the key links between the external world and internal cognition is imitation.

23. Neuron webs, from EyeWire Project. Sebastian Seung, Massachusetts Institute of Technology.

John Onians suggests that when the painters at Chauvet began their work they were first inspired by bears that scratched paw marks on the walls of the caves. The earliest drawings appear to have been made over the marks made by bears. Another imitative impulse may have come from the particular environment around the caves. Nearby there is a distinctive stone arch that forms a natural bridge over the river. Upper Paleolithic humans in this area saw animals cross this bridge during seasonal migrations, and were undoubtedly struck by the wonder of thousands of beasts walking over the narrow span. An arch-like apse in the Chauvet Cave became the "setting" for a painting re-creating this migration.[8] [22]

When Aristotle wrote his famous definition, "tragedy is the representation [imitation] of an action," he was describing one of the fundamental activities of both art and the brain.[9] *Imitation* is how humans activate inherent and acquired brain schemata, whether by drawing them, expressing them in words, singing them, or otherwise demonstrating their content. It is becoming clear that even athletes benefit from a type of imaging or mimesis when they prepare for their performances. Various Olympic champions close their eyes and imagine their routines before actually performing them, employing a form of visualization to lock in their motor functions before a race, game, or performance. Mirror and canonical neurons facilitate this kind of pre-cognitive priming of the action centers of the brain.

Since the brain has a capacity for empathic signal reception, the repetition of artistic patterns that are admired or emotionally charged creates stronger neuron circuits than mere rote exercises. The more an artist identifies with a hero or master, and practices in his/her style, the better he or she will be at creating new art. This is not because she is a copycat or slave to fashion (though for a time that may be the case), but because good mental habits are strengthened by emotional connections with other human beings. Recent experiments have shown that emotions play a vital role in both memory and perceptual processing.[10] Every perceived object or setting has a corresponding affect recorded in memory, and each time a similar object is seen (smelled, felt, represented in a symbol such as a word, or heard) the feelings associated with it in memory recur. Likewise, when an artist draws a human form, or any object, memories and feelings rise to the surface of consciousness.

Repetition of both mental and motor exercises builds competency into expertise in many human endeavors. Muscle exercises and workout routines are based upon repetition of specific actions over long periods of time—days, weeks, and months. Likewise, musicians practice scales, fingering patterns, and muscle development exercises for years before attaining a level of mastery that allows them to perform in public. The brain, with its dense bundles of neurons, is not unlike a muscle in that strands become more densely developed over time, but only with repeated activation. When synapses fire, connections become stronger, and more adjacent neurons are activated. So repetition makes conscious brain patterns into unconscious ones—we form mental habits that make all sorts of repeatable activities easier, less mentally enervating.

Recent scientific attempts at brain "mapping" also emphasize the density of neuronal tissue in bundles or nests, using thin slices of tissue to construct 3-D models. Sebastian Seung's *Eye Wire* computer game enlists the aid of laypersons in producing a huge model of a mouse retina. [23] The results are being collected in his laboratory at MIT. The Human Connectome Project, funded by NIH, is building a larger model of a human brain using more conventional means.[11] What both show is that each memory or other stimulus builds connections between neurons—brainwork produces not more brain cells but more complex wiring between cells. Repeated activity in specific areas of the brain produce larger and more resilient networks, or circuits.

Some of this may seem counterintuitive. Because novelty and originality have come to be regarded as the ultimate tests of aesthetic worth in the modern era, imitation has lost the currency that it once had among artists, who at various times have relied upon it in all sorts of training. Emulation of masters has long been the core of any kind of apprenticeship, yet somehow "fine artists" came to be excluded from the canon of craft teaching during the Romantic Era. We will discuss why this was the case in Chapter 9.

Leonardo himself, as an artist imbued with embodied or experiential knowledge of the natural world, was indeed demonstrating the human capacity for invention using mimetic representation. By exploring natural phenomena through both drawing and writing, he reinforced networks in his own brain that unlocked secrets about physical sciences, biological development in fauna, and the lives of plants, among other areas of natural history. He even made startling discoveries about visual perception that anticipated the work of today's neuroscientists.[12] When doing creative work, such as making a painting, this great artist applied embodied knowledge to virtually every detail of a composition, from the haze in the middle distance to the flowing water in far off rivers. He did not "reproduce" scenes via mimicry or copying, and would not understand our contemporary dilemmas over realism or imitative forms of artistic "production."[13]

It is therefore a trap to assume the posture of a Hegelian dialectician when discussing the work of artists and architects who followed humanistic or classical methods of painting, sculpture, or poetry. Walter Benjamin, who wrote persuasively about the human capacity for mimetic "similarities" in both literature and the visual arts, was nevertheless constrained by his early twentieth-century intellectual milieu. Perched on the threshold of Modernism, he reveled in the potential of film and dynamic art movements that promised to supplant dying forms of imitative representation. His famous analysis of the mechanical replication of all forms of visual art, and the necessary alienation from religious, mythic, and monarchic cultures that he found amidst the ruins of Europe after the Great War, cannot be applied carte blanche to the current moment.[14] There is no dialectical struggle between producers and consumers of visual culture in the world today, though their relationship is increasingly complex and inscrutable. At the level of human cognition, there is no curtain barring any artist from accessing embodied forms of knowledge, culled through sensory experience, and applying this knowledge through an external symbolic field to create an original work, whatever may be its similarity to previous examples of its genre or type.

Concept Formation vs. Problem Definition

When architects indulge in creative thinking they trigger action-perception loops that are based on habits learned during their school years and apprenticeships. As we saw in Chapter 2, all drawings and other external mnemonic forms stimulate the brain during these sequences. The three step processes are very like the fundamental signal loops that accompany visual perception on a neuronal level (input-processing-output). They are also analogous to computer processes and information theory. The industrial engineer Morris Asimow was among the first to identify such loops during the 1950s.[15] I will argue that concept formation differs from problem solving in much the way sketching differs from working out the technical and programmatic aspects of an architectural design.[16]

It is common to think of problem definition and computation as aspects of "design" that computer scientists and other engineers use in finding "solutions" to all sorts of practical problems, from packaging, to propulsion, to analysis of economic activity among nations. In current parlance, these creative problem solvers are looking for "algorithms" that will perform the functions needed to solve a given problem. It is even fashionable to speak of algorithms in evolutionary theory, as if organisms are computing the probability that one phenotype will prevail over another in a particular environment.[17] We will discuss why this is the case in Chapter 9.

There are certainly analogies in this way of thinking to the processes followed by architects and craftspeople in their work, but artists are not, by and large, looking for such solutions in their creative endeavors.[18] They are looking, broadly, for ideas that can be defined in the cognitive space of short-term memory and refined via representational modes such as those discussed in Chapter 4. We noted in Chapter 2 that the brain uses images or concepts in all kinds of cognitive and sensory motor operations. At the procedural level, there are cognitive action-perception cycles that direct the organism in the desired activity. At the perceptual level there are places in memory and in the sensory cortices where images can be stored and manipulated.

Conception differs from abstract cognition in that sensory information becomes an intermediary in the loop between eye, brain, and object.[19] The imagined space of the concept, which can be expressed in some external representation, is captured in memory but not fully formed in the mind until realized in its medium (sound, light, paint, physical materials, etc.). An artistic concept thus differs from a purely symbolic one in that a physical form is the desired outcome, but that form cannot be apprehended all at once in the mind's eye.[20] More importantly, when bringing images into the conscious mind, the brain cannot hold complex ideas in working memory for the duration of design sessions, and must rely on drawings or other external memory aids to retain critical information. Drawing is in this way a medium of thought. Drawings hold vital information while the mind contemplates how to mold that information into a useful concept.[21]

Buildings and other occupied spaces are particularly difficult to conceive without sophisticated understanding, through training and experience, of multiple means of representation and physical construction. Cognitive models related to problem solving (algorithmic and logical) do not fully account for the internal and external embodiment of knowledge related to the object that is to be designed and constructed in the physical environment. Precedents must be part of the discussion in the designer's mind.[22] Knowledge of the site, place, and environment is critical. And social/psychological factors are essential in evaluating how successful a building will be when occupied by a particular user group. It is therefore critical for any theory of cognition that pertains to the building arts to recognize the nature of visual ideation, and its fundamental difference from linguistic symbolization. I will argue that since building design is a visual discipline, its fundamental cognitive problem is one of ideation rather than solution generation. There is a teleological distinction between the two. There is also a distinction in the type of cognitive activity used in making conceptual judgments: analogical and abductive reasoning are both engaged, whereas pure inductive or deductive reasoning may not be part of the process.

Designers depend upon analogies with existing artifacts in their work, for in the marketplace users are conditioned to the forms of existing buildings, tools, machines, and designed environments of all kinds. Drawing information from these artifacts creates a comparison by analogy in the mind of a designer.[23] In addition, architects may use what Charles Sanders Pierce called "abductive reasoning" when working on a design, for "abduction suggests that something *may* be" in an Aristotelian ontology.[24] Such thinking depends upon references to places, experiences, and activities spent with useful artifacts in everyday life. These things often constitute a pattern of thinking, or of inference, in the minds of trained artists or artisans.

Memory: Internal and External

Hominids developed larger brains with greater internal memory capacity about 150,000 years ago, and began creating pictorial images about 50,000 years ago. With this development came an ability to employ short and long-term memories in the conceptualization of language, culture, social relations, art, and myths.[25] More importantly, the brain of *Homo sapiens* was primed to use external representations of the world in solving problems, creating objects for use, and understanding the environment. Some of these representations were primarily narrative or linguistic. However, even before written language emerged, humans had the capacity to stimulate creativity with pictures.

Pictorial representation is an external device for recording not only the visual field apprehended by the eye and brain, but also abstract concepts about the world, especially spatial concepts.[26] The psychologist Merlin Donald suggests that the limited capacity of biological memory, used for storing language symbols, sensory motor functions, cognitive functions, and the like, did not allow early humans to think theoretically or conceptually. The development of many kinds of external symbolic forms such as writing and drawing opened up the capacity for conceptual thinking at all levels. The brain thus had an adjunct to its internal neuronal storage.

Donald's analysis of the emergence of the modern mind constructs a persuasive argument that virtually all pedagogic, artisanal, theoretical, and even narrative systems are bound together by what he calls the EXMF or External Memory Field, a part of the brain that processes loops between biological memory and external memory. Though speculative, this theory tracks beautifully with the observations of many scholars about design schools and methods, from the attributes of Greek sculpture to the functional designs of

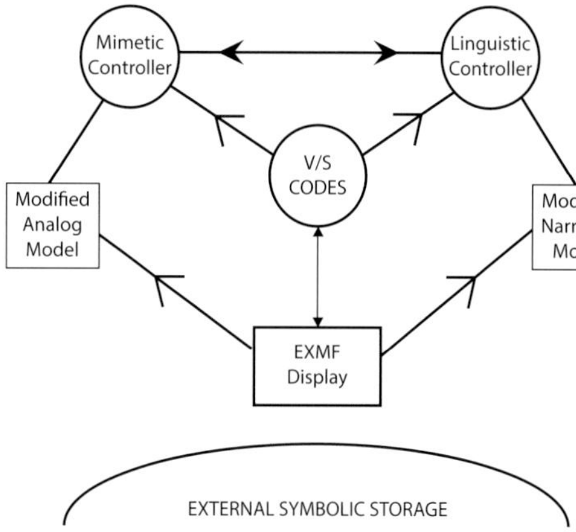

24. Merlin Donald's diagram of the EXMF or external memory display. (Author)

the Bauhaus. "The EXMF is essentially an external loop for organized thought process, and there is no substitute for it in biological memory," Donald argues.[27] He suggests that the brain uses exograms for external memory storage, which may be things like models, drawings, or even entire buildings. Trevor Watkins has suggested that Neolithic humans found that their dwelling and settlements were potent vehicles for keeping cultural memories alive.[28] The buildings were primitive exograms. One emerging theory that points to the potential "loop" that occurs when architects make a diagrammatic drawing and stare at it, hoping to find a further idea or wrinkle in its vague lines, suggests that the brain anticipates the formation of a new image or idea in the sketch at hand, which is an "affordance" for new thoughts. Daniel Dennett has proposed a theory that uses AI models to account for the use of working memory in such cognitive loops. In such sequences the drawing/image is more than a physical or symbolic representation; it is actually attached to thoughts occurring during conception.[29] The cognitive image and the drawn exogram are melded as a concept takes shape, but always as what Dennett calls a "rough draft." For this reason, it is critical to understand how, and in what sequence, an architect makes his or her initial drawings. [24]

Scientists studying the relationship between cognition and memory (long term, working, procedural) have recently divided their experiments into verbal, spatial, and visual modalities. Though the basic operations are similar, it has been convenient to look first at verbal/linguistic/semantic problem solving. For instance, when parsing the digits in a phone number, how does the brain hold portions of the number in working memory while assembling the entire figure?[30] Unfortunately, multi-sensory cognition is not precisely analogous to symbolic cognition, so some work remains to be done to sort out the kinds of memory required for each. Spatial memory, it seems, can sometimes bridge the gap between the two in problem solving.[31]

A particularly significant avenue of research concerns the relationship between memories that have occurred in sequence or in close proximity to one another in time. The brain seems to store these memories in similar areas, making it more likely that when one is recalled, a "nearby" memory may also come up. The chemical used to direct the brain to allocate space in a certain area is called CREB. Experiments have shown that this chemical has a powerful effect on monkeys and mice, directing them to associate memories with each other and even improving the memories of aging animals. Scientists have postulated an "allocate-to-link" hypothesis that may explain why memories formed within hours or perhaps a day of each other tend to be associated, though their content may differ significantly. As we shall see in Chapter 9, it appears that neurons in close proximity to each other may even contribute to the creation of "original" concepts and thoughts.[32]

Idea Sketches and Mental Images

Neuroscientists have studied the ways in which the brain constructs images, observing that there is a kind of "equivalency" between perceived visual stimuli from the visual cortex and analogous visual concepts.[33] It is therefore essential to compare perceptual processes with representational processes when studying how architects design. Moreover, generating visual concepts engages the visual cortex in concert with other areas of the brain, but may not necessarily engage the language and number processing areas of the left hemisphere. Images captured in long-term memory are commonly seen as mental "pictures" that help to generate artistic compositions. As we saw in the previous chapter, embodiment redefines these pictures as sensory-motor, cognitive images that do not exist merely in visual cortex, but in proprioceptive areas generally.

In studying design of various genres (mechanical, architectural, industrial, environmental), cognitive science has until recently focused almost entirely on what is called the Computational Theory of Mind—a theory entirely dependent on language as a symbol system and symbolic logic as an operator. Vinod Goel, at York University, has examined art and design cognition using experiments that track closely with other scientists, but attempted to contest the so-called CTM view. He argues persuasively that designers, confronting physical problems, must employ a wide range of symbolic representations in order to solve cognitive puzzles such as the design of buildings. Moreover, "different symbol systems correlate with different cognitive phases which in turn are associated with different cognitive processes." If an architect is exploring alternative plan diagrams, he or she will employ plan sketches and draw on examples from memory of previous plans. It is unlikely that the designer will engage left cortical areas concerned with language processing while doing this. He or she won't need logical operators or subject-predicate pairs in this type of thinking.[34]

Goel argues that "sketches" are perfect for the kind of ill-defined, indeterminate cognition that architects use when conceptualizing their buildings. The so-called "problem space" in these instances is open and requires ambiguous or multivalent images. Later, during design development and construction drawing phases, architects are drawn to more conventional linguistic modes of thought, because the information that must be assimilated into designs is often coded in written language or mathematics.

Architects, painters, composers, and poets "conceive" their works as holistically as possible, despite the potential complexity of seizing the concept for a complete work in one cognitive flash. This is one reason why most cognitive science theories do not adequately encompass artistic processes in their diagrams. Many experiments with architects and designers use common forms of logical and linguistic charts, and are often quite persuasive in their attempts to capture the design process. But, like most systems of this type, they err in one crucial respect: the loops used in conceptualization do not readily conform to logical operations. One emerging theory is that visual thinking may be broken into temporal "segments" that are readily definable in all design tasks. These segments can be "forward oriented" or "reflective—backward oriented." Even this "Creative Segment" theory does not adequately consider the role of habits and internalized modes of thought in most design problems.[35]

As we shall learn in Chapter 4, humans have used many forms of representation as aids in designing and building their houses and cities. The idea sketch is a relatively recent drawing type that emerged after 1600, when Renaissance painters-architects translated their studio practices into architectural design methodologies.[36] However, sketches provide the most direct evidence of the kinds of mental images that psychologists believe are generated when the brain forms visual concepts of any kind. The habitual sequence of generating sketches, assessing them cognitively, and proceeding to more detailed schematic images on paper generally emerges as architects reach a level of mastery or competence after their schooling and apprenticeships are complete. These routines are largely subconscious once an architect enters professional practice, just as they are following an apprenticeship in, say, violin making, woodworking, or timber framing.[37]

Habits and the Habitus

Habits are the means by which humans and other organisms connect memories with actions in the pursuit of homeostasis. Without deeply embedded habitual memories we could not find our way home from a walk, tie our shoes, or perform in a sport or game. We develop habits by bringing memories from the cortex into other areas of the brain in order to store the motor actions that are required to perform whatever activity is required for everyday tasks. Working memory allows only a few iterations of a task, whereas memories in the hippocampus are more or less permanently available. Recent research in both memory and perception has begun to pinpoint areas of the brain in which memories and actions are integrated. As one might guess, it turns out that memories related to particular perceptual and motor activities are often located in nearby areas of the cortex—i.e., running memories near motor neurons controlling leg muscles.

Most architects understand that they work according to habits acquired during their student years, often under powerful mentors. This master-pupil relationship is similar to that of many artists, particularly to painters with traditional academic training, such as Chinese classical, Renaissance, or French academic masters. Mentorship is essential to the development of any successful architect, which is why internships are required for licensure in most countries.[38]

Pierre Bourdieu has developed the concept of "habitus" to explain the social and cultural dimensions of professional relationships. He locates the practicing professional in his social milieu by analyzing the ways in which he acquires or loses credibility and status *vis à vis* both his peers and his patrons, a quality he calls "distinction." By using the term *habitus* he also indicates the degree to which the day to day experience of professional practice is colored by routines, repeated encounters, and codified relations between persons of lower or higher social rank.[39] Evolutionary psychology has proved that humans, like their nearest relatives, are compelled to seek higher levels of status in order to ensure mating with equal or higher level individuals within a social group or clan.

The habitus directs individuals in a field through their unconscious socializing behaviors. In a professional field, one "plays" according to one's habitus and the "capital" one accrues by reputation and exposure. In contemporary architecture, pursuit of fame or reputational capital supersedes normative practices in the habitus, leading to the unhealthy imbalance all architects experience in their daily lives.[40] Bourdieu's theory of competitive social practice contrasts with cooperative models that are closer to brain-based research than to post-structuralist, Marxist positions.

Acquiring good design habits does not come naturally to any architect, for the complexity of most design problems will test a brain that is not prepared to think in three dimensions. When architects were also builders they could reproduce common building types without sophisticated representational techniques. In terms of the embodied, active and empathetic qualities of our brains, such direct translation of the experience of building to the design of buildings was advantageous. But once they began to rely on drawings as their most important design tool, routines for teaching drawing and design emerged, as we shall see in Chapter 5. In these routines are the keys to conceptual processes, some followed by all architects, others by groups, and some by individuals. Architects often lock onto a particular drawing or modeling type in order to further their investigation into new forms, as we shall see in the following chapter.

In the 1990s scientists began to pinpoint areas of the brain that seemed to contain "programs" that directed humans to repeat learned routines, or habits, that were often subconscious and in many ways bypassed cortical or rational thinking.[41] Habits are necessary for many human activities, from brushing one's teeth to skiing down a slalom course without hitting gates. There are so many that it is often difficult to distinguish habitual behavior from daily living; yet, a distinction has now been made that may revolutionize behavioral science. Many habitual behaviors are buried so deep in the unconscious—in the amygdala and hippocampus—that they are immune to conscious interference or control. David Hume, William James, and Edmund Husserl all held that habits were fundamental to our experience of the world, so it is not surprising that cognitive research has proven their hypotheses.[42]

Habits often begin as rational thought sequences that require hard work in the medial frontal lobes of the cortex. Repetition, which we saw as necessary for the building of neuronal connections, is also critical to anyone who wishes to form a new habit. Without it, the brain spends too much energy re-learning patterns that the organism must execute on a regular basis, using as little brain energy as possible. Since the cortex needs to conserve energy for higher cognition, it shifts the mental activity acquired in perfecting say, the way to ski slalom gates, to another part of the brain called the *basal ganglia*. This small area is actually a portion of the "old brain," and one that was just as necessary to our primate ancestors as to us.

Experiments show that once laboratory animals learn a habitual routine like running a maze to find chocolate, their brains settle down and run on autopilot. As might be expected, their first trips through the maze are marked by false starts, frustration, and lots of mental effort. Later, they rely on their *basal ganglia* to direct their routes. Moreover, it appears that humans, even ones with brain lesions in key cortical areas of short-term memory, can learn new habits like combing their hair, or brushing their teeth.[43] The famous brain research patient H. M. (Henry Moliason) was able to remember paths between his home and the grocery store without being able to describe the route in words. The old brain has control of habitual behavior that can mean the difference between life and death.[44]

Since habits are learned from normal daily routines, it would make sense that factory workers, seamstresses, chefs, and artists share similar experiences not only in learning their craft, but also in the way they use habits in their working lives. Like most young architects in the drafting room, I was quite inept at proper lettering when I began working for a living. Though I had learned to draft with a pencil during my studies at Penn in the 1970s, and thus outstripped most of my apprentice mates in that skill (few programs still taught pencil drawing then, relying instead on leaky ink rapidographs), I was no better than they at quickly putting notes on working drawings. I learned fast enough by developing a technique, or habit, that is with me to this day. Every architect develops a storehouse of such habits and uses them in virtually every design problem.

These ingrained ways of pursuing an art, craft, or profession are well-known to historians and sociologists. However, the mental activity that drives humans to repeat both healthy and unhealthy habits have not been well understood until very recently. Because of this, it has been difficult to distinguish predilections that are idiosyncratic from those that prevail in social groups, cultures, or larger societies. Moreover, many researchers on design thinking have ascribed too much power to the cortex, or reasoning, area of the brain, ignoring the more intuitive and sense-based areas such as the hippocampus, amygdala, and basal ganglia.[45] We shall see in Chapter 8 that habitual, rule-based practices in design began to lose favor as "artists" rose in stature during the eighteenth and nineteenth centuries, leading to some unfortunate consequences for the modern architect.

Visual Culture and the Period Eye

Among early psychoanalytic studies, Carl Jung established theories of the relationship between the individual artist and the "collective unconscious" that are still employed in the social sciences. Anthropologists have explored cultural archetypes in myths and among primitive cultures. Jung's typology errs in dividing the "rational" and "irrational" but it does account for the relationship between individual and group consciousness that is critical to conceptual modalities. As he wrote about literature as an art:

> Art is a kind of innate drive that seizes a human being and makes him its instrument. The artist is not a person endowed with free will who seeks his own ends, but one who allows art to realize its purposes through him. As a human being he may have moods and a will and personal aims, but as an artist he is "man" in a higher sense—he is "collective man," a vehicle and moulder of the unconscious psychic life of mankind.[46]

Styles, folkways, and vernacular patterns in material culture are well-established modalities that tie groups of builders, craftsmen, and artists together in temporal, ethnographic, and physical contexts. During the last century an explosion of new scholarship focused attention on such social and cultural modes of production, freeing historians from the limitations of *kunstwissenschaft* (high art studies), dominant since the first years of the twentieth century.

Hegelian historicism has given way to wider lenses under which to examine all kinds of design, and the concept of art itself has broadened to include more vernacular production.[47]

Neuroscience has just begun to examine these vernacular practices as examples of the human need for art as a means of understanding the world. With respect to architectural design, the field of inquiry is vast, but much work has already been done to identify key historical milestones and shifts in modalities, such as the Renaissance pictorial revolution. Focusing more directly on modes of conception that emerged among groups of architects will be one essential area of inquiry.[48] During the mid-twentieth century, Ernst Gombrich and his students paved the way by linking the psychology of vision to leading artists, comparing their working habits throughout the centuries. Recently, John Onians, a Gombrich student, has written extensively on the neuroscience of art.[49]

Because of his serious attention to perception and memory, the late English art historian Michael Baxandall was a pathbreaker in this area of study. Like Gombrich, Baxandall was drawn to schools of artists that demonstrated consistent methods of making their art, such as the limewood sculptors of the German Baroque period. His relative youth allowed him to make use of mid-century psychological studies, and he seemed unconstrained by the conventional purview of his peers. He was a pioneer in establishing modes of conception among artists within regional cultures at circumscribed periods—a phenomenon he called the "period eye."

Baxandall's contribution to the understanding of how groups of artists perceive, conceive, and create art was revolutionary. He was the first scholar to study the role of shared experiences, memories, and practices in art. Among 15th-century Italian painters, he found that not only artists but also their patrons were obsessed with certain mathematical and formal schemata that could be shared when viewing a picture. Gauging, or calculating shape and volume, was one such skill among most educated Florentines; another was an awareness of proportional relationships. Likewise, painters might be judged according to commonly held standards and criteria, many borrowed from classical rhetoric and Christian theology. Even the language used to discuss the value of art was highly specific to the minds of Renaissance Italians from particular city-states or regions. Together, this web of common experience, learning, cultural values, and taste make it possible for us to identify a modality that resided in the brains of these famous Italians.[50]

He distinguished the apparatus of perception—the eye—from individual experience by sorting linguistic terms and modes of representation such as perspective and geometry, social rank and position, artisanship, and even contractual expectations. Thus, a fifteenth-century Italian might view a modern painting with little understanding of its nuances, whereas a work of his own time would immediately strike a chord. “That [mode of understanding art] which we tend toward will depend upon … the interpreting skills one happens to possess, the categories, the model patterns and habits of inference and analogy: in short, what we may call one’s *cognitive style*.” What he describes here is a mode of conception common to both artist and patron during a fixed period, such as that of his subjects, the Italians of the Quattrocento and Cinquecento.[51] He proved, with close analysis of human relationships, not just aesthetics, that shared experience resulted in distinctive mental patterns that allowed artists, patrons, and the viewers at large to engage in the enjoyment of the products of a new culture—humanism. Building upon a dense scholarly edifice constructed by giants such as Burkhardt, Panofsky, Wofflin, and Wittkower, he moved beyond the literary and aesthetic tradition of humanistic research to one based upon the recorded activities of his protagonists, not on their assumed intellectual influences. To the vast edifice of books on architectural theory and written treatises, he added a powerful tonic that may prove crucial to unlocking secrets about the brains of architects.

The more common approaches to establishing cultural biases among artisans, especially vernacular makers, emerged in material culture, anthropology, and sociology during the last half-century. Acceptance of such research among art historians has been slow and reluctant. Henry Glassie wrote convincingly about distinctive formal patterns in the plans of vernacular houses in Middle Virginia during the seventeenth through nineteenth centuries.[52] He was influential in establishing a school of vernacular historians who did pioneering work on log cabins, folk houses, barns, slave quarters, and other elements of the cultural landscape of North America. Jules Prown brought some of their techniques to the study of fine art such as painting.[53] Herbert Gans observed some similar cultural patterns when researching the residents of twentieth century suburbs such as Levittown, Pennsylvania. He subsequently referred to these biases as “taste cultures” residing among social groups, classes, and castes in contemporary American society.[54] All of these scholars influenced architectural historians and theorists to a greater or lesser degree.

Enacted Cognition

Habits of mind, as we noted above, emerge as a result of months and years of practice, often under mentorship or in collaboration with others. Since perception is itself enacted, and conceptual thinking engages action-perception loops, scientists are beginning to look at a mind-body-environment concept of human cognition. Two schools of thought have engaged the attention of both philosophers and neuroscientists: the *extended mind* hypothesis, and the *enacted mind* hypothesis.[55] Though both are persuasive in some measure, the latter offers the most cogent explanation for how designers think. This is because the process of conceiving buildings must necessarily involve a kind of acting out of the building task, whether it be in the use of tools in fabrication, the use of drawings in visualization, or the use of a model in mocking up the thing one wants to create. In the next chapter we will discuss three ways in which this can happen, in the context of human evolution and history.

It is essential to recognize that much of the brain's activity during design sessions is unconscious or subconscious; patterns of thinking become ingrained once an architect completes an educational or apprenticeship program. Many books on "design thinking" elaborate on the conscious portions of studio work to the exclusion of habitual modes of thought. It isn't hard to see why—interviewers and observers can easily record their subjects' conscious thoughts and actions. Social scientists use what is called "protocol analysis" to assess the mental activities followed by subjects in design sessions. Either the subject explains his/her thinking to a lab assistant while drawing, or is asked to relate thoughts following the session while viewing a videotape.[56] Since as much as 87% of cognitive activity is subconscious, such experiments are not likely to expose the most fundamental mechanisms of design thinking. "Mindwandering" is a persistent activity in conscious periods of brain activity; laypeople call it daydreaming.

In previous studies I have advanced the idea that architects use "modes of conception," habitual routines in cognition, to imagine their works prior to building them.[57] These modes are ingrained, or subconscious. They are also shared with like-minded colleagues, or with members of a cultural group. Richard Dawkins and Daniel Dennett have argued that one explanation for the persistence and cultural coherence of unselfconscious design traditions is implicit in Darwin's theory of natural selection. In a recent book Dennett develops the hypothesis that memes are brain-based cultural schemata that developed because humans required mental tools for manipulating objects in their particular environments. This notion is persuasive because it explains the brain's capacity for holding cultural practices that can be transmitted to other brains, and that persist over generations in all traditional societies. Again, Jung's notion of the collective unconscious seems to bear scrutiny from empirical science.[58]

In other fields the work of psychologists and cognitive scientists has produced discoveries that subsume or even contradict well-established theories and beliefs. Some of the most provocative have come in economics. Nobel laureates Daniel Kahnemann and Amos Tversky became world renowned when they overturned some of the most tenaciously held beliefs about rational economic behavior, utility, and the role of intuition in decision making. By recognizing the role of unconscious, evolved cognitive processes in all humans, and contrasting these with conscious "reasoning," they were able to rewrite entire theories of judgment, decision making, and even statistical analysis that influences not only daily routines but our political and social system as a whole.[59] Little research of this kind has been done on designers.

If individual and group-based modes of conception are evident in the historical record, and new means of experimental testing can establish their presence in the brains of contemporary designers, what would these thought patterns look like? The diagram below gives a basic picture of the loop-like sequence of thoughts, using both internal working memory and exograms (external memory prompts). A designer begins with an inchoate "concept" that is largely subconscious. The concept may or may not constitute a pictorial or visual notion. Only when represented in a drawing or series of drawings, perhaps connected with words, does the concept emerge as something concrete. More important, these cognitive patters are not linear, nor are they logically derived—a loop can proceed in a forward or backward direction, winding and unwinding.

Each iteration of the action-perception loop advances the design, though the process may not be linear. Some experiments have tracked the forward, backward, and lateral progress of drawing sequences in contemporary designers and students, but patterns used today may not reflect those used historically by the most successful architects and designers. The most recent memory research indicates that working memory holds images for sufficient periods of interaction with external "codes" linked to language, vision, and hearing. More specifically, there are neural pathways that divide the processing of spatial from object information. There is always a feedback loop at the end of the processing sequence.[60] In embodiment, the external environment is at the far side of this loop. The exogram allows designers to link the particular object to its environment in a conceptual way.

As the philosopher Alva Noe observes, "Neural activity during perception is not a one way thing. Neural activity is characterized by loops and two-directionality. There are neural pathways heading back into the brain from the senses, but there are even more neural pathways heading back out again. This should not be surprising. Consider how much easier it is to hear a sound that you are expecting than one that you are not expecting."[61] He notes that cues in our environment constantly interact with our sensory-motor systems and brain, providing information about how to understand the world. In design, drawings and other forms or representation trigger memories and responses to visual problems—they are part of our embodied environment. When drawing, the brain brings experiential aspects of an environment into working memory in order to construct the image on the page.[62] By seeing in the mind's eye, the object is connected to its place, time, and circumstance.

My diagram of the *mode of conception* used by most architects [24] indicates a sequence of greater and greater definition, achieved by the re-representation of image/ideas over an extended period. This period need not be temporally continuous (as in a single session). However, the mode will follow a similar habituated series of inward/outward loops until a satisfactory schematic design is reached. There is evidence for this methodology in the work of architects throughout the modern period, from Leonardo da Vinci to the present day. Moreover, the evidence is compelling that the forms of representation used influence the inner process of conceptualization, and this is consistent with new theories of grounded cognition. [25]

As we shall see in the following chapters, evidence exists that "schools" of architects acquired and transmitted memes among their cohorts, and that some of these habits of mind even persisted over generations. By examining design practices, representational modes, training, and written theories, architectural historians have already documented the existence of distinct modes of thinking among masters, pupils, and disciples during many periods. By using the enacted mind hypothesis and embodiment, the case studies of key transitions in architecture in this book uncover significant mental schemata. Such unconscious patterns of thought and preferences for certain representational types occur throughout the "modern" period, and may even exist prior to the emergence of professional architects after 1600. The only way to corroborate brain-based theories of cognition with artistic practice is to look critically at history. So, we will begin by surveying the representational types used by our ancestors and then proceed to examine the history of representation during the Renaissance, Baroque, Neoclassical, and Romantic eras, and proceeding up to the present day.

25. Diagram of a mode of conception used by many contemporary architects. Exograms proceed from sketches to diagrams to plans, elevations, models, and finally to the presentation stage (Author).

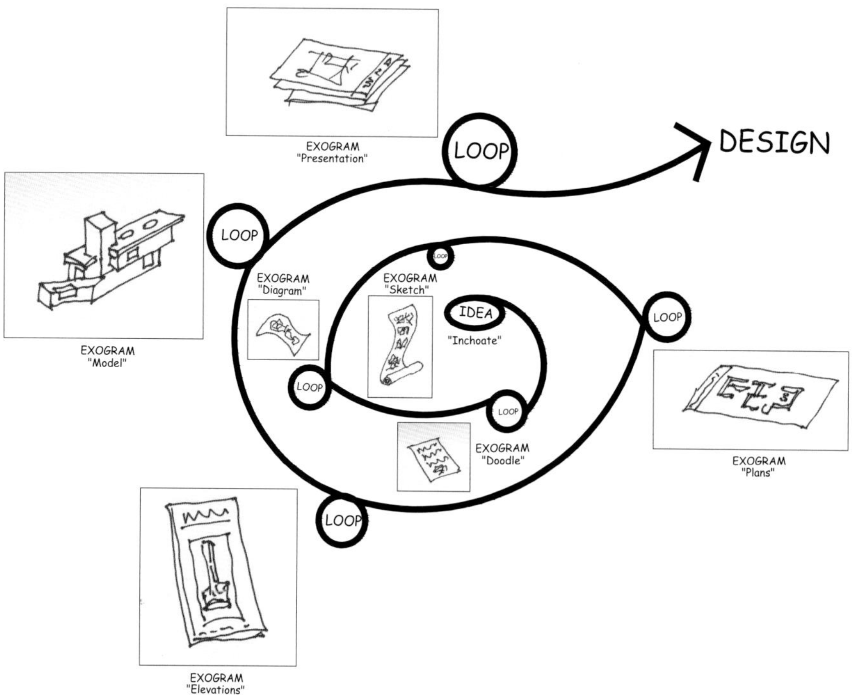

Notes

1 The critical papers include: Rizzolatti, G. & Craighero, L. The mirror neuron system. *Annual Rev. Neuroscience*. 27, 169–92. 2004; Rizzolatti, G., Fadiga, L., Gallese, V. & Fogassi, L. Premotor cortex and the recognition of motor actions. *Cogn. Brain Res.* 3, 131–41. 1996; Rizzolatti, G., Fadiga, L., Fogassi, L. & Gallese, V. The space around us. *Science* 277, 190–191. 1997; Rizzolatti, G., Fogassi, L. & Gallese, V. Neurophysiological mechanisms underlying the understanding and imitation of action. *Nat. Neurosci. Rev.* 2, 661–70. 2001

2 The pertinent research is in: Gallese, V. The manifold nature of interpersonal relations: the quest for a common mechanism. *Phil. Trans. R. Soc.* B 358, 517–28. 2003; Gallese, V. The roots of empathy: the shared manifold hypothesis and the neural basis of intersubjectivity. *Psychopathology* 36, 171–80. 2003; Gallese, V. A neuroscientific grasp of concepts: from control to representation. *Phil. Trans. R. Soc. B* 358, 1231–1240. 2003; Gallese, V. Embodied simulation: from neurons to phenomenal experience. *Phenomenol. Cogn. Sci.* 4, 23–48. 2005 Gallese, Vittorio, "Being like me": self-other identity, mirror neurons and empathy; in S. Hurley & N. Chater, Eds. *Perspectives on imitation: from cognitive neuroscience to social science*, vol. 1 (Cambridge, MA: MIT Press. 2005): 101–18.

3 Merlin Donald, *Origins of the Modern Mind*, 168–69.

4 Aristotle, *Nichomachean Ethics*, 1140b, 25–30, *Selections*, 402–03. See also David Roochnik, "What is Theoria: Nicomachean Ethics 10.7-8," *Classical Philology*, Vol. 104, No. 1 (January 2009), 69–82.

5 Arbib, discusses the progression from gesture, to protolanguage, to spoken language in *How The Brain Got Language*. His book is not the only study to advance an embodied theory of language development, but I find it persuasive.

6 Gombrich, *Art & Illusion*, 66–74 on making before matching.

7 The standard account of its discovery is Jean-Marie Chauvet et al., *Chauvet Cave: the discovery of the world's oldest paintings* (London, Thames & Hudson: 2001).

8 John Onians, *European Art: A Neuroarthistory* (New Haven, Yale University Press: 2016).

9 *The Poetics of Aristotle*, Ed. Stephen Halliwell (Chapel Hill, Univ. North Carolina Press: 1987): 37, or Chapter 5. "Tragedy, then, is a representation of an action which is serious, complete, and of a certain magnitude ..."

10 L.F. Barrett and Moshe Bar, "See It with Feeling: Affective Predictions during Object Perception," *Philosophical Transactions: Biological Sciences*, Vol. 364, No. 1521, Predictions in the Brain: Using Our Past to Prepare for the Future (May 12, 2009): 1,325–334.

11 See the official website at http://www.humanconnectomeproject.org/

12 For a perceptive overview, see James S. Ackerman, "Leonardo's Eye," in *Distance Points: Essays in Theory and Renaissance Art and Architecture* (Cambridge, MIT Press: 1991): 97–150.

13 According to historian J.R. Hale, "Leonardo's scientific desire to understand was rooted in his artistic impulse to copy." in J.R. Hale, *The Renaissance in Europe* (London, Folio Society: 2001): 291.

14 Walter Benjamin, "On The Mimetic Faculty," in Peter Demetz, Ed., *Reflections* (New York, Schocken Books: 2007): 333–36.

15 Peter Rowe has characterized Asimow's model as "deterministic" and tied to behaviorism, though the latter's observations now seem prescient. See Rowe, *Design Thinking*: 47. For the original concepts, see Morris Asimow, *Introduction to Design* (Englewood Cliffs, NJ, Prentice Hall: 1962).

16 For an excellent discussion of the role of sketching see Kendra Schank Smith, *Architects' Sketches: dialogue and design* (Oxford, Architectural Press: 2008).

17 See Vinod Goel and Peter Pirolli, "The Structure of Design Problem Spaces," *Cognitive Science* 16 (1992): 395–429.

18 On these distinctions, see Nigel Cross, *Design Thinking: Understanding How Designers Think and Work* (London, Bloomsbury Academic: 2011): 19–30.

19 For studies of sequences and interactions between drawings and the brain, see Rivka Oxman, "Design by re-representation: a model of visual reasoning in design," *Design Studies* 18 (1997): 329–47. The author also cites earlier srudies that postulate a "conversation" between internal and external representations. Bryan Lawson, *How Designers Think* (London, Architectural Press: 1980); and Omer Akin, *A Psychology of Architectural Design* (London, Pioneer: 1986).

20 This has been tested experimentally in psychological studies. See Sarah M. Shuwairi, Rebecca Bainbridge, and Gregory L. Murphy, "Concept formation and categorization of complex, asymmetric and impossible figures," *Attention Perception Psychophysics* 76 (2014): 1,789–1,802. The study also indicated that only through artistic experience could subjects identify "impossible" or illusionistic figures, indicating how important it is for designers to create drawings as part of the learning process.

21 See Herbert, *Architectural Study Drawings*, for an extended analysis of conceptual design loops, Chapter 5, 75–85.

22 The most persuasive case for this interactive thought process is in Donald Schon, *The Reflective Practitioner: How Professionals Think in Action* (New York, Basic Books: 1984).

23 One fascinating study of two types of analogy is Willemien Visser, "Two functions of analogical reasoning in design: a cognitive-psychology approach," *Design Studies* 17 (1996): 417–34. Among the observations in it are that retrieval of analogous models from memory is not "cued" but is subconscious; and that in Case-Based-Reasoning the brain depends not only on analogy for generating potential solutions to sub-tasks, but also in "cognitive economy"—husbanding brain capacity.

24 Charles Sanders Pierce, *Collected Papers of Charles Sanders Pierce, Vols. 1–6*, 1,931–935 (Cambridge, Harvard Univ. Press: 1958).

25 Colin Renfrew notes that the "speciation phase" of evolution occurred between 150,000 and 100,000 years ago. See Renfrew, *Prehistory: The Making of the Human Mind* (London: Folio Society: 2013): 94.

26 Groh, Jennifer. *Making Space: how the brain knows where things are* (Cambridge: Belknap Press, Harvard: 2014).

27 Donald, 331.

28 Trevor Watkins, "Building houses, framing concepts, constructing worlds," *Paléorient*, Vol. 30, No. 1 (2004), 5–23. As the author writes, "Architecture as a external symbolic storage allowed Neolithic communities to become the first humans to inhabit a "built environment" rich in symbolic meaning, making possible the emergence of new, larger and more complex forms of social organization."

29 Dennett, 264.

30 C.f. Matthew L. Shapiro and Howard Eichenbaum, "Learning and memory: computational principles and neural mechanisms," in Rugg, Ed. *Cognitive Neuroscience:* 77–130.

31 On basic working memory function, see Nelson Cowan, "The Magical Mystery Four: How is working memory capacity limited, and why," *Current Directions in Psychological Science* 19:1 (Feb. 2010): 51-57.

32 Alcino J. Silva, "Memory's Intricate Web," *Scientific American* 27:3 (Fall 2018): 82–89.

33 Farah, Martha J. 1988. "Is visual imagery really visual? Overlooked evidence from neuropsychology," *Psychological Review* 95, 307–17. Finke, R A. 1980. "Levels of equivalence in imagery and perception" *Psychological Review* 87, 113–32.

34 Vinod Goel, *Sketches of Thought* (Cambridge, MIT Press: 1995).

35 Sun, L., Xiang, W., Chai, C., Yang, Z., and Zhang, K. 2014. "Designers' perception during sketching: An examination of Creative Segment theory using eye movements," *Design Studies* 35: 593–613.

36 James S. Ackerman, "The Origins of Sketching," in *Origins, Invention, Revision* (New Haven, Yale University Press: 2017): 1–20, cautions us not to view the earliest sketches simply as vehicles for creative thinking, but rather as tools for "a variety of functions" needed by artists during the Renaissance.

37 C.f. Alan Chimacoff, "Figure, System and Memory: The Process of Design," in Akin & Weinel, *Representation and Architecture*, pp. 139–57.

38 See Garry Stevens, *The Favored Circle: The Social Foundations of Architectural Distinction* (Cambridge, MIT Press: 1998).

39 Pierre Bourdieu, *The Field of Cultural Production* (New York, Columbia Univ. Press: 1993).

40 For a cogent explanation of the theory, see Stevens, *Favored Circle*, 50–79. For a more extensive explication, See Pierre Bourdieu, *Distinction: A Social Critique of the Judgment ofTaste* (Cambridge, Harvard Univ. Press: 1984): 169–226.

41 Research in economic behavior was particularly striking. See Daniel Kahnemann, *Thinking Fast and Slow* (New York: Penguin Books: 2011. Kahnemann and his research partner, Amos Tversky, were awarded the Nobel Prize for their work on Prospect Theory, Framing, Availability and other new theories of decision making.

42 This is discussed in a wider philosophical context by both Evan Thompson and Alva Noe. See Thompson, *Mind in Life: Biology, Phenomenology and the Sciences of Mind* (Cambridge, Belkapp Harvard: 2007): 28–33; and Noe, *Out of Our Heads: Why You Are Not Your Brain and Other Lessons from the Biology of Consciousness* (New York, Hill & Wang: 2009): 97–128.

43 Charles Duhigg, *The Power of Habit* (New York, Random House: 2014): 1–30.

44 Arbib, How *The Brain Got Language*, presents a good summary of the H.M. research, 102–05.

45 See, for instance, Mitchell, *The Logic of Architecture*, op. cit.; Peter G. Rowe, *Design Thinking* (Cambridge, MIT Press: 1987); Christopher Alexander, *Notes on the Synthesis of Form* (Cambridge, MIT Press: 1969).

46 Carl Jung: "Psychology and Literature," in *The Spirit in Man, Art and Literature* (Princeton, Bollingen Series, 1966): 101.

47 One of the leading journals in the field, *Buildings and Landscapes*, is published quarterly by the Vernacular Architecture Forum.

48 A number of papers in Robinson & Pallasmaa, *Mind in Architecture*, suggest directions for this research. The Academy for Neuroscience in Architecture, attached to the Salk Center in San Diego, is actively pursuing these avenues.

49 Onians, *Neuroarthistory*, op. cit.

50 Michael Baxandall, *Painting and Experience in Fifteenth Century Italy*, 2nd Edition (Oxford, Oxford Univ. Press: 1988): 29–108.

51 Ibid., 29–30.

52 See Henry Glassie, *Folk Housing in Middle Virginia* (Knoxville, Univ. of Tennesse Press: 1975).

53 Jules Prown, "Mind in Matter," in Upton & Vlach, Ed. *Common Places: Readings in American Vernacular Architecture*, (Athens, Univ. of Georgia Press: 1986).

54 Herbert Gans, *The Levittowners: How People Live and Politic in Suburbs* (New York, Pantheon: 1967). One of the things Gans found among the residents of this new suburb of Philadelphia was that members of ethnic and religious groups at first "transferred" their old world or urban culture to Levittown. Gradually their habits morphed into new cultural patterns. 45–80.

55 A brilliant exposition of the two positions can be found in Christian Tewes, "Embodied Habitual Memory Formation: Enacted or Extended?" in Tewes, ed., *Embodiment in Evolution and Culture* (Cologne, Mohr Siebeck GambH & Co.: 2016).

56 C.f. Gero, J. and McNeil, T. "An approach to the analysis of design protocols," *Design Studies* 19:1 (1998): 21–61.

57 Hewitt, "Architectural Drawings and Modes of Conception," *JAE*, 1985, op. cit.

58 Daniel Dennett, *From Bacteria to Bach and Back: The Evolution of Minds* (New York, Norton: 2017).

59 See Amos Tversky and Daniel Kahneman, "Judgment Under Uncertainty: heuristics and biases," *Science*: 185, 1974; entire article reprinted in Kahneman, *Thinking Fast and Slow*. Appendix A.

60 John Jonides and Edward E. Smith, "The architecture of working memory," in Michael D. Rugg, Ed., *Cognitive Neuroscience* (Hove, UK, Taylor & Francis: 1997): 261–65.

61 Alva Noe, *Out of Our Heads*: 22.

62 This has been shown in several studies, including Huette S, Kello CT, Rhodes T, Spivey, M., "Drawing from Memory: Hand-Eye Coordination at Multiple Scales," *PLoS ONE* 8(3): e58464, Published March 15, 2013.

[4]

CRAFTING, DEPICTING, AND ASSEMBLING

26. Paleolithic cave signs, after Van Petzinger, *The First Signs*. (Author)

Humans have been making images that represent their perceptions of the world since the first artists scratched marks on stone walls during the Ice Age.[1] Recently, scientists have proved that the earliest cave paintings were made not by *Homo sapiens* but by Neanderthals, in caves located in modern Spain, at least 64,000 years ago. There is also new evidence that various abstract signs were used by *Homo sapiens*, based upon cave drawings.[2] [26] What happened once humans began to make buildings as shelter is not clear since we have little information about their representations in the archaeological record. It is tempting to believe that the first builders found it necessary to depict their creations using pictures, though nothing in the record suggests that this was the case. Given the primitive propensity for representation in stone and clay, it is more reasonable to posit that small models or effigies were made. Notably, among hundreds of examples of primitive sculpture and decorative art, all are either figures of humans and other living things, or useful objects such as tools and vessels.

Of course, humans also represent the world in modes such as narratives, films, and physical models. Today we understand that capturing personal sensations and intellectual concepts using just one mode or system is impossible, given the complexity of the human organism and the environment, let alone the meanings that we may wish to convey. Artists in all media are aware of this, and often exploit the paradoxes that attend their efforts to depict, or mimic, reality. Even centuries ago builders and artisans had a firm concept of how their models of the constructed world, and their sense of their bodies in relation to it, could be used to create representations of one sort or another.

Yet, for decades scholars and social scientists have focused most of their energies on describing and analyzing external mechanisms that have their limitations as analogs of the perceptual and cognitive faculties of the central nervous system. By focusing on he communicative (external) role of representation, many theorists have ignored the fundamental connection between cognition and internal haptic awareness, often assuming that designers simply "draw" what is in their brains.[3] Though the conventions of graphic representation have been well described, the revolution in brain science offers twenty-first century investigators a new means of exploring questions of how closely conventions of drawings, models, and other forms track with the activities of perception itself.[4] It may also be possible to ascertain how the presence of different types and scales of buildings changed the way humans represented their world. If the environment changed around them, human builders may have been forced to change with it, altering even their internal models of experience.

In this book I will suggest that three large schemas have governed humans' understanding of the built world: *crafting, depicting, and assembling.* In prehistory humans were compelled to think about their dwellings and monumental buildings as fully constructed entities, and remember the materials, sequences, and techniques required for their fabrication by looking at the artifacts in their environments. As craftsmen, they had a fully embodied understanding of the artifacts required for everyday life, including small tools, containers, vehicles, and houses. The practice of crafting buildings persisted through the ancient civilizations of Mesopotamia, Persia, Egypt, Greece, and Rome, lasting through the Middle Ages. Though drawings and models were employed to give a more detailed view of creating buildings, the broad practice of crafted buildings did not change until around 1400, when Renaissance artists began to look at depiction as a means of designing artifacts. Both crafting and depicting were enacted, cognitive modes that gave builders the necessary cognitive capacity to envision their work prior to construction. However, when more abstract concepts of building emerged during the industrial revolution, a new cognitive model emerged—assembly lines allowed production to occur much more rapidly, and builders began to look at their work as disembodied assemblies. In this chapter we shall discuss the various representational modes that existed prior to the industrial revolution, and the current information age.

Physical Forms of Representation: Templates to Models

As Vitruvius tells us in his famous accounting of the origins of building in the primitive hut, humans must have initially made their dwellings without the aid

27. Model of a Greek megaron, from Lawrence, *Greek and Roman Architecture*. (Author collection).

of abstract verbal or pictorial representation of any kind. They built by imitating the natural world, leaving caves and constructing their domiciles from the materials at hand. Some of the earliest representations of their buildings, including early Greek megarons and Roman huts, are preserved in clay models.[5] [27] Like the buildings they depicted, models were crafted out of one or more readily available materials, such as clay, wood, or soft stone.

As direct analogies to larger buildings, models have the advantage of showing any viewer an overall depiction of the artifact to be constructed, available from all vantage points. Models have the disadvantage of not always depicting details, interiors, or structural components of buildings, especially if not fabricated of similar materials to the building in question. So, as the Greeks surmised, a model of the overall structure might need to be augmented by templates and models of specific components before a complete description could be conveyed to contractors. Physical effigies of things in the outer world also had the advantage of relating, in miniature, to the objects to be constructed in direct relationship to the bodies of those who sought to build. When buildings were relatively small, little more than a basic image was necessary for humans to recreate them in larger form. Physical intimacy with any object facilitates understanding its qualities, especially when contemplating its replication. Were it possible for designers of massive structures such as bridges or warehouses to touch every part of their creations before conceiving them many problems might be more easily resolved. Today, the scale of the undertaking usually prevents this. In primitive eras scale was not a significant deterrent to conceptualization, and thus very basic representational types served well.[6]

Traditional craft knowledge is passed down through folklore and folkways, often through repeated experience with artisan masters. If an apprenticeship exposes the student to a range of critical processes and provides "hands on" exposure to materials and methods, then physical artifacts can represent their essence effectively. Take, for instance, the steps in basket weaving. A student is first shown an example of a simple vessel. Next she is introduced to the materials she must master—reeds, cane, wooden components, perhaps string or leather. Finally, she is instructed in techniques of weaving, often complex and not easily grasped on the first try.

Baskets are lightweight, easily rotated, and perfectly consistent inside and out. There are few hidden details in their construction. A beginner can not only concentrate on perfecting skills in weaving reeds to form the bottom and walls of the vessel, but also has a reference to look at while doing so. Greek stonecutters and Chinese carpenters were likewise able to look at full-scale models of the components they were making while on the building site. Such close observation and interaction with the "workpiece" (as furniture makers refer to their creations) provided all the necessary knowledge to these master builders right on the building site or in the workshop. The brain acquired memories and images from which to re-enact the process of fabrication.

The rock cut caves created by skilled artisans in ancient India provide one example of design knowledge passed from generation to generation. In the Buddhist *chaiyta* (house of worship) a stone sculpture called a *stupa* would be placed in the apse of a rock cut cave. Carving of the vaulted chamber would begin at the top and follow a conventional form taken from earlier wooden buildings. Thus models either extant or remembered could be used as templates for the carvers, who would later replicate them in more complex spaces, joining the main chamber with smaller antechambers, much as we see in Egyptian religious buildings[7] and tombs. The largest example, at Karli (32 BCE), rivals Gothic cathedrals in size, at 146 feet in length and 46 feet in height. The arched entrance suggests that it is being supported by elephants, and the interior columns are remarkably detailed.[8] [28]

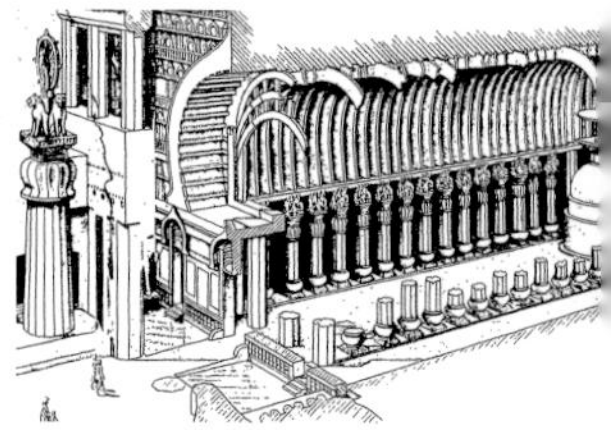

28. Karli, India, rock cave and stupa, hewn out of solid rock. Archaeologists now believe that many early pit houses in Southwest Asia were dug out of the earth before being roofed. (Wikimedia Commons)

One of the most significant challenges to builders in the ancient world was the increasing scale of their intended structures. The Romans were the first builders to create complex, and colossal structures out of masonry, wood, and concrete. By contrast, Egyptian, Persian, and Babylonian buildings were simple monoliths. We have few records of the process by which Roman structures were created, but surmise that basic geometry served as a planning tool for even the largest cities, camps, viaducts, and buildings. Scholars also underline the logistical genius of Roman builders, who were capable of organizing teams of skilled artisans and laborers to construct huge buildings efficiently. This system was not lost during the Middle Ages, and some of its methods were passed down to early Christian and Romanesque builders. One of the reasons for this is that, while texts and drawings may have vanished, building techniques could be transferred by demonstration, from one generation to the next, without being recorded in books or drawings.[9]

The intricate system of "geometric" design known to European medieval masons seems to have been passed down from masters to apprentices through physical demonstrations on the building site. The masonic guilds were generally attached to large cathedral or monastic building programs, fully functioning by the late Carolingian era. Robert O. Bork and Lon R. Shelby studied guild practices by looking closely at pattern books, reading building accounts, and measuring parts of cathedrals. They and many previous scholars have postulated that some of the world's largest stone buildings were designed and constructed primarily using templates, full-scale models, and stone mock-ups. This is not to say that visual and graphic tools were not employed, but their use was far more limited than in today's design and construction processes.[10] When a medieval draftsman sat down to draw a façade or portion of a cathedral he was constrained by the paucity of parchment available for such a drawing, and thus made fewer design studies than a Renaissance draftsman.

29. Nave of the cathedral of Reims, the subject of Villard's drawings. (Wikimedia Commons)

From the approximately 600 medieval drawings that have come down to us, scholars have re-constructed the sequences of geometric transformations necessary for a mason to create the building blocks of a cathedral.[11] This system was based on the method of "progressive squaring." Using only a compass, a square, and simple integers, these designers developed ways of producing templates and rudimentary plans for stonecutters to follow when fabricating each piece of a wall, buttress, pier, or pinnacle. Full drawings of pinnacles show that by drawing a square, dividing it into smaller squares, circles, and triangles, and projecting the figures upward in tiers, the mason could devise many variations on a standard pinnacle form. Similarly, when making full-scale templates and models for moldings and columns, masons used geometry to draw the shapes required for fabrication. [29, 30] Some of their full-scale drawings were inscribed into the walls of the buildings themselves.

For centuries the mason's template has been an essential aid in fabricating stone buildings. Generally made of wood or metal (copper is easily cut for this purpose), a template replicates the shape of a molding or portion of a pier so that a stonecutter may easily transfer the design to his or her workpiece. Carpenters also employ templates for similar molded components. At the building site teams of stonecutters would lay out their daily work using such templates, and proceed to rough out the pieces, later using the templates to hone in on the precise shape in the final dressing of the stone.[12]

These small templates were designed in the same manner as the larger elements in the cathedral, using either a square, a circle, or a triangle as a basis. [31] Likewise, the master mason could employ basic geometry to lay out plans and sections of a building. The system of designing "ad quadratum" and "ad triangulum" became conventional enough that by the late Gothic period in Europe experts from several workshops could advise the builders of Milan cathedral on the advantages of each during the fourteenth and fifteenth centuries, as we shall see in the following chapter. James Ackerman, who studied this and many other Italian building campaigns, suggests that the northern European masons, trained in the system described above, were unsuccessful in converting Italians, trained more as painters than engineers, to their way of thinking. He postulates that the Italian obsession with drawing, particularly perspective drawing, may have been a deep cultural bias even during the later Middle Ages. Architects who depended on models and looked at building from a "hands-on" point of view were suspect because they did not create pictures before undertaking their work.[13]

Thus, during eras in which crafting was a primary schema for building, physical representation served well for the builders, designers and patrons erecting everything from small huts to whole cities. In the post-Enlightenment period both drawings and physical models were used in design, often together, and in a wide range of modes such as perspectives, obliques, and various section cuts though a building or buildings. Modalities of representation brought with them advantages and biases that might well influence the formal development of building types or characteristics of a style.

30. Villard d'Honnecourt sketchbook, elevation and section, Reims cathedral. On the following page, Fig. 30, Villard shows the templates used to make the moldings in the elevation. (Wikimedia Commons)

Non-figural Representation in Architecture

Pictorial or graphic representation systems were not available to primitive humans until the Paleolithic Era, yet our species managed to construct dwellings, temples, and even stone monuments without any means of drawing them. The brains of early humans were well-equipped with a capacity to represent objects in other ways, and apparently used them effectively prior to the advent of geometric and orthogonal drawings. Some representation was simply supplied by constructed dwellings themselves, but narrative modes appear to have played a part as well.

In ancient Mesopotamia the Hittites, Babylonians, and Persians used clay and wax tablets to record numbers, early written symbols, and pictures of their buildings. Only rudimentary examples of these records have come down to us. After the Archaic Era, Greek architects appear to have functioned quite well with only the most rudimentary types of plan drawing, and perhaps without any plans at all, for two centuries or more. Attic Greek thinkers possessed a sophisticated aesthetic, moral, and philosophical worldview and a brilliant command

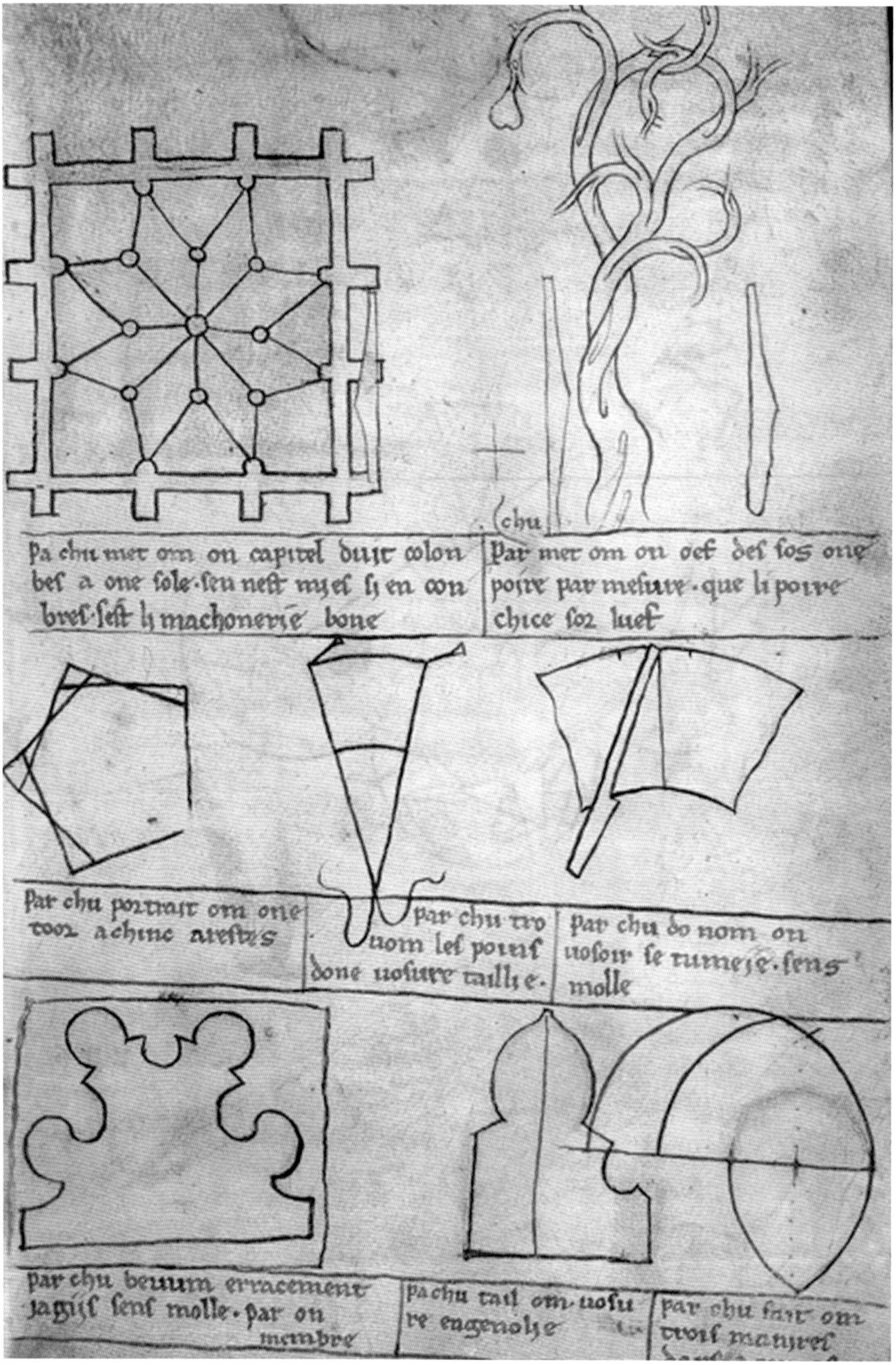
chu
Pa chu met om on capitel duit colon
bes a one sole·sen nest myes si en con
bres·sest li machonerie bone
Par met om on oef des sos one
poire par mesure· que li poire
chice sor luef
Par chu portrait om one
toor a chine arestes
par chu tro
uom les poins
done vosure tailli e.
par chu do nom on
vosoir se turneie· sens
molle
par chu devum erracement
iagisis sens molle· par on
membre
pa chu tail om vosu
re engenolhe
par chu fait om
trois manires

31. Drawings showing the construction of some molding templates used in the building of Reims cathedral, by Villard. The middle drawings show the use of a square, chalk line, and ruler to lay out the figures in the third row. (Wikimedia Commons)

of written and spoken language. Centuries in development, the literary achievements of Classical Greek culture were matched only by the perfection of their architecture and the depth of their scientific inquiry. It may not be surprising to learn that these three cultural predilections were intimately related.

The Cambridge archaeologist J. J. Coulton studied the practices of Greek architects, sculptors and builders intensively for decades, finally surmising that the stunning lack of extant evidence for drawing as a means of design was not a result of lost artifacts but rather a bias toward using words to describe the most intricate aspects of temple building and design. All who teach the terminology of Classical architecture are aware of the precision by which each component part of the three Greek Orders can be defined and named. Indeed, virtually everything in a typical temple, including the relationships between sub-parts, could be described in technical language.[14]

Many historians believe that Classical Greek architects designed with words, even in their most elaborate stone buildings. Because their buildings were highly conventionalized by the fifth century BCE they could forego drawings and instruct contractors by using what we now call specifications, or *syngraphai,* as contract documents. Though plan drawings were well known to them, they had little need for drawings of the overall building—representation was verbal as well as visual. [32]

These written instructions were so detailed and so rich with terminology in common usage that even the elaborate Greek system of "optical correction" known to archaeologists from measurements of temples could be aptly conveyed on the building site. One such document, the Prostoon Incription from Eleusis, described a method for stone construction from quarry to site:

> Stones to by quarried of Pentelic marble, length 17 feet, width 2 feet, thickness 1 ½ feet, to be trimmed straight and square on all sides and handed over for loading sound white and unstained, and with appropriate surplus, number 8. Stones to be transported from Pentelikon to Eleusis. … Eight Pentelic stones to be dressed, … straight and square on all sides, and to be rubbed smooth, and to be hoisted and set on both of the thresholds, the joints being made tight and unbroken, and to be clamped and leaded, and to be dressed true on top.[15]

An architect and contractor today would be perfectly familiar with such an instruction, but the likely form of communication between them would be a drawing and a short specification. There is no economy in a written description of this kind if drawings are available. It is telling that in Greek the word *graphos*, from which the English *graphic* is described, can mean either "draw" or "write." Conceptually, the architects of Greek temples and other high-style buildings were using the most logical, straightforward means of describing their designs to those who would fabricate the parts and the whole. And, as we know from their emphasis on symmetry, proportion, and consistency of part to whole (in Greek, *taxis*) these architects had linguistic analogies in mind for virtually everything.[16]

Design began with a simple description: *octastyle* referred to a temple of eight front columns and 17 flank columns. Temples were laid out on a stylobate made of square stones, and the basic peripteral plan (a cella encircled by columns) could be laid out directly on the ground using this matrix or grid. The diameter, and therefore height, of each column determined the height of the overall building. The three parts of each order—shaft, capital and entablature—could be determined by the scale of the column and its diameter. Thus, a small temple was in principle simply a scaled down version of a large one, though refinements were made on visual principles. Measurements needed to be very precise, as stone construction demanded close tolerances.

Because of this, Greek builders relied on two other forms of non-graphic representation: the *anagrapheus* and *paradeigma*. Both were types of models, but the former resembled a wooden template and was used for moldings. The latter was often a full scale mock up of a portion of the building. Using this system of verbal documents and demonstration models gave the architect explicit control of all aspects of the construction.

Chinese architects, as might be expected given their language and its character symbols, were also adept at using words in the design of their temples and other buildings. In the famous *Yingzao Fashi* (c. 1103) by the Tang dynasty building official Li Jie, line drawings occupy pages in concert with vertical columns of characters that are often more like specifications than captions.[17] Unfortunately, archaeologists have not yet chronicled the relationship between Li's treatise and the buildings he describes, so our knowledge of the graphic/verbal system is not as comprehensive as that of the Greeks. The layout and proportional systems of wooden temples were nonetheless just as tightly related and codified in Chinese architecture as in Greek, and words were important in making this system clear to carpenters and builders for centuries. [33]

In fact, a recent study of Song dynasty street design by Jin Xie argues that "the narrative of Premodern Chinese architecture is literary rather than physical." Illustrations from his article show the depiction of the Imperial Street in Hangzhou as neither a map nor a plan, but a pictogram using words and lines to represent buildings on a thoroughfare. Dating from the Southern Song period (1127–1279 CE), one drawing was mainly a record of commerce, as hundreds of merchants and stalls crowded both sides of the wide main street. Of course, Chinese graphic art was by this time quite sophisticated, so that the choice of textual/character description was deliberate on the part of the drawing's author.[18] [34]

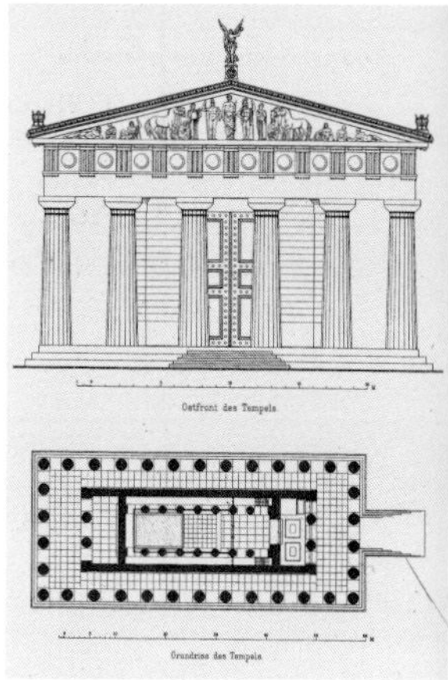

32. Temple of Zeus, Olympia, from Mauch and Normand. A conventional 6 x 13 columnar peristyle is used. From the stylobate to the columns and entablature, every part is related to the diameter of these columns, and each element has a specific name that can be used in a written specification. (Author collection)

33. Rules of carpentry, from Ying tsao fa shih, *Pictorial History of Chinese Architecture* (1984). The Chinese system is as balanced and harmonious as the Greek, and also based on part-whole relationships. (Author collection)

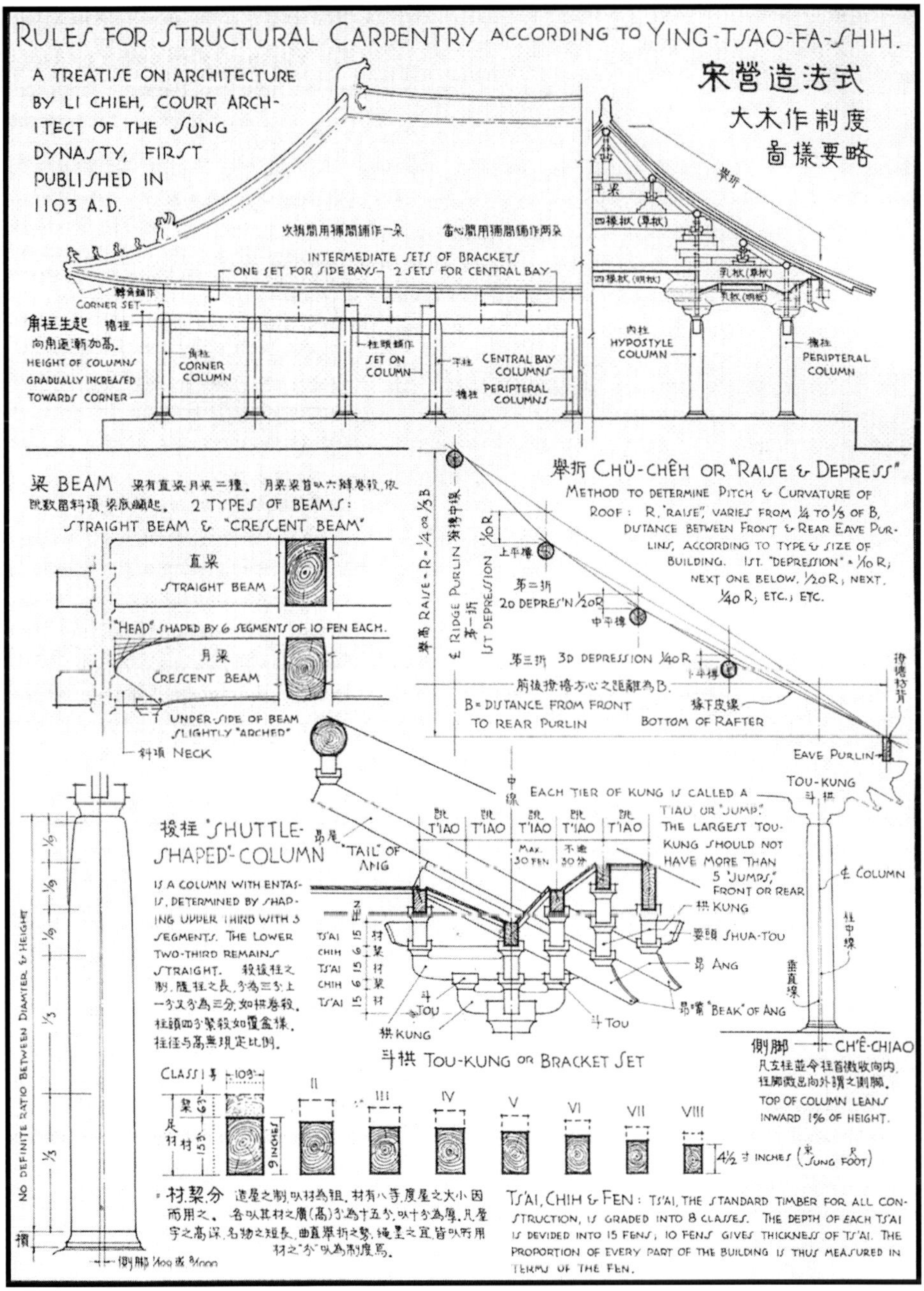
RULES FOR STRUCTURAL CARPENTRY ACCORDING TO YING-TSAO-FA-SHIH.
宋營造法式
大木作制度
圖樣要略
A TREATISE ON ARCHITECTURE BY LI CHIEH, COURT ARCHITECT OF THE SUNG DYNASTY, FIRST PUBLISHED IN 1103 A.D.
INTERMEDIATE SETS OF BRACKETS
ONE SET FOR SIDE BAYS
2 SETS FOR CENTRAL BAY
CORNER SET
HEIGHT OF COLUMNS GRADUALLY INCREASED TOWARDS CORNER
CORNER COLUMN
SET ON COLUMN
CENTRAL BAY COLUMNS
PERIPTERAL COLUMNS
HYPOSTYLE COLUMN
PERIPTERAL COLUMN
梁 BEAM
2 TYPES OF BEAMS: STRAIGHT BEAM & "CRESCENT BEAM"
直梁 STRAIGHT BEAM
"HEAD" SHAPED BY 6 SEGMENTS OF 10 FEN EACH.
月梁 CRESCENT BEAM
UNDER-SIDE OF BEAM SLIGHTLY "ARCHED"
斜項 NECK
舉折 CHÜ-CHÊH OR "RAISE & DEPRESS"
METHOD TO DETERMINE PITCH & CURVATURE OF ROOF: R, "RAISE", VARIES FROM 1/4 TO 1/3 OF B, DISTANCE BETWEEN FRONT & REAR EAVE PURLINS, ACCORDING TO TYPE & SIZE OF BUILDING. 1ST "DEPRESSION" = 1/10 R; NEXT ONE BELOW, 1/20 R; NEXT, 1/40 R; ETC.; ETC.
℄ RIDGE PURLIN
RAISE - R = 1/4 OR 1/3 B
1ST DEPRESSION 1/10 R
2D DEPRES'N 1/20 R
3D DEPRESSION 1/40 R
B = DISTANCE FROM FRONT TO REAR PURLIN
BOTTOM OF RAFTER
EAVE PURLIN
梭柱 "SHUTTLE-SHAPED" COLUMN
IS A COLUMN WITH ENTASIS, DETERMINED BY SHAPING UPPER THIRD WITH 3 SEGMENTS. THE LOWER TWO-THIRD REMAINS STRAIGHT.
NO DEFINITE RATIO BETWEEN DIAMETER & HEIGHT
EACH TIER OF KUNG IS CALLED A T'IAO OR "JUMP." THE LARGEST TOU-KUNG SHOULD NOT HAVE MORE THAN 5 "JUMPS," FRONT OR REAR
T'IAO T'IAO T'IAO T'IAO T'IAO
"TAIL" OF ANG
MAX. 30 FEN
拱 KUNG
耍頭 SHUA-TOU
昂 ANG
昂嘴 "BEAK" OF ANG
斗 TOU
TOU-KUNG
COLUMN
側脚 CH'Ê-CHIAO
TOP OF COLUMN LEANS INWARD 1% OF HEIGHT.
斗栱 TOU-KUNG OR BRACKET SET
CLASS I II III IV V VI VII VIII
9 INCHES
4½ 寸 INCHES (宋尺 SUNG FOOT)
材栔分
TS'AI, CHIH & FEN: TS'AI, THE STANDARD TIMBER FOR ALL CONSTRUCTION, IS GRADED INTO 8 CLASSES. THE DEPTH OF EACH TS'AI IS DEVIDED INTO 15 FENS; 10 FENS GIVES THICKNESS OF TS'AI. THE PROPORTION OF EVERY PART OF THE BUILDING IS THUS MEASURED IN TERMS OF THE FEN.

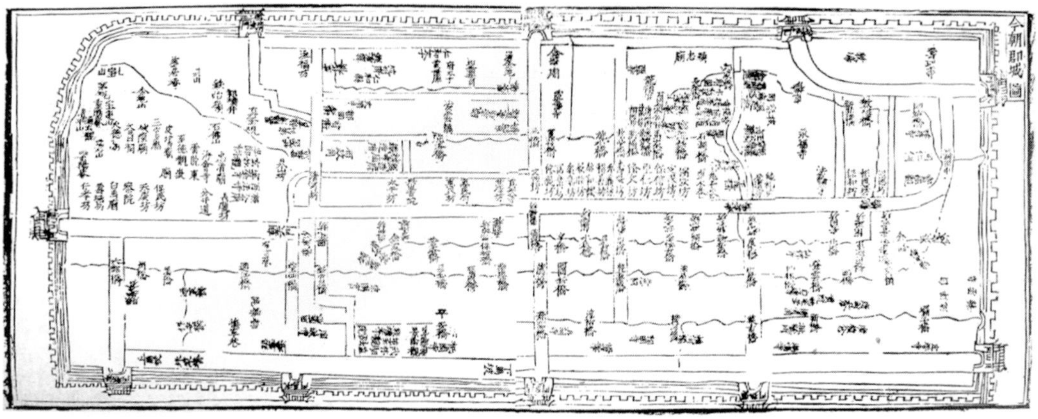

It is also important to recognize the role of numbers, both in proportion and in logistics, in the making of pre-modern buildings.[19] Roman and Gothic building workshops were organized on economic models that allowed materials to be quantified before production, so that materials were neither wasted nor delayed in reaching the building site. During the building of Hadrian's Pantheon the massive front portico columns had to be reduced in size so that they could be transported more easily via rivers and land routes to Rome. Roman contracts were often quite precise about costs, demanding fixed price bids, but not specific as to timing. Such bids required good accounting and cost estimating, things that the Roman system of numbering provided.[20]

34. Schematic plan of Hangzhou (Song Lin'an) from the Song period. Instead of drawing pictures of the buildings and shops, the artist uses standard characters to describe them. (Author collection)

Archbishop Gervase of Canterbury was able to provide one of the most detailed descriptions of a cathedral building process because he had been the sacristan or monastic accountant during most of its early construction. By going back over his records he could be precise about when portions of the building were put up in the first through fourth years under the master mason William of Sens. He tells us why a French mason was chosen for the construction of a church commemorating Thomas Becket, the man murdered in the now burnt previous cathedral—Becket had spent time at Sens and knew of his abilities. Gervase could also provide precise accounts of the numbers of windows, columns, and blocks of stone used in each early campaign simply by looking over his record of payments for the work completed.[21]

The Geometries of Projection

Countless books have been written about the systems of representation that architects and other visual artists use to do their work, and nearly all focus on what may be called modes of representation.[22] Most of these modes are "projections" that abstractly depict two- and three-dimensional figures on some sort of plane or array of points in space. Only recently have scholars begun to

35. Fresco, Room of the Masks, Domus Aurea, Rome. The perspective used is not "scientifically" constructed as in later Renaissance examples, but is effective at presenting an illusion of depth. (Wikimedia Commons)

recognize that these systems carry ideas that can bias designers toward specific forms and spatial schemata. The most famous work in this regard has been done on perspective, notably by Erwin Panofsky, Alberto Perez-Gomez, Martin Kemp, and Massimo Scolari.[73] Renaissance artists rediscovered this mode of representation and used it to revolutionize both painting and architecture beginning in the early 1400s. For a time it was thought that perspective held the secret to optical science, and was capable of rationalizing space and form in buildings, prejudicing designers for almost four centuries. It is fair to say that the digital computer, with its powerful capacity to represent three-dimensions on visual displays, will have an equally profound impact on the way we create all sorts of images.

Builders have used geometric projection since antiquity. Vitruvius discussed modes of architectural representation in his *Ten Books on Architecture* in the late first century BCE, during the reign of Julius Caesar. In "Book I" he described the common means used by Romans to represent their designs, establishing a taxonomy that his readers took to be canonical after the Renaissance: *Ichnographia, Orthographia, and Scaenographia.* Ichnographia, according to this Roman author, corresponds to our understanding of plans; orthographia to building elevations. Both of these orthogonal projections could have been derived not only from visual stimuli but also from the pragmatic experience of building construction. Both were used by Egyptian builders, though no parchment drawings of the type survive, only wall paintings. Whatever their origins,

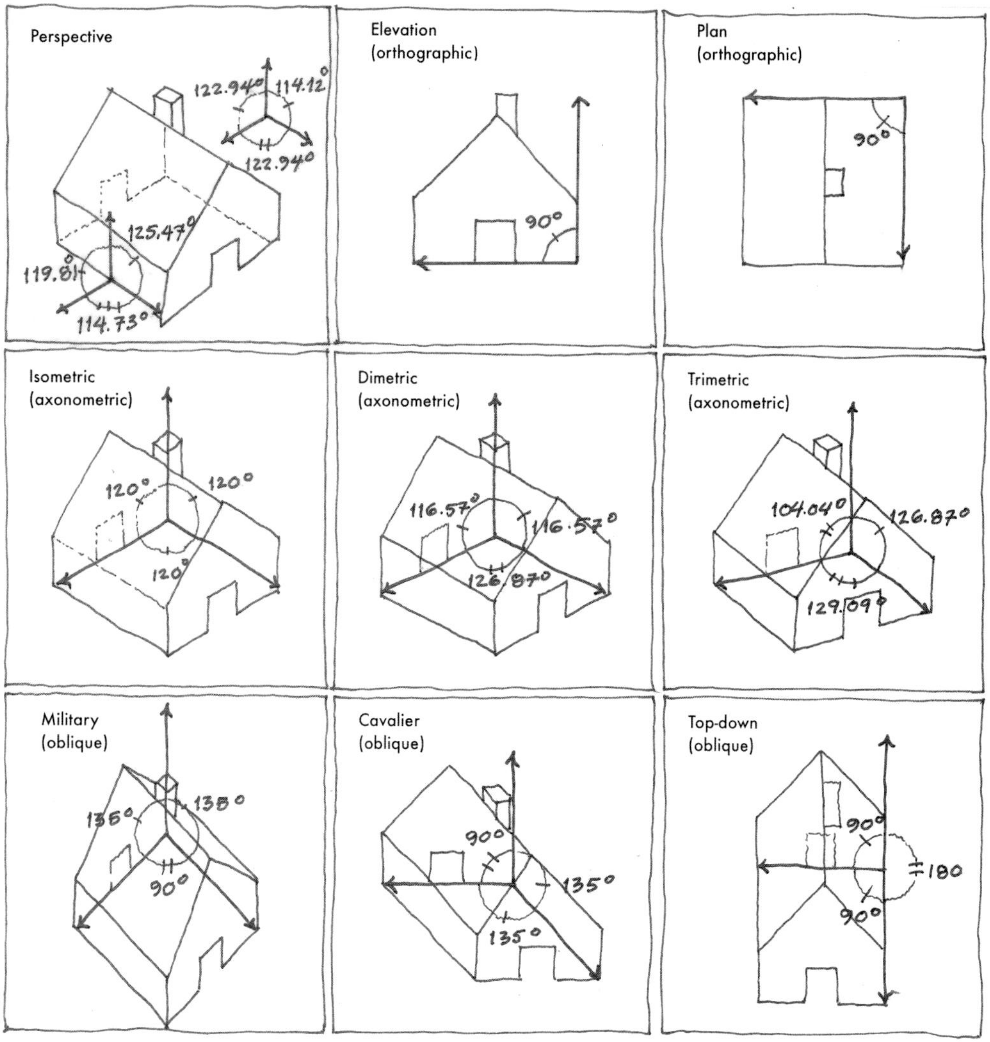

36. Principal types of projections used in architectural, engineering and surveying drawings. (Author)

they were established modes by the time Roman architects began building their most astonishing structures in the Common Era.[24]

Scaenographia seems to have described a form of perspective that did not capture foreshortening the way a camera does. Apparently such a view involved projecting an elevation in the foreground of the painting and rendering one or more sides in a kind of foreshortening "toward the center of a circle." Shading was applied to give a more three dimensional appearance. Roman wall paintings from Pompeii and elsewhere depict spaces and forms in a kind of perspective/axonometric projection that so far has not yielded to geometric analysis. Since Roman artists had limited knowledge of the science of optics, their depictions of the physical world must have more directly reflected their perception of inanimate objects and living things. This is not to say, however, that painted pictures of some sort were not part of their experience.[25] [35]

One of the most common observations about such pictures is that the distance between things often corresponds more to their perceived relationship in space than to their position when foreshortened in an optical projection. If one considers our neuroscientific understanding of the visual cortex, such a re-presentation of a scene is less puzzling. With no cultural bias, painters may well have placed objects in a room or bounded space according to their understanding of their relative positions in the space itself, accounting for their phenomenal size or placement. Since the eye and brain do not seize on a single pictorial projection of any scene, the various "snapshots" in the artist's mind were synthesized into these primitive fresco views.

37. Auguste Choisy. Axonometric drawing of Roman vaulting, Palatine palace. Note the measured layout lines in the lower right portion of the drawing. (Author collection)

A recent study of Third Style landscape paintings speculates that maps and dances were also important to those creating the frescos in town houses and villas.[26] The author compares mural paintings at places such as the Villa of the Mysteries in Pompeii to the ancient cartographic genre of *chorographia*, in which pictures or processions were employed as maps. A famous example is the Nile Mosaic at Palestrina (dated speculatively at 100 CE), a linear depiction of places along the famous river. Such experiential representation of places in time follows the way the brain apprehends such information, placing it in memory for later retrieval.

Another observation that has become canonical in Roman archaeology concerns the influence of such perspectives on the design of later Roman buildings. The Third and Fourth Styles in wall painting have often been cited as large influences on Hadrianic and later Roman structures, which employed split pediments, discontinuous cornices, and illusionistic motifs. Research has also indicated that wall paintings often "quoted" scenes from noted dramas and comedies. Stage illusionism, a lasting influence on architecture, has been with us since antiquity. We will discuss it in more detail in Chapter 6.

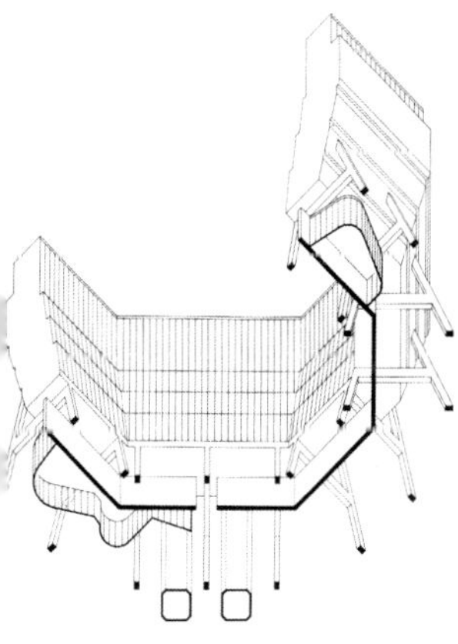

38. James Stirling, axonometric drawing of Olivetti building. Non-western viewers cannot "see" the building in axonometric drawings. (Author collection)

Orthographic projections, such as plans, sections, and elevations, capture objects by casting their outline on planes at right angles to their faces, in a Cartesian grid. The x, y, and z axes from Euclidian geometry constrain such projections from presenting a "picture" of the object as captured by eye and brain, but render the dimensions and form of the object commensurable, or measurable in Euclidian terms. The three conventional drawings of this kind are the plan, the elevation, and the section. Descriptive geometry, invented by Gaspard Monge (1746–1818), provided a mathematical basis for constructing the projection of an object on a plane or surface from any point in space.[27]

If the object is rotated so that the angle of projection is non-orthogonal, other modes of projection come into play, often described as *obliques* in languages other than English. (In English oblique drawings are specific as to

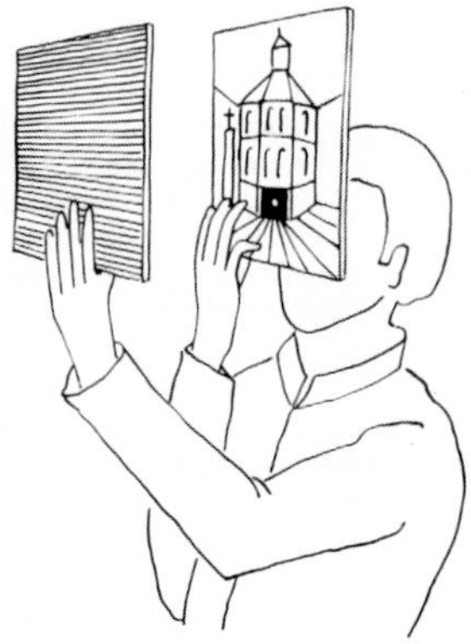

angles). The most common is an axonometric view that abstractly represents the external facets of a solid from a non-orthogonal angle such as 45 degrees. Thus, more than one plane is projected on the drawing or computer screen, and the object can be rotated to reveal all of its surfaces in a series of views. The conventional description of these projections as views derives from historical assumptions about visual perception that are no longer widely held. Experiments with so-called primitive societies revealed that not all humans understood such projections in the same way.[28] [36]

The most common variant of the axonometric drawing is one in which the right angles of one plane are preserved (say, the y and z axis), while the third projection is acute. The sum of the angles is always 360 degrees, and a 60-30-90 triangle is often used to create the drawing. Architects favored the axonometric during the last century because it offered a simple means of depicting buildings in space without resorting to more complex perspective construction. Some theorists even advocated for the adoption of the axonometric as a standard representational mode as an alternative to perspectives, owing to its presumably more "rational" and culturally neutral qualities. One of the tantalizing aspects of axonometric drawing is that the three sides of the object are represented without foreshortening, in a quasi-perspective projection. In an isometric projection the three angles are each 120 degrees, resulting in a more even rendering of the sides, very useful for engineers. Any three sides may be presented—in some drawings the bottom or plan view is incident to the paper plane, in others the top, or roof. Sometimes the incident plane is a section through the object. If the same views are presented with foreshortening the drawing becomes a perspective.[29]

Two masters of axonometric projection in the twentieth century were August Choisy, a French architect and historian, and Peter Eisenman, an

39. Samuel Edgerton. Florentine Baptistry, photo, with camera at station point and two vanishing points on horizon line. (Author collection)

40. Samuel Edgerton. Reconstruction of Brunelleschi's Florentine "pin hole" perspective panel. The facing surfaces are mirrors. (Author collection)

41. Sebastiano Serlio, antique details, from Book III. These elevation drawings use a form of perspective foreshortening when none is strictly necessary. It was probably engraved from a sketchbook drawing like the ones in Figs. 57-60. (Author collection)

American with some of the same penchants. Choisy found the method particularly useful for representing the construction of historic buildings such as Roman vaulted rooms and Gothic cathedral bays. [37] His innovative book on architectural history has been admired for its analytical incisiveness but criticized for its often confusing "up" axonometric views. The English architect James Stirling used such views to depict many of his buildings during the late twentieth century because they could capture both interior and exterior features. [38]

Eisenman, also the target of much criticism, has maintained that his oblique drawings allowed him to design buildings that stretched the boundaries of common experience. His "House III" in Connecticut set walls and other objects in grids that were perceivable only through drawings, much as Baroque architects set points of view in their buildings based upon ideal linear perspective projections. A number of his later residential projects, such as "House X," used axonometrics as compositional devices, often producing skewed and barely functional spaces in the built realization of the drawings.[30]

42. Clay tablet showing plan of the city of Nippur, Babylon. Maps preceded building plans as means of depicting the environment. (Wikimedia Commons)

Linear perspective, established definitively by painters and architects in Florence in the 1410s, is also a form of projection, and one that obsessed designers for more than a century afterward. Its representation of forms in space corresponds roughly to the way in which light is refracted through a pinhole or lens, casting an upside down image on a picture plane.[31] Brunelleschi conducted a famous experiment in 1425 in which he used a primitive camera obscura to make a projected image of the Florentine Baptistry on a wooden plate that was then traced to create a picture. Samuel Edgerton and other scholars determined the probable influences on the Florentines, including map projections, that placed the discovery of perspective earlier than 1400, but there is little doubt that Brunelleschi and his contemporary, Leon Battista Alberti, introduced it to architects.[32] [39, 40] Various types of perspectives influenced architects for several centuries afterward. Indeed, as Marin Kemp has demonstrated, perspective projection changed the way humans saw the world, and with it the way architects conceived their buildings.[33] Yet even the many masters of this scientific means of representation were aware that their constructions on paper could not describe what they perceived with their eyes.[34]

Wolfgang Lotz, in a famous essay from the post-WWII period, discovered a significant number of Italian drawings made during the period following Brunelleschi's panel that recorded standard building sections (well-known to medieval masons) as perspective drawings. Such a fixation on one mode of representation was unusual enough to pique the interest of Lotz, a German intellectual who was conversant with Gestalt psychology. He concluded that most of the painter-architects who made such drawings were compelled to use the latest discovery in drawing to render interior spaces even though such views distorted the spaces enough to produce useless drawings. Something similar is happening today with many computer-generated views.[35] [41] The seductive power of such projections has often gone unrecognized by architects and designers, as have their limitations. In his book, *The Projective Cast*, Robin Evans discussed several alternative kinds of projection, including stereotomy for stonecutting, which proved useful throughout the modern period as means of representing buildings and their component parts. Though his approach did not favor perceptual studies, he acknowledged the myopia of many previous

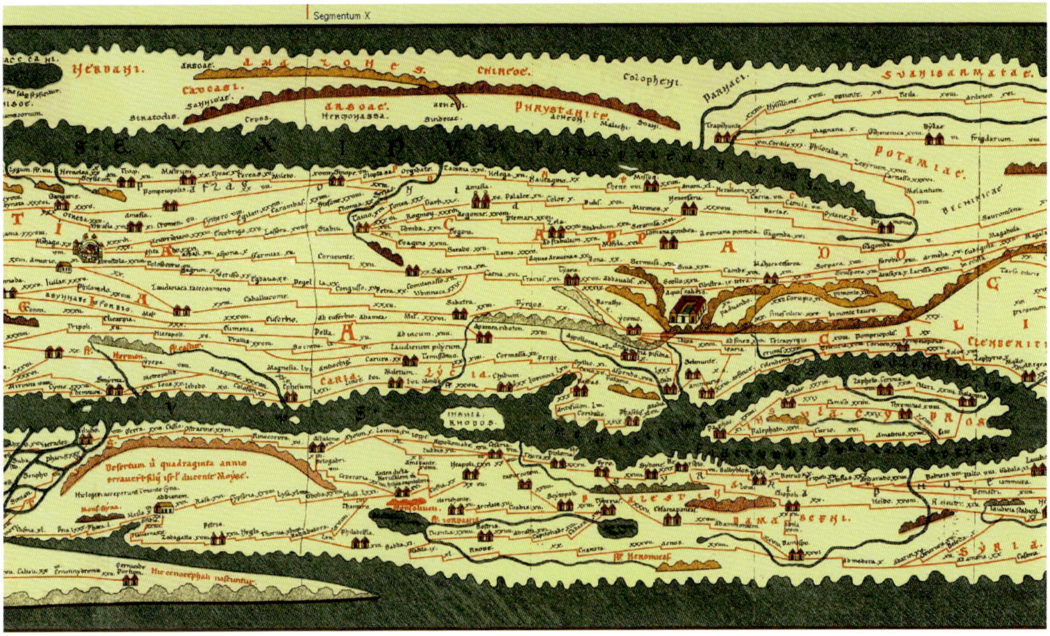

43. Peutinger map of the known world, c. 500 CE. (Wikimedia Commons)

scholars who limited their purview to *costruzione legittima* (Alberti's perspective method) and engineering drawing according to various kinds of axonometric projections. Evans, a practicing architect and scholar, brilliantly pointed to the wide range of experience and knowledge that building designers must use in their work, experience that no single form or representation can capture. I shall return to this topic later, but the issue of projected images pre-dated the Renaissance and Enlightenment periods that most scholars have written about since the early twentieth century.[36]

For now, we may observe that for any mode of pictorial representation, both cultural (learned) schemata and purely sensory (visual, haptic, aural) data play a role in the final product of the artist or designer. Much as Semir Zeki has observed, the brain seems to build its imaging capacity along the nature/nurture continuum. It is experience with the world that leads humans to be able to function competently in it.

Maps, Movement, and Location

Most buildings are designed for particular places or locations on the earth. When considering the placement of any structure, designers use maps of the surface of the earth to guide their decision making. From the beginning of civilization cartography has been essential to the movement of people, vehicles, and ships, and no less so for building settlements. In order to create maps, humans had to understand the placement and orientation of their bodies in space and time.

Recent research has suggested that the brain also creates "maps" that help us locate our bodies in space and provide needed cues for moving around in the environment. "When the brain makes maps it informs itself," writes Antonio Damasio; "The information contained in the maps can be used [unconsciously] to guide motor behavior efficaciously, a most desirable consequence considering that survival depends on taking the right action. But when brains make maps, they are also creating images, the main currency of our minds."[37] Maps depicting the stars have been found in the caves at Lascaux, drawn about 16,000 years ago.

Not surprisingly, the earliest ground maps date from collective civilization in the Fertile Triangle, some 7,000 years ago. A cuneiform tablet, probably from Sipper in Babylon, depicts a circular disk purported to show the known world. A clay tablet from the Babylonian Kassite period (1400–1200 BCE) depicts a ground plan of buildings in the city of Nippur. [42] When people began to live in large settlements, the need to organize dwellings demanded a new form of representation. A surface such as clay or stone could be used as a surrogate for the earth itself. Transferring a brain image to a drawing surface was a simple matter—a stylus did the trick. Cognitive maps of territories and river valleys emerged as people traveled farther distances in order to trade with their neighbors.[38]

As noted above, the Greeks were adept at mapping their world in literature—Homer and Hesiod both provide detailed descriptions of the Aegean world and beyond. Anaximander of Miletus (610–546 BCE) was the first recorded mapmaker in Greek texts, but his legendary world map has never been found or substantiated by archaeology. The technique of projections, using meridians and parallels, was first expounded by Marinos of Tyre in the first-century BCE, and later critiqued by Ptolemy of Alexandria (87–150 CE) in his *Geographia*. This major treatise on map making was as sophisticated as the 10 books of Vitruvius, written just decades before in Rome. Like his Roman predecessor, Ptolemy was a theoretician who incorporated many kinds of knowledge into his text, including philosophy, natural history, and mathematics.[39] Like his Greek predecessors, he believed that the Earth was a sphere.

Some scholars believe that Ptolemy employed a kind of perspective projection in his method of mapping. He was able to calculate the circumference of the Earth at its equator using a sundial and some tricky mathematics.[40] There are no preserved illustrations from his original text, but Byzantine copies (labeled A and B) show examples intended to conform with his theories. These, divided into eight books, are remarkable in that the Indian, Caspian, and Mediterranean seas are depicted, along with parts of Europe, Africa, and Asia. All were based upon a circle, with a prime vertical meridian passing through the Fortunate Islands. Significantly, the Ptolemaic maps provide temporal

44. Forma Urbis Romae, stone fragment (Stanford University, Digital Forma Urbis Romae Project). The most detailed known map of Rome was inscribed onto the walls of a building.

notation to describe the distances between parallels and meridians. If journeys between points were recorded in days, hours, and minutes, it would be logical to see this information on early maps. The time-based measurements are consistent with what we know about the brain—humans equate distance with the time taken to travel between places.[41]

Though the Romans presumably made accurate maps to guide their conquests, we only know of one notable example: the so-called "Tabula Peutingerianda" preserved at Augsburg. Though information on it may date from the 1st century CE, its final form probably dates from 500 CE. [43] In pictorial form it corresponds closely to modern maps: a figure ground composition using white for land and dark for seas. In this table map the Earth was presumed to be a flat disc.

In keeping with their ordered civilization the Romans surveyed cities using a Cardo and Decumanus along the cardinal axes, and also drew ground plans of their settlements. The most famous of these is the Forma Urbis Romae, a large plan of the city of Rome dating from 203–211 CE during the reign of Septimus Severus. Carved on a block of marble, individual building and streets are visible, and the depiction is to scale, 1:240. The original was carved on a wall of the Templum Pacis, and many parts were lost to time and vandalism. Remarkably, 10–15%, or 1,186 pieces, survive and are being digitized by scholars at Stanford

University. David Reynolds believes that the map was a copy of papyrus plans, drawn from surveys, providing a cadastral reference for Roman administrators recording land ownership. The plan conventions are quite modern, as openings for doors correspond to those in excavated buildings from the time.[42] [44]

In such a map, Roman engineers and planners had a fully commensurable depiction of all major buildings in the city, with the width and direction of streets also rendered in scale. Unquestionably, by this date much of the technical knowledge in surveying, plotting, and drawing structures, even on varied topography like that of Rome, was in the brains of these officials. Like much that vanished with the Roman Empire after about 400 CE, this knowledge was not to resurface for almost a thousand years.

The earliest modes of representation available to builders were all aids to memory, means of communication, and embodied constructs of our sensory-motor systems. If these modes also helped to understand the world within our grasp, humans could make good use of them in constructing dwellings and other forms of shelter, to be placed in particular environments. Humans are not the only creatures that build dens or shelters to protect themselves from the environment. Birds, crustaceans, and insects are also capable builders; none use pictorial representation as far as we know. It is important, as we look at the history of representation, to understand how builders and designers differed in their use of models, non-pictorial tools, and drawings when conceiving their works. Current scholarship strongly suggests that, prior to the fifteenth century, those charged with producing buildings did not fully depict their creations before building them. Modern architectural design began when full depiction became necessary for the production of monumental structures such as cathedrals, fortifications, and palaces.

Notes

1 See Jean Clottes, *What Is Paleolithic Art* (Chicago, Univ. of Chicago Press: 2016). He points out that only about half of the known paintings of Paleolithic people are in caves. Many are on individual rocks, cliff faces, or rock shelters. This suggests that hominids may have decorated their dwelling places much earlier than previously thought.

2 See Genevieve von Pettinger, *The First Signs: unlocking the mysteries of the world's oldest symbols* (New York: Simon & Schuster: 2016). The author has identified 32 signs, some with affinities to architectural decoration,that exist in large numbers throughout Upper Paleolithic sites.

3 In an otherwise brilliant book, the late Dalibor Vesely presents a critique of the current crisis in architecture on this basis, without any reference to brain science or perception. See Vesely, *Architecture in the Age of Divided Representation: The Question of Creativity in the Shadow of Production* (Cambridge, MIT Press: 2004). This a common problem with the current literature.

4 The best research has been presented in the journal *Design Studies*, founded in 1979 and published in London. Neuroscientists, social scientists, and designers all contribute research to this interdisciplinary publication.

5 A wonderful exploration of the ramifications behind the Vitruvian myth of origins is in Joseph Rykwert, *On Adam's House in Paradise* (New York, Museum of Modern Art: 1972).

6 Pinker describes several experiments that demonstrate the brain's capacity to rotate objects in both two and three dimensions. *How the Mind Works*, 279–84.

7 C.f. Binda Thapar, *Introduction to Indian Architecture* (Singapore, Periplus Editions: 2004): 30–35.

8 Benjamin Rowland, *The Art and Architecture of India: Buddhist, Hindu, Jain* (Harmondsworth and Baltimore, Pelican History of Art: 1981): 113–20. Scholars generally believe that Indian carvers were influenced by Greek examples following the conquests of Alexander the Great.

9 See Richard Krautheimer, "The Afterlife of Vitruvius in the Middle Ages," in *Studies in Early Christian, Medieval and Renaissance Art* (New York, NYU Press: 1969).

10 Robert O. Bork, *The Geometry of Creation: architectural drawing and the dynamics of Gothic design* (Farnham, UK, Ashgate: 2011); and Lon R. Shelby *Gothic Design Techniques* (Carbondale, IL, Southern Illinois Univ. Press: 1977).

11 Bork notes that of the approximately 600 Gothic-era drawings that have come down to us, some 480 come from the former Holy Roman Empire, and 420 are in Vienna, 21.

12 On the template as well as other tools and methods of Gothic workshop production see Stephen Murray, *Plotting Gothic* (Chicago: Univ. of Chicago Press: 2014): 113–29.

13 James S. Ackerman, "Origins of Architectural Drawing in the Late Middle Ages and Renaissance," in *Origins*, 60–61. Ackerman points to the "linear" emphasis on late medieval drawings from northern workshops, as contrasted to an emphasis on mass in early perspective drawings by Filarete, Francesco di Giorgio Martini, and Baldassare Peruzzi.

14 J.J. Coulton, *Ancient Greek Architects at Work: Problems of Structure and Design* (Ithaca, New York, Cornell Univ. Press: 1977).

15 Coulton, 53.

16 For an excellent explication of these systems see Alexander Tzonis and Lianne Lefevre, *Classical Architecture: The Peotics of Order* (Cambridge, MIT Press: 1986); and George L. Hersey *The Lost Meaning of Classical Architecture: speculations on ornament from Vitruvius to Venturi* (Cambridge, MIT Press: 1988).

17 The best English analysis is in Quingua Guo, "Yingzao Fashi: Twelfth Century Chinese Building Manual," in *Architectural History* (V. 41:1998): 1–13.

18 Jin Xie, "Disembodied Historicity: Southern Song Imperial Street in Hangzhou," *JSAH* 75: 2 (June 2016): 182–200. Quote on 198.

19 For an interesting history of their use in the Renaissance, see Mario Carpo, "Drawing with Numbers: Geometry and Numeracy in Early Modern Architectural Design," *Journal of the Society of Architectural Historians*, Vol. 62, No. 4 (Dec., 2003), 448–69.

20 For an excellent account of the Roman design and building process see Rabun Taylor, *Roman Builders: A Study in Architectural Process* (Cambridge, Cambridge Univ. Press: 2003): 21–92.

21 On Gervase and his Tractatus, see Murray, *Plotting Gothic*, 47–71.

22 C.f. Helen Powell & David Leatherbarrow, Eds., *Masterpieces of Architectural Drawing* (London, Orbis: 1982); Frank Ching: *Architectural Graphics* (New York, Van Nostrand Reinhold, 1985); David Dernie, *Architectural Drawing* (London, Lawrence King: 2010). Iain Frasier and Rod Henmi, *Envisioning Architecture: an analysis of drawing* (New York, Van Nostrand Reinhold: 1993); C. Leslie Martin, *Architectural Graphics* (New York, Macmillan: 1970).

23 Erwin Panofsky, *Perspective as Symbolic Form* (New York, Institute of Fine Arts: 1991/97): 1,924–25, paperback MIT Press. Alberto Perez-Gomez and Louise Pelletier, *Architectural Representation and the Perspective Hinge* (Cambridge, MIT Press: 1997). Massimo Scolari, *Oblique Drawing: A History of Anti-Perspective* (Cambridge, MIT Press: 2012).

24 For a cogent explanation of their meaning, see Ingrid Rowland and Thomas N. Howe, *Vitruvius: Ten Books on Architecture* (Cambridge, Cambridge Univ. Press: 24–25; 143–50. For a discussion of the influence of these terms in the Renaissance see James S. Ackerman, "The Conventions of Late Gothic and Early Renaissance Drawing," in *Origins, Imitation, Conventions: Representation in the Visual Arts* (Cambridge, MIT Press: 2002): 45–65.

25 The nature and meaning of these paintings continues to be a source of speculation. Some observers of the Third and Fourth Styles of Roman mural painting see the dominant influence of theater design in all "framed" views, while others look at the flat frontality of the foreground elements as attempts to represent the threshold between interior and exterior. See Diana Kleiner, *Roman Architecture: A Visual Guide* (New Haven, Yale Univ. Press: 2014).

26 J. Keith Doherty, "Roman Landscape Painting and Ancient Cartography," PhD. dissertation, Boston University, 2010.

27 See W.H. Roever, "Some Overlooked Mathematical Principles of Descriptive Geometry," *The American Mathematical Monthly*, Vol. 41, No. 3 (Mar., 1934): 142–59.

28 John Onians, *Neuroarthistory: From Aristotle and Pliny to Baxandall and Zeki* (New Haven, Yale Univ. Press, 2007): 150–58.

29 The most far-reaching exploration of these drawing types to date is in Scolari, *Oblique Drawing*, Op. cit.

30 An interesting study of this can be found in Kristina Luce, "The Collision of Process and Form: Drawing's Imprint on Peter Eisenman's House VI." *Getty Research Journal*, no. 2, 2010, 125–37.

31 Michael Kubovy calls this "Brunelleschi's peepshow" and explains in detail why Renaissance artists developed more sophisticated means of representing space in their paintings after the 1420s. See Kubovy, *The Psychology of Perspective in Renaissance Art* (Cambridge, Cambridge Univ. Press: 1986): 32–64.

32 Samuel Y. Edgerton, *The Renaissance Re-discovery of Linear Perspective* (New York, Harper & Row: 1976).

33 See Martin Kemp, *The Science of Art: Optical themes in western art from Brunelleschi to Seurat* (New Haven and London, Yale University Press: 1990): 7–99.

34 For an interesting recent critique of linear perspective see Michael Sciarillo and Scott Aker, "Drawing as an Insight Into Wholeness," *Journal for Geometry and Graphics* 12:1 (2008): 87–98. The authors use what they call Reverse Engineered Perspective, something commonly employed in computer visual reconstruction, but fail to account for cultural knowledge of conventional perspective systems in their research.

35 Wolfgang Lotz, "The Rendering of the Interior in Architectural Drawings of the Renaissance," in *Studies in Italian Renaissance Architecture* (Cambridge, MIT Press: 1977): 1–65. This famous paper has colored all subsequent attempts to discuss the impact of perspective on Renaissance architects. Citing a letter from Raphael to Pope Leo X, he presents a decisive break in the practice of drawing interior sections using perspective at 1519, 20–21.

36 Robin Evans, *The Projective Cast: Architecture and Its Three Geometries* (Cambridge, MIT Press: 1995).

37 Antonio Damasio, *Self Comes to Mind: Constructing the Conscious Brain* (New York, Pantheon Books: 2010): 63.

38 On Babylonian maps see Leo Bagrow, *History of Cartography* revised and enlarged by R. A. Shelton (Reprint of 1985 edition, New Brunswick, Transaction Publishers: 2010): 31 ff.

39 Bagrow, 33–36.

40 See John Noble Wilford, *The Mapmakers* (London: Folio Society: 2016).

41 Bagrow, 34–35. One hour would equal fifteen degrees, and one minute was a fifteenth of a degree. This system, in nomenclature, is preserved in modern geography.

42 The extensive digital archive is at formaurbis.stanford.edu and is available to anyone, free of charge.

[5]

THE DISCOVERY OF DEPICTION

Donato Bramante, who came to Rome in 1501 to oversee the "fabbrica" or workshop charged with rebuilding the basilica of St. Peter's, would not have been called an "architect" a century earlier. In fact, the profession of architecture did not exist anywhere in Europe prior to Bramante's time.[1] Buildings were designed and built by teams of craftsmen in the guilds or trades that produced various parts of the whole edifice: masons, carpenters, painters, metalsmiths, glaziers, plasters, and sculptors. The "maestro" of the workshop was most often a member of the mason's guild if the building was to be built of stone, or of the carpenter's guild if wood was used. Giuliano da Sangallo, Bramante's predecessor at St. Peter's, and Antonio da Sangallo, one of his key successors, were trained as carpenters. Because of their training, close knit working relationships, and understanding of materials, these artisans did not need to have a full representation of every component in order to make it.

Building conventions throughout Europe varied somewhat, but generally followed the process described in Chapter 4. Using a combination of folk wisdom, mason's rules of thumb, and specific requirements of an individual owner, a contractor prepared drawings, models, and specifications for areas of construction requiring pre-planning. If a façade or plan followed well-known conventions, a drawing might not be necessary. Likewise, if models of building details were made of wood or plaster, carvers could simply work from these templates, fabricating multiple pieces of say, a cornice or string course. It was not necessary for the workmen to see all parts of the building before or during construction, because their work was highly conventional and supervised.[2]

Builders and designers were from the same class, had the same education, and knew the same "folkways" as their fathers and grandfathers. Moreover, they did not feel the need to document every piece or every procedure used in producing the building described in their contract with an owner. This well-established system of creating buildings flourished throughout the Middle Ages, but was to change radically in the mid-1400s in Italy, as patrons demanded larger, more elaborate palaces, churches, and government buildings.

From Guild to Studio

The pioneering Italian designers of the Quattrocento were largely trained in workshops that followed an apprentice-journeyman-master system of education. If we think of design as a cognitive process, it is reasonable to expect the brain of an apprentice to absorb information about the production of useful objects very effectively by imitating the example of better-trained journeymen and masters. As we saw in Chapter 3, mimesis, or imitation, is one of the most powerful ways of passing on knowledge between individuals or groups. Mirror neurons facilitate the replication of gestures, hand skills, and coordinated activities.[3]

The guild system, as practiced by masons, metalsmiths, glass blowers, wood carvers, joiners, and other trades, depended for success on proprietary skills, methods, and knowledge. Thus, most of the techniques and educational tenets imbued in apprentices were passed along via word of mouth and demonstration, and were kept secret outside of the guild.[4] Today one of the few professions to maintain this practice is magic—masters do not reveal their secrets to anyone outside the fold.

The structure of the medieval cathedral workshop was hierarchical, with a *magister* or master at the top, assisted by an *appareilleur or apratator* as second in command of all stonecutters and other workmen. All those involved in this construction learned their skills via apprenticeship. Scholars generally believe that, prior to the thirteenth century, these workshops functioned without elaborate pictorial or written directives. When a contract was prepared, the workshop promised to execute a building under cost, design, and time guidelines. One such contract survives for the Palazzo Sanredoni in Siena, begun in the 1340s under Agostino di Giovanni and two other masters. Franklin Toker's analysis of the existing building in comparison to this document provides some fascinating information on aspects of design, aesthetics, and technics.[5]

It appears that Sienese builders were quite familiar with the forms, materials, and construction methods needed to erect an infill building in the dense fabric of the medieval town. A *maestro* would often ally himself with masters of building technics as a team when competing for a project. In the case of the Palazzo

Sanredoni, Signor di Giovanni and his father were joined by Agostino di Rosso and Ceccio di Cassino, both experienced building contractors. The full contract document is a large sheet of paper with a drawing of the proposed elevation of the palace at its top. No other drawing appears, but the remainder of the sheet contains an extensive specification for the work. How could builders construct the palace without a plan?

The contract contains the answer to this question in the form of detailed instructions on the rooms, functions, and furnishings of the building—a kind of inventory that would be familiar to any nobleman wishing to keep track of his household and belongings. Since the palace would be constructed next to an existing house with a party wall, and stretch along the Piazza del Campo for *34 3/4 braccia*, its plan dimensions were fixed by the site. The number and thickness of the interior walls were given in the specifications. The height of the elevation, and that of each story, would be determined by using two proportional systems based the "art of geometry" as known to all master masons in Europe. Toker lays out the process by which an equilateral triangle and a square, both based on the façade width, were employed in the elevation drawing to give several key vertical dimensions to the designers. Window and door dimensions, as well as distances between windows, were given in scale on the drawing, with dimensions provided as numbers on the drawing as well. Toker believes that the scale elevation was used to determine floor heights based only on geometric/graphic lines, whereas other key dimensions such as the width of an opening, were determined numerically. Some of these were dictated by law and convention, as the Palazzo Pubblico on the square was to be the model for subsequent infill buildings.[6]

Toker argues that combined textual and graphic contracts such as this one were common during the thirteenth and fourteenth centuries in Italy. He correctly points out that such a document contains all of the elements in modern "construction documents"—a contract between the owner and the builder, the architect's duties during construction, and specifications. He also points to the highly conventional building trades in Siena at the time of construction: each tradesman was fully familiar with the paradigms guiding the overall project as well as his role in its successful completion. With such a closely-knit brotherhood, facilitated by the guilds, more extensive graphic depiction of the completed building was unnecessary.[7] The kinds of drawings made by painters in preparation for their work were not part of any building contract.

Few aspects of mason's guild practices survived into the modern era once design passed into other hands. One of the advantages of a secret bond between masters in a trade was that anyone requiring their skills was compelled to use the guild's members, much in the way that trade unions have worked in the modern era. A clear disadvantage was that some knowledge could wane and disappear. Another was that innovation might well be quelled by the desire for masters to exercise their prerogatives and power over new members.

Such was the case in the building of the Duomo of Milan over several centuries. In a process documented and analyzed by James Ackerman, the masons in charge of construction quarreled over many aspects of its design over a period of more than two centuries. One faction, consisting of traditionally trained Lombard masons, argued that "ars" or craft knowledge should guide the decision for determining the height of the nave of the huge building. But the clergy overseeing the building sought advice from masons in the provinces to the north, who had more theoretical knowledge—"scientia." The two groups disagreed vehemently about whether the nave should be proportioned "ad quadratum" or "ad triangulum" using either a square or a triangle as a guide.[8] [45]

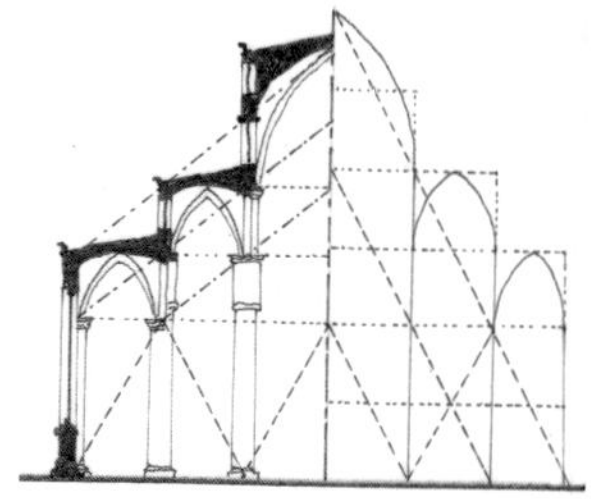

45. Comparative sections through the nave of Milan Cathedral, showing ad Quadratum, (right) and ad Triangulum, (left) configurations. (Author, after Ackerman)

Ackerman's study wasn't just a chronicle of obscure disputes about design methods that long ago ceased to matter. He underlined an important dichotomy in the way architects look at their discipline: as the craft of building versus the art of visual design. This dichotomy divides pre-Renaissance from post-Renaissance architecture in Europe and the West. It forms the basis for analyzing the history of architectural design in terms of the brain-body-environment theories that we have traced in the first three chapters of this book. When projects were conventional and problems well-understood, the artisanal craft methods for building design served well, but when a building project exceeded the conventional knowledge of guild masters something more speculative, and visionary, was needed.

The most celebrated church building project of the early Renaissance was the erection of the "cupola" or dome of Santa Maria del Fiore, the cathedral of Florence. Guild masters from all over the city were consulted—painters, silversmiths, and masons. All recognized that spanning the huge octagonal crossing would require innovative construction and design. Only Rome's Pantheon, built centuries before with methods no longer understood, provided a model. Many doubted that a solution could be found for erecting the dome without "centering" or massive wood supports for the arched ribs of the masonry.[9]

The triumph of Filippo Brunelleschi in solving the problem of the dome has been well documented in both scholarly and popular literature. His feat was almost miraculous, given the knowledge, materials, and machinery available in 1420. What has often been lost in histories of the Duomo of Florence is the critical battle between hide-bound guild practices (the Ghiberti team) and more innovate designs based upon drawings, models, and empirical study (the Brunelleschi team). [46]

Brunelleschi employed a range of drawing types to depict not only the dome and its construction, but also machines for hoisting men and materials, geometric diagrams, and details of brickwork. His skill as a draughtsman was key to his success. Had he relied only on the conventional tools of masonic

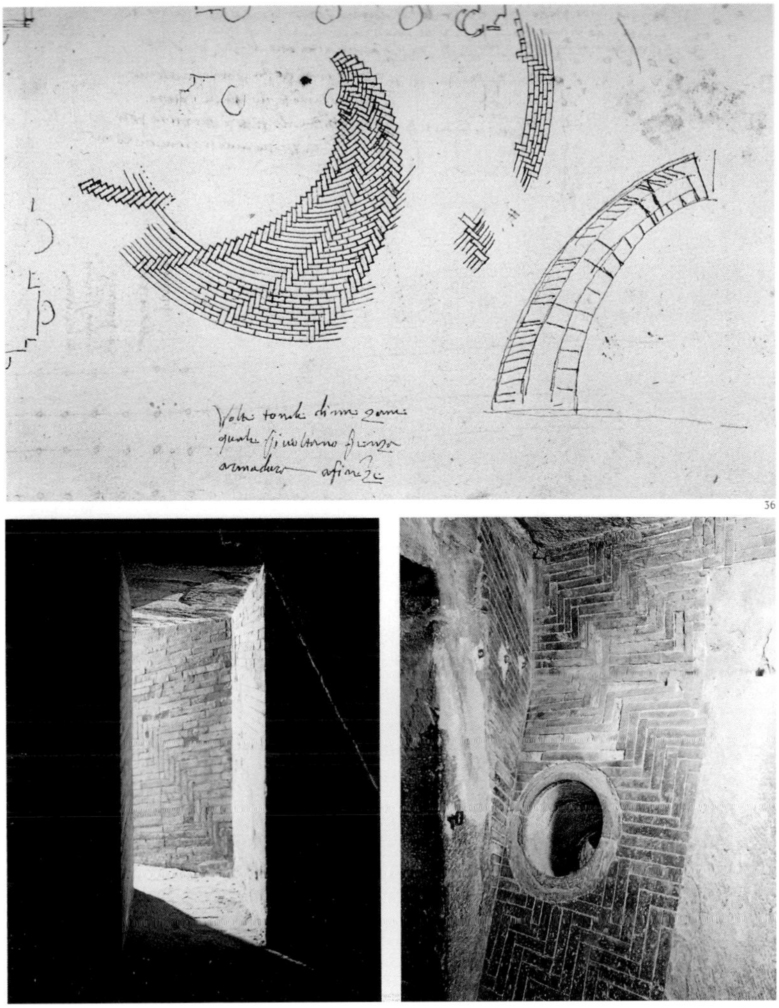

46. Drawing by Antonio da Sangallo of herringbone brick in Brunelleschi's dome, Florence cathedral. Below, photos of brickwork from inside the dome. (Author collection)

practice, he could not have understood the scale and complexity of the problem. Moreover, his curiosity about Roman classical architecture drove him to visit the Pantheon and other ancient buildings to study their construction. He extended his inquiry beyond the abstract to embrace bricklaying techniques that were shared with the workmen on the job. Massimo Ricci, a Florentine architect, and his colleagues discovered the secrets behind the brickwork only during the past decade by building a one-fifth-scale model of the dome's masonry. In so doing they refuted decades of interpretations of the dome's design and construction by some of the world's most eminent scholars and engineers.[10]

Despite the manifest progress made in building the dome according to his methods, Brunelleschi was forced to share responsibility with Lorenzo Ghiberti, his rival, because Florentine officials clung to traditional beliefs about the efficacy of guild practices. His final elevation to hero's status did not come for decades, but, when achieved, paved the way for profound changes in the way architecture was conceived. The younger Leonardo da Vinci and Michelangelo Buonarroti witnessed some of the construction. The former assisted with the design and fabrication of the gold orb that tops the lantern of Santa Maria del Fiore while an apprentice to Andrea del Verrochio. The latter used the cupola's design in his conception of the dome of New St. Peter's. [47]

47. Santa Maria del Fiore, Florence, with Brunelleschi's dome (Wikimedia Commons).

Brunelleschi and his contemporary, Leone Battista Alberti, were largely responsible for the change in attitude among patrons such as the Medici, Sforza, and Gonzaga families that pressed artists (painters and sculptors) into architecture, circumventing the mason's guilds in many key building campaigns. Alberti's important book, *De re aedificatoria* (*Ten Books of Architecture,* 1450) argued that the architect should be a man of learning and experience, not just a member of a specialized trade such as carpentry. In order to "profess" his expertise, the architect should have not only practical experience but also learning in literature and the arts.[11] Moreover, the artist was an intellectual, whose knowledge and expertise far exceeded that of any tradesman, a distinction that obsessed Giorgio Vasari and other sixteenth-century advocates for academic training in the visual arts.[12]

The workspace of the professional architect also changed from the stoneyard or building site to a studio in which elaborate drawings and models could be prepared by a master and his assistants. When Leonardo created designs for the massive equestrian monument to Ludovico Sforza in the 1490s, he was given space in the old castello near Milan's cathedral. It was here that he not only drew countless schemes for the horse and rider, but also created the celebrated clay model that advanced his reputation as an artistic genius.[13] Such a studio was not only home to Leonardo and his assistants; it was also a place where he collaborated with other masters in projects for the Sforza court. This included a trip to Pavia Cathedral with the Sienese engineer and architect, Francesco di Giorgio Martini, and many discussions with Donato Bramante about the ideal form for churches.[14] [43]

The Crucible of New St. Peter's

The transition from guild to studio would not have progressed rapidly without the patronage of popes, cardinals, and the nobility in the Italian city states. Large building projects such as the new basilica of St. Peter's, urban ensembles in Pienza, Urbino, and Vigevano, and military infrastructure protecting cities such as Milan and Siena, pressed painter-architects into service, often in collaboration with tradesmen.

When painters began to experiment with the powerful representational techniques made possible by papermaking, perspective construction, and harmonic proportions, they introduced a layer of knowledge and technical prowess that expanded the capacity of designers exponentially.[15] Scholars have often observed that the construction of New St. Peter's after the decision of Nicholas V to demolish the early Christian basilica would not have been possible without such an expanded "professional" habitus. Though records are too scant to prove close communication and trace individual relationships, it is clear that the workshop begun under Giuliano da Sangallo and fully developed under Bramante at the turn of the fifteenth century was a laboratory for the development of architectural practice in the modern sense. Virtually everyone connected with the enterprise was to influence architecture during the succeeding two centuries.[16]

Such was the feverish atmosphere at the fabbrica of St. Peters during the pontificates of Julius II and Clement X that talented painter architects developed remarkable means of studying and depicting the massive basilica that continue to fascinate scholars and the public.[17] Models, large-scale elevation and section drawings, perspective studies and sketches, and even measured notebooks with annotated details from antiquity have been associated with the workshop.[18] However, the fabbrica was by no means stripped of the knowledge and talents of master masons and sculptors—they worked hand in hand with visual designers to create the basilica. Their efforts have heretofore not been accorded the same status as those of famed artists such as Raphael and Giulio Romano, but artisanal modes of approaching design were still considered during these transitional years. Without them we would not have the fully developed Renaissance architect, nor any modern practitioner.

The group of painter-architects who came to Rome to work for Bramante formed a unique "collaborative circle" that shared knowledge and discussed their progress as building progressed on the new basilica.[19] Though there were disagreements about the design, structure, and construction of the building, and changes made by successive popes after the death of Julius II, a consistent design methodology can be traced to the remarkable group of drawings now preserved in the Vatican Library and at the Uffizzi in Florence. Scholars have studied these drawings for nearly a century, and distinct patterns emerged from their analysis that we can use today in looking at design as a discipline.[20]

Meg Licht, in a close examination of the early drawings, suggests that discussions or *ragionamenti* occurred during 1505 and 1506, as Bramante directed the construction of the four crossing piers of the new cathedral. His assistants included some of Italy's most talented draftsmen: Baldassare Peruzzi, Antonio da Sangallo, and eventually, Raphael. According to Giorgio Vasari, Giuliano da Sangallo lost his position as head of the cathedral workshop to Bramante after the latter created more persuasive designs for an entirely new church rather than accepting earlier plans to renovate the old basilica. These designs emerged from a series of "brainstorming" sessions in which competing schemes were discussed, new schemes drawn, and revisions made. The process was similar in many respects to that followed by teams of architects on a large project today.[21]

Bramante's status as a genius rests on both his executed works and the opinions of his fellow artists. Those in the St. Peter's workshop saw him as an inspiring leader and man of ideas, despite the fact that he no longer had the steady hand of a painter due to arthritis. He was aware, through his encounters with Leonardo and Francesco di Giorgio, of the power of visual representation available to contemporary artists, and its fundamental usefulness in building design. He encouraged the most talented members of his studio to produce speculative drawings showing problematic aspects of the design and construction of the new church. Many of the best were the work of Baldassare Peruzzi (1481–1536) of Siena.

48. Baldassare Peruzzi, design for St. Peter's, Uff. A107r. Four different variations on the corner niches appear in the plan, but one is represented in the center sketch. (Courtesy, Prints & Drawings Department, Galleria degli Uffizi, Florence).

One group of plans, from early 1505 (Uff. 529A verso), shows the influence of Francesco di Giorgio, with whom Peruzzi worked and studied in Siena. The geometric figures are drawn in single lines, much as Francesco did in his first treatise. On the other side of the same sheet are more detailed plan drawings and perspectives (Uff. 529A recto). The plans are close to one subsequently associated with Bramante's accepted design for the new cathedral. Remarkably, Peruzzi also sketches thumbnail perspectives of the interior of the crossing to illustrate how the plan would appear in three dimensions. Before Leonardo, architects did not make such drawings. Equally remarkable are the small sketches at the upper right of the drawing, and below. None are directly connected to the larger plan, but instead offer related ideas taken from Roman precedents.

Licht points out that a plan and perspective appear to refer to the famous Piazza d' Oro at Hadrian's Villa in Tivoli, near Rome. The perspective below seems to depict a Roman basilica or an early Christian church, as if to indicate the precedents that were being considered in the workshop. The large plan seems to refer to dimensions taken from the building site, since rows of numbers are added in several places as if calculating heights, widths, or other proportional relationships. One of the columns is labeled "Ionico" or Ionic. Uffizzi 107, verso and recto, depict plans and perspective that refer directly to the design of the crossing of the new cathedral. [48]

This sheet contains two of the most revolutionary drawings in the entire corpus of designs for the basilica. The recto sets the stage by creating a hybrid of speculative figures for the four crossing piers and the diagonal spaces behind them, usually interpreted as domed chapels. Instead of drawing a bilaterally symmetrical plan of a central dome, as appears in the central perspective sketch, Peruzzi surrounds the pantheon-like circle with four different variations on a theme. The lower half has two apsidal chapels that may (or may not) open directly into the drum of the dome above. The upper half depicts one complete pier on the right, and half a pier on the left, suggesting that the niche-like

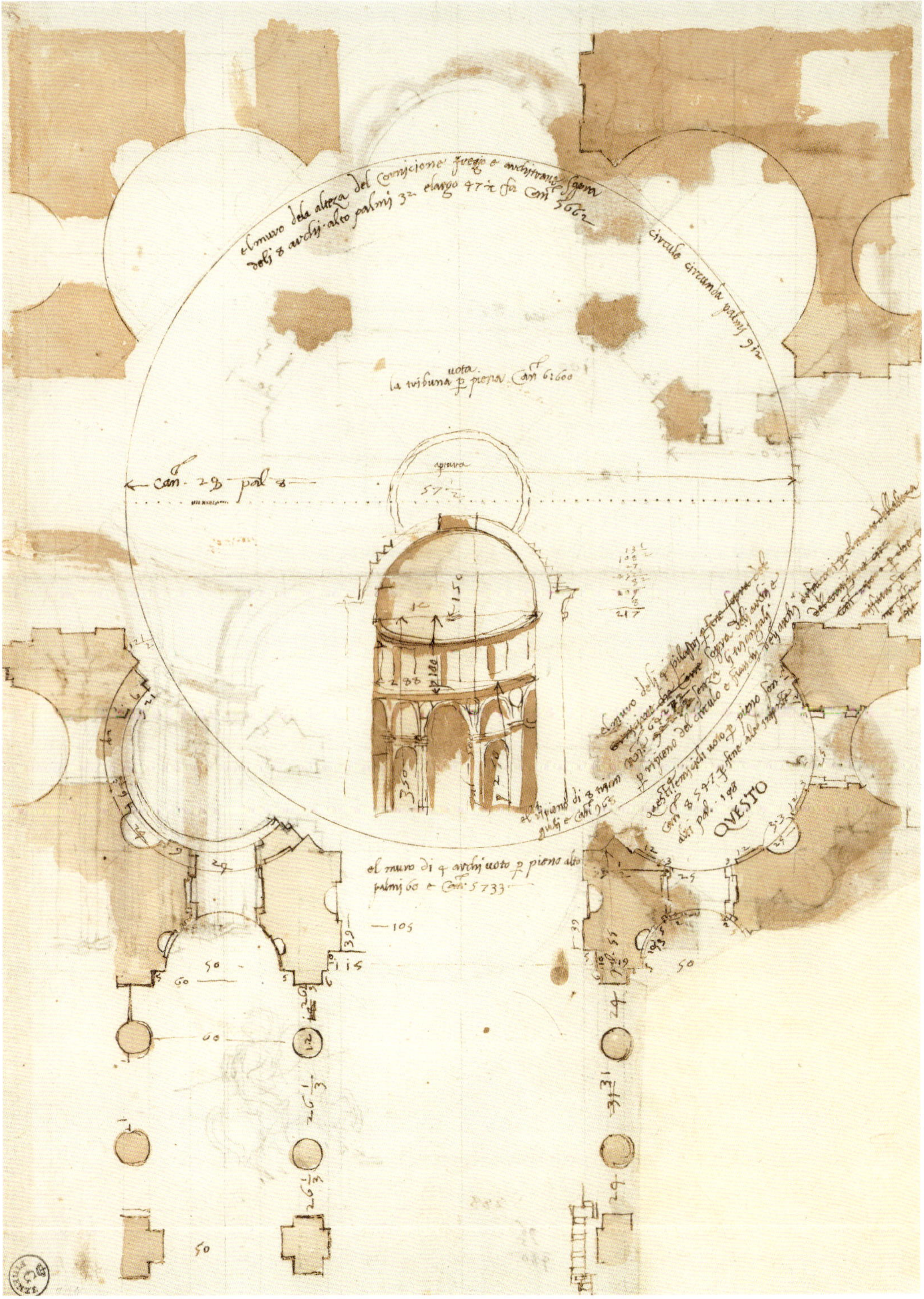

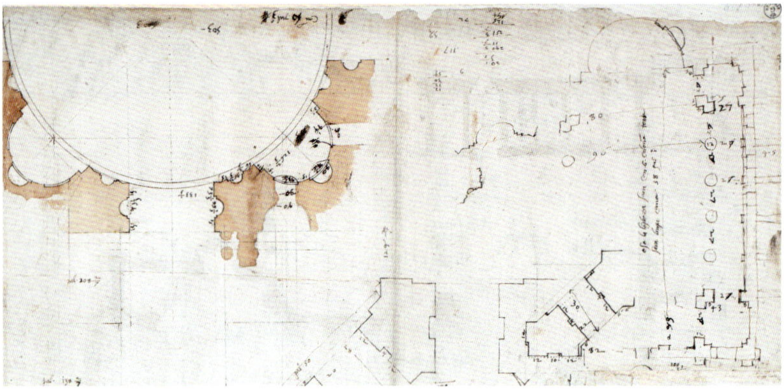

49. Peruzzi, design for St. Peter's, Uff. 109A. Both the crossing and the outer nave are part of the same drawing. (Courtesy, Prints & Drawings Department, Galleria degli Uffizi, Florence).

chapels below have morphed into fully separate spaces behind the piers. The drawing is incomplete, but notes on the lower right indicate that some decision has been made—"Questo" or "this one" suggests that the architects wish to explore one of the variations further. The exploration continues on another sheet, (Uff. 109A) [49]. Discarding the apsidal designs completely, the draftsman depicts two octagonal chapels behind the central piers, which are split into two parts or perforated—a highly unusual and structurally daring conceit. Adjacent to the half plan, on the right, is a perspective of the split pier showing how it might meet the pendentive above, and divide the group of pilasters depicted on the plan. Nothing of this kind will be constructed until the next century, as we will see in the following chapter, but Peruzzi has opened the door to new thinking about the design of the chapels. According to Ann Huppert he often drew a perspective next to a plan in order to visualize possible massing before continuing his studies, probably in imitation of Bramante.[22] This cognitive habit was not common in the previous century. It is interesting to note that below the perspective Peruzzi draws a faint sketch of a horse and rider that cannot but refer to Leonardo's Sforza monument, as no other source was known to artists in 1506, when the clay model was still standing in Milan.

One of the most difficult problems Bramante and his team faced was the reconciliation of the vaulted nave and chapels with the columns and pilasters that ensured their classical pedigree. No building since antiquity contained such a complex array of spatial figures, or a comparable collection of classical elements syntactically ordered.

In order to envision the three-dimensional mass of the piers around the nave, Peruzzi produced two extraordinary drawings, both based on the Bramante plan. One is a very modern cutaway perspective drawing (Uffizi A2) that is published in virtually every description of St. Peter's. [50] No scholar has explained how this drawing may have been laid out and conceived, as nothing like it appears for at least two centuries. Like the smaller perspectives, and Uffizi 27A, it appears

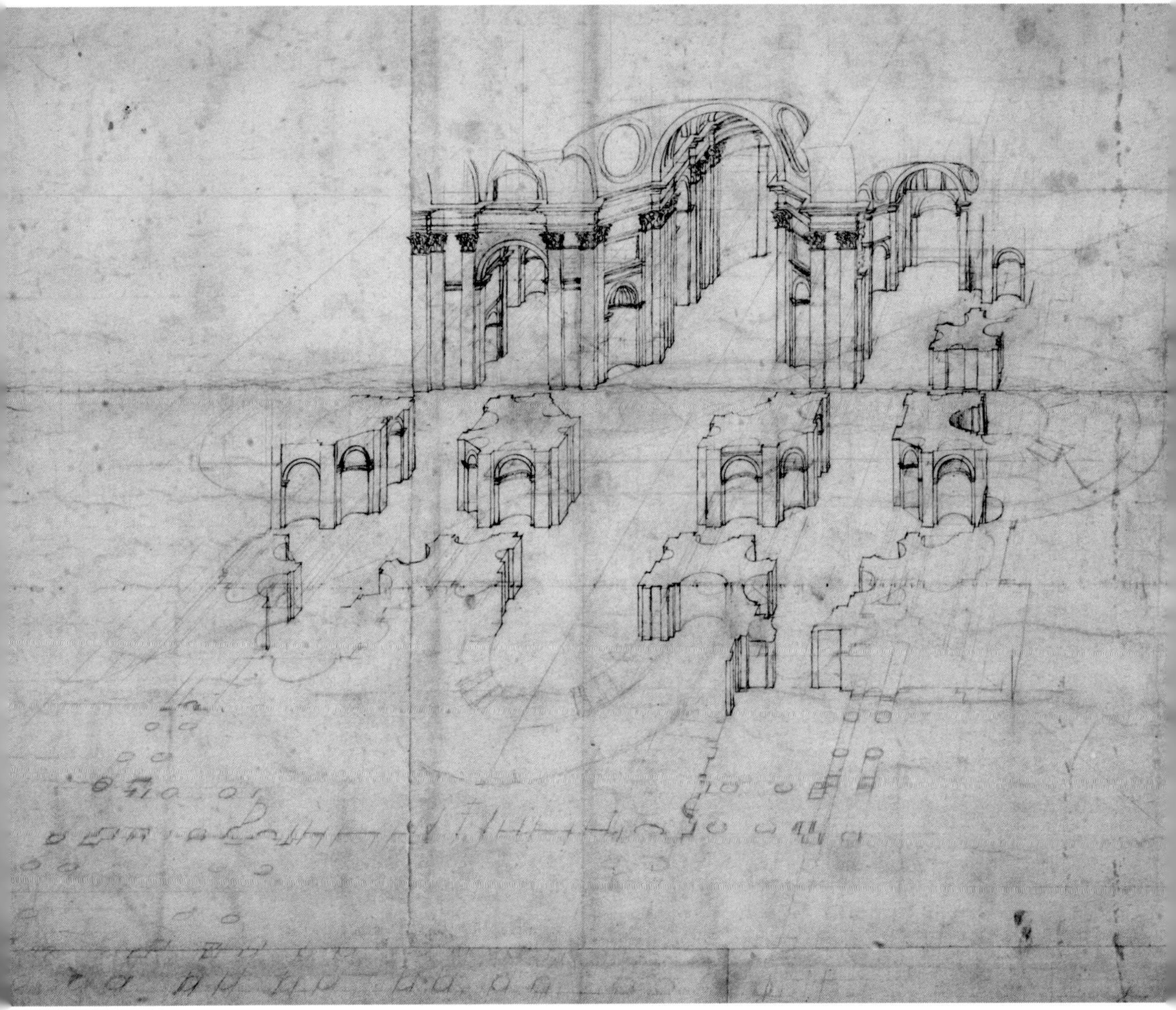

50. Peruzzi, cut away perspective, possibly based upon Bramante's preliminary plan, Uff. A2r. One of the most extraordinary drawings ever made during the sixteenth century, anticipating techniques that were not to emerge for 250 years. (Courtesy, Prints & Drawings Department, Galleria degli Uffizzi, Florence)

to have been made to extend the wall masses conceived by Bramante upward and thereby visualize their realization in three dimensions. In this respect they are speculative sketches that anticipate modern study drawings. Uff. 27A is an exterior view, perhaps intended to complement the interior perspective in Uff. 2A. [51] Licht has dated both of these drawings to 1506, though there is no consensus that Peruzzi made them at that early date.[23] We do know that a medal by Carradosso was struck in 1506 to commemorate the laying of the cornerstone of Julius II's new church. Julius died in 1513.

Bramante's drawings contain the armature of what would become the "plan on parchment" (Uff 1A) that we believe became the basis for his final, unfinished project. [52] The sheets created by his assistants are fascinating because each, in its way, interprets the master's ideas differently, and according to the training of its author. When Bramante died in 1514 his position was given to Raphael Sanzio (1483–1520), another painter-architect, rather than to Antonio da Sangallo (1484–1546), a carpenter-architect. Raphael changed the structure of the workshop somewhat, and instituted a stricter regimen of drawing and model making standards. When Raphael died in 1520, Sangallo assumed the mantle, and made both a large model and a vast number of study drawings for the basilica. Though often criticized for his unpainterly draftsmanship, Sangallo made many extraordinary orthogonal and perspective studies of the basilica, and advanced its formal development significantly. He faded into the background on his death in 1546, after which Michelangelo took over his project and created an entirely new, and superior design.

51. Peruzzi, perspective based on Bramante's approved scheme, Uff. A27r. The drawing is quite small, very like Leonardo's central plan church sketches. (Courtesy, Prints & Drawings Department, Galleria degli Uffizzi, Florence)

Since the mid-twentieth century there has been scholarly quarrel about the relationship between drawings done in orthogonal projections versus those using some form of formal or informal perspective.[24] Though the debate was necessary, the two positions have now been found to be less important than the historical record of the drawings and written accounts themselves. In the circle of architect-builders who worked on the project for nearly a century, all contributors added necessary knowledge and technical analysis to the final work. Perhaps more importantly, the "school" of St. Peter's made possible not only the standardization of architectural design practice, but also the educational system that is the basis for the Ecole des Beaux Arts.[25]

Michelangelo

Because of his long life and final association with the St. Peter's workshop, it is convenient to see Michelangelo as the most persuasive candidate for not only the universal genius of the Renaissance, but also ultimate example of a new kind of architect, a master of depiction. A recent book by Cammy Brothers argues for such a reading, but underestimates the evidence giving Leonardo and Bramante precedence as the inventors of new modes of conception in architectural drawing.[26] If considered as the inheritor of a "habitus" developed over almost 100 years since the time of Ghiberti, it is far more reasonable to place Michelangelo at the beginning of a persistent "cognitive style" or conceptual bias during the late sixteenth century, one that finally resided with Carlo Maderno at the beginning of the baroque era. He is undoubtedly the best exemplar of a powerful conceptual mode shared by two generations of Italian architects who created some of the most extraordinary buildings in history. Most importantly, his life spanned the years most critical to the establishment of the architectural profession as we know it today.

Unlike Francesco di Giorgio, Bramante, Raphael, or the Sangallos, Michelangelo insisted upon the comprehensive representation of the buildings

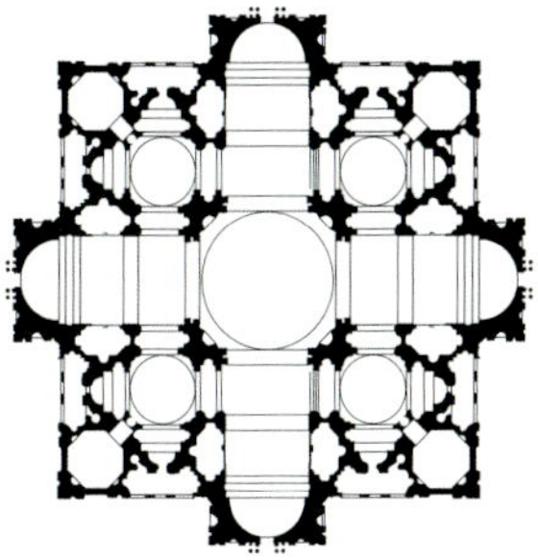

52. Bramante, the so-called plan on parchment of 1505, (Wikimedia Commons).

53. Michelangelo, modello of a door, Casa Buonarroti, Florence. Author photo from Metropolitan Museum exhibition. The artist would often explore several variations on a single sheet, even as it was presented in "finished" form for the patron.

and objects under his design supervision. He needed to know every detail, every surface, and every material prior to the construction of a project. The great German scholar, Christof Frommel, observed that he limited his use of perspective to conceptual sketches of interiors, using only the "triad" of plan, section, and elevation in all of his projects. "The most illustrious designers of the years to come [following Michelangelo] … inherited a highly developed patrimony which needed no fundamental improvement. In the same way in which every step in the development of graphic representation was linked to

54. Michelangelo, study for a double tomb. (Courtesy, British Museum, London). The speed of the drawing is evident in the lines that are built up on the sheet.

transformations in the history of architecture, so the retention of these methods of designing also implied a continuity of architectural form."[27] [53]

As Semir Zeki has said, Michelangelo was an artist with a unique understanding of his own creative, intellectual, and visual processes—quite literally a man offering a clear view of the brain of a genius. By consulting his poetry when looking at unfinished drawings and sculpture, Zeki has postulated that the great artist left his art works in a state of "becoming" until quite late in the fabrication, thereby leaving open a range of possible endgames or resolutions. In his architecture this modality worked particularly well, as several of his projects

55. Michelangelo, study for the Laurentian Library. (Courtesy, British Museum, London). A true demonstration of "thinking on paper," as the artist explores the wall in both two and three dimensions.

were inherited from previous designers in an unresolved state—the most famous case being the completion of St. Peter's basilica.[28] More importantly, he extended the process of conceptualization so that more "unseen" aspects of the building could be studied prior to construction. In this way he avoided the most serious problem with Bramante's designs—their ultimate failure to anticipate spatial and structural contradictions as the masonry piers arose from a ground plan that was inadequately represented and understood by its author.

Michelangelo was a master of many arts, but he considered himself a sculptor foremost. As he modeled his figures he engaged mirror neurons, and empathy, to connect with living humans. His uncanny haptic connection with the body has struck virtually every critic from de Tolnay to Pater. We know that while acting as an architect he had assistants build elaborate models of works in design, and also cut examples of key moldings before final stones were quarried and cut. The dome of St. Peters was designed using these tools.[29] Years working with quarrymen at Carrara taught him that three-dimensional design required three-dimensional material representation. This experience, though not unique among

architects (Sansovino and Palladio were stonecutters) of his time, gave him a fuller understanding of the complex relationships between mass and space.[30]

His sculptures also demonstrate an understanding of what Paul Frankl has called "corporeal form," and such physical and psychological sensitivity to the human body was central to his architecture.[31] As James Ackerman has written, "his association with architecture with the human form was no longer a philosophical abstraction, a mathematical metaphor. By thinking of buildings as organisms, he changed the concept of architectural design from the static one produced by a system of predetermined proportions to a dynamic one in which members would be integrated by the suggestion of muscular power." Ackerman also observes that his sketches were fluid, and that he continued to explore new ideas even as his buildings were under construction.[32] [54]

The 1950s analysis of the Laurentian Library in Florence by Rudolph Wittkower remains a paradigm among studies of Renaissance architecture. [55] The author showed that key drawings during the development of the design gave Michelangelo visual cues for a transformation that produced the masterpiece we know today. A key group of sketches proved that the artist was able to change course following decisions by Pope Clement VII (Medici) about the height of the vestibule as well as the staircase. [56] Unlike Leonardo, whose sketches were purely speculative, these drawings were used to refine ideas according to conditions in the field, as issues of structure and scale forced reconsideration of early design ideas. Though they are now lost, we know that he constructed a number of models of the library to assist in making key decisions. He also drew full-scale outlines of key moldings. [57] In this way Michelangelo acted in much the same capacity as a modern design-builder, who must continuously react to exigencies outside his control to guide a project to completion. Significantly, many of Michelangelo's projects were additions to or completions of buildings already standing.[33]

These individual habits of design mark Michelangelo as an innovator during one of the most revolutionary periods in the history of art. They are keys to workings of the brain in a unique genius. Yet Michelangelo worked in a milieu in which all visual artists defined their craft in similar ways. He shared with other painter-architects a bias toward *disegno* as the mother of all arts, and therefore created appropriate drawings at various stages in a project, some for his own edification, some for presentation to clients, others as instruction for other artisans. Like Leonardo, he explored the formal and material aspects of architecture using study drawings that were often left in unfinished or barely legible states. Often he would create an ambiguous sketch or even presentation drawing (modello) and then use a variation on it to produce the final design, perhaps in a model.[34] As Vinod Goel has noted, this use of a "sketch" allows for maximum creative freedom, something Michelangelo valued greatly.

56. Michelangelo, ricetto or vestibule, Laurentian Library, Florence. See also Fig. 146. (Author collection)

57. Michelangelo, drawing of stairway and moldings, Laurentian Library, from Casa Buonarotti, Florence. Note the cutaway in the upper right corner of the sheet, a full scale molding profile. Author photo from Metropolitan Museum exhibition.

58. Bramante's Tempietto, Rome, from Codex Mellon. 1513. (Courtesy, Morgan Museum and Library)

Teaching Disegno with Illustrated Manuals

As artist-architects acquired more prestige and professional status in the courts and cathedrals, they drew increasing numbers of aspiring designers to their studios. It wasn't long before the most esteemed painters, sculptors, and architects formalized the training given to these apprentices. A large workshop such as the one at the Vatican in Rome required standardized training in drawing, model-making, and rudimentary engineering. Generally the new workshops were organized by leading patrons, such as the Medici in Florence or the Gonzaga in

59. Bramante's Tempietto, Rome, from the Codex Coner. Though the use of perspective is similar, it does not have a "cutaway" section. (Courtesy, Trustees of Sir John Soane's Museum)

Mantua. Some patrons, such as Giangiorgio Trissino in Venice, trained gifted young designers such as Andrea del Pietro (Palladio), even sending them to Rome to study ancient monuments.

There were no established standards for drawing classical details and ornament in the early Renaissance, so Quattrocento architects experimented with various means of depicting say, cornices, by reading Vitruvius and sketching Roman ruins in situ. They kept collections of such drawings in sketchbooks that could serve as pattern references in the studio. Several created "treatises" that were annotated collections of their drawings, while others, such as Alberti,

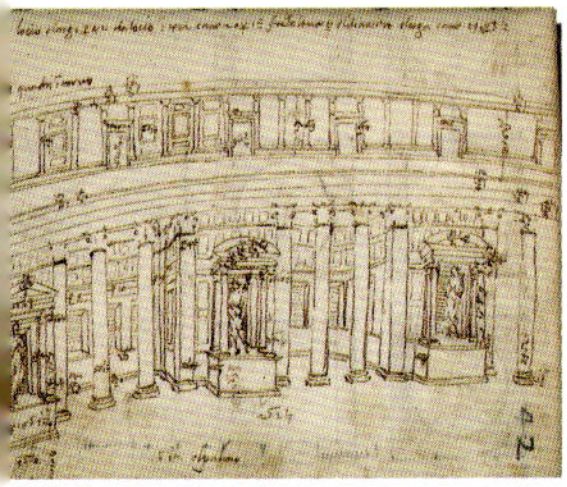

60. Interior of the Pantheon, Rome, from Codex Mellon, showing the obsession with perspective. Note the measurements in the drawing. It was done in situ. (Courtesy, Morgan Museum and Library)

61. Interior of the Pantheon, from Codex Coner, using the common section /perspective technique noted by Lotz. Here no measurements are included. (Courtesy, Trustees of Sir John Soane's Museum)

wrote without including groups of illustrations. Leonardo's notebooks are something of a hybrid between the two types of loosely bound collections of text and illustrations. It is important to note the he, like many painters, had a sketchbook strapped to his belt so that he could record his visual impressions of anything that caught his fancy. The earlier pattern books were literally source material for paintings.

Two unsigned architectural sketchbooks have been analyzed by scholars: The Menicantonio Sketchbook (or Codex Mellon) and the Codex Coner in the collection of Sir John Soane. Each explains some of the problems faced by designers wishing to impart professional standards of drawing and model making on their associates. Each demonstrates the mnemonic purpose of the illustrations. Amazingly, they are almost identical in the types of drawings used to study Roman buildings, and the subjects they present. [58]

The Codex Coner, in Sir John Soane's Museum, has been available for study since the nineteenth century, but has only recently been interpreted in connection with St. Peter's and Bramante's circle.[35] [59] Many of its exemplar drawings of Roman and Renaissance monuments are drawn using perspective, even conventional elevation and section projections. There is no question that the architects making the drawings were directed to measure and draw every key example of a Roman building that could readily be accessed, including the Pantheon, the Colisseum, and the extant Roman baths.

The "Menicantonio Sketchbook" (Codex Mellon) was discovered only in the 1970s and has not received a similarly informed reading by contemporary scholars. Both, it has been argued, were intended as reference and teaching manuals for Renaissance architects, and were made in connection with the building of the new basilica of St. Peter's. Unlike didactic treatises, such reference books helped student architects, and practicing experts, understand the proportions, ornamental syntax, and contours of the classical orders. Without them these systems could not be passed on from apprentices to masters, nor could extant Roman buildings be compared with Vitruvian rules and ideals.

The Codex Mellon, now in the Morgan Library, is wide ranging and fascinating in its subject matter, a fact that puzzled the great Rudolf Wittkower when he first saw it.[36] Like the Codex Coner, it contains both ancient Roman and contemporary classical buildings such as Bramante's Tempietto. Its date, 1513, places it during the late tenure of Bramante and the transition to Raphael in the St. Peter's workshop. At least two drawings among its more that 80 pages depict Raphael's designs for the cathedral; the originals are lost. Though Wittkower thought it the preliminary draft of a treatise, he was mislead by his fixation on theory in Renaissance books. The drawings, probably made by an assistant to Raphael, were not a homage to Vitruvius or Alberti. They were references, memory prompts, for use in the studio by practicing designers. [60, 61, 62]

Organized by subject matter, the 80-page book does not have large text portions, but rather contains captioned illustrations of building plans, sections, elevations, and details. By comparing ancient examples with "modern" works by esteemed masters such as Bramante, Raphael, Sangallo, and their contemporaries the book was an ideal teaching tool for apprentices. It is likely that Raphael himself supervised its creation, and directed draftsmen under his charge to measure Roman buildings. Unquestionably, a man charged with the biggest building project in Europe would look to organize reference material for his assistants, and to document the finest extant buildings in the region. It is also reasonable to expect that Raphael would develop some sort of educational program for exposing new architects to new methods of design, building materials, and building technics.

62. Perspective sections of Baths of Diocletian and Basilica of Maxentius, Rome, from Codex Mellon. (Courtesy, Morgan Museum and Library)

We do not know how many books such as the two above were created during the sixteenth century, though others have been discovered and examined. It is quite likely that such sketchbooks passed into the hands of Sebastiano Serlio, through his teacher Peruzzi, and became the basis of his epochal treatises published in the 1530s and after. There are, however, manuals for teaching painting, sculpture, and decorative arts from that period, and many were associated with teachers or founders of art academies. Alberti wrote two treatises devoted to those subjects: *De Pictura* and *De Statua*. Luca Pacioli created a book in concert with Leonardo, entitled *De divina proportione* (1509) as a manual for the visual arts, with some references to music. Most influential were works illustrating the training of visual artists that could be used in studio instruction.[37]

The inventor of *disegno* as an academic principle, and one of Michelangelo's key exponents, was Federico Zuccaro (or Zucchari) (1540–1609). Though the "divine" master did not write a treatise, his admirers were quick to appropriate the artist's ideas in advancing their own careers. None was more powerful than the notion that visual art be taught through the vehicle of design. Federico's brother Taddeo was a student of Michelangelo's and made a sterling start as an artist prior to his untimely death. After a colorful career as a historical painter both in Italy and abroad, Federico returned to Rome in 1568 and helped to found the most important art academy in Italy, the Accademia di San Lucca.[38] Its sister academy, the Accademia del Disegno, had honored Michelangelo as a founder shortly before his death in 1564. Federico created a fascinating painting of the artist in the pose of his sculpture of Moses for the tomb of Julius II, and made a number of drawings depicting events in the life of his brother. The portrait, from around 1600, shows Michelangelo holding a role of paper, and surrounded by drawing instruments, a testament to his visual abilities, his "giudizio dell'occhio." One of his drawings depicts art students in the New Sacristy at San Lorenzo, intently drawing the sculpture and architectural ornament designed by Michelangelo.[39]

B
A

When Zuccaro wrote about *disegno* he presented a strong case for drawing as a means of cultivating such "judgment of the eye." *Disegno interno* concerns the inner logic, geometric structure, and Platonic essence of the object to be drawn or designed. *Disegno esterno* is the final realization of that intellectual and spiritual idea, in physical form. When he developed the curriculum at the Accademia di San Lucca he inserted these concepts into all aspects of teaching and learning. Moreover, he extended the achievements and theories of Michelangelo through the vehicle of the academy in Rome, and gained the patronage of the church by doing so. In his *L'idea de pittori, scultori ed architetti* of 1607, Zuccaro outlined his comprehensive system of design education along the lines of the dominant philosophical and theological tenets of the Roman church.[40]

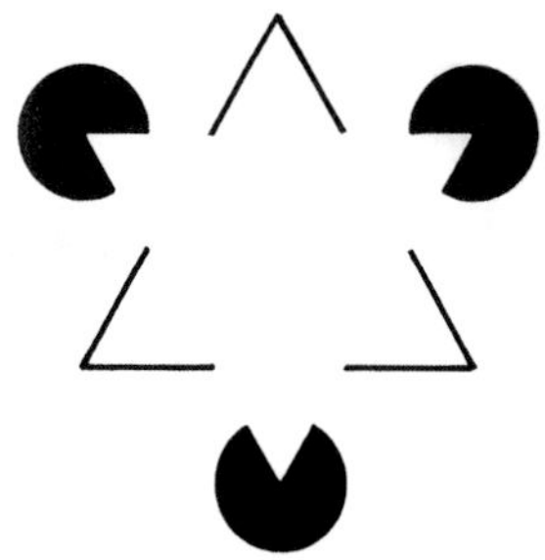

63. Kanisza illusion. One of the two triangles is not "present" in the drawing. Many such illusions result from the visual system's favoring one type of information over another.

Theologians and artists had shared an attachment to Neoplatonic philosophy since the early Middle Ages. Plotinus, Thomas Acquinas, and Marsilio Ficino synthesized aspects of Platonic, Aristotelian, and Stoic philosophy in their work over several centuries. Plotinus began teaching in the third century, and it was he who first associated Plato's *eidos* with proof of the divine in the works of humankind and nature.[41] Aquinas extended the intellectual and spiritual breadth of Neoplatonic philosophy into Scholastic doctrine, as Erwin Panofsky demonstrated in several important publications during the mid-twentieth century.[42] Ficino taught artists and clergy at the Florentine Academy during the Quattrocento, using precepts from the two earlier philosophers. From this tradition, painters and sculptors learned that a "spark of the divine" could inspire their creative endeavors, and saw its apotheosis in the work of Michelangelo, "il divino."[43]

Zuccaro and his fellow artists used pattern books such as the Codex Coner, examples from the sketchbooks of renowned painters such as Michelangelo and Raphael, and drawing exercises common in the *botteghe* (studios) of such masters as tools for teaching *disegno*. The studio curriculum stressed fundamentals in early years, later encouraging students to specialize in painting, sculpture, decorative art, or architecture. The Accademia di San Lucca became the first "Beaux Arts" design school in Europe, and the model for dozens of others during succeeding centuries. Its roots were in the notebooks of Leonardo and the *ragionamenti* of the St. Peter's workshop, but its influence extended into the architectural practice of the Baroque era, as we shall see in the following chapter.[44]

Drawings and Perceptual Presence

The discovery of depiction by Renaissance artists and architects stimulated the creation of larger and more complex buildings and urban ensembles. Today, psychologists understand that humans use various contextual affordances in order to make sense of their environment, but have seldom explored how the things we make require similar kinds of external aids to perception and memory.

What they call "perceptual presence" is a phenomenon in which non-sensory, non-veridical (non-real) images occur in the mind to "fill in" aspects of object or scene perception that cannot be provided by the senses themselves. The most famous example of this is the existence of two triangles in the Kanisza illusion. [63] Neither is illustrated by the lines or "Pacman" figures in the corners of the drawing. The triangles are present only in the mind of the observer. In hearing, a similar phenomenon occurs when listeners see the lips form a consonant and "hear" the sound even though the ears/auditory cortex are not receiving the same signal—the McGurk Effect.

Were Renaissance architects to look at Uffizzi A2, the Peruzzi perspective illustrated above, those trained to see in perspective would fill in the vaults and domes above the truncated walls. This is not unlike the picture of a cat behind a picket fence described by Alva Noe in his book, *Action in Perception.* As he puts it, "The strictly unseen environmental detail [parts of the cat occluded by the fence] seems perceptually present even though we do not see it all at once. We do not merely *think* that these features are present."[45] Architects, of course needed to acquire a knowledge of drawing in order to make a picture like the one in Peruzzi's sketch, but once enlightened, could establish a "presence" for unseen aspects of their designs on paper.

This comes at the issue from a different direction than that normally explored by philosophers like Noe, or psychologists like James Gibson, who was one of the first researchers to discuss the way the environment constrains and reveals the world to the senses and brain. I think, however, that historians must continue to grapple with the problem of how humans began to change their environment (buildings and landscapes) as a result of new means of depicting it. Before the widespread use of study sketches and linear perspective, builders could not explore complex spatial characteristics of buildings because they had no means of imagining them outside their minds, their immediate understanding of the artifacts themselves. Once a wider range of drawings and models became available, they were able to generate perceptually "present" ideas. Pattern books and canonical drawing methods gave designers a clear reference not only for use in design sessions, but also as a means of imagining complex forms and spaces to generate a presence in their minds. We can understand Steven Pinker's comment about the pastiche or slide show of views available to an educated "looker" when we see such Renaissance drawings.

Once a new pattern of design emerged in post-Renaissance practice, architects were delirious about the potential of new modes of representation. They began a two-century period of experimentation with radically plastic and dynamic form that we call the Baroque and Rococo. The new materials, craft practices, and trades that emerged during this same period spurred their inventions to heights that continue to impress us today.

Notes

1 For an excellent treatment of Quattrocento practice, see Leopold D. Ettlinger, "The Emergence of the Italian Architect during the Fifteenth Century," in Kostof, *The Architect: Chapters in the History of the Profession* (New York, Oxford Univ. Press: 1977): 96–123.

2 On terminology in the middle Ages see Nikolas Pevsner, "Terms of Architectural Planning in the Middle Ages," *Journal of the Warburg and Courtauld Institutes* 5 (1942): 232–37.

3 For an excellent explication of gestures and other kinds of activities stimulated by the mirror neuron system, see Arbib, *How the Brain Got Language*, 72–147.

4 See Paul Frankl, "The Secrets of the Mediaeval Masons," *Art Bulletin* 27:1 (1945): 46–65.

5 Franklin B. Toker, "Gothic Architecture by Remote Control: An Illustrated Design Contract of 1340," *Art Bulletin LXVII: 1* (March 1985): 67–94.

6 Toker, 80–83.

7 Toker, 84.

8 James S. Ackerman,"Ars Sine Scientia Nihil Est: Gothic Theory of Architecture at the Cathedral of Milan," *The Art Bulletin*, 31 (1949): 84–111.

9 For an up to date, well-written account see Ross King, *Brunelleschi's Dome: How a Renaissance Genius Reinvented Architecture* (London, Bloomsbury Books: 2000).

10 See Barry Jones, Andrea Sereni, and Massimo Ricci, "Building Brunelleschi's Dome: A Practical Methodology Verified by Experiment," *JSAH* 69:1 (March 2010): 39–61. The authors demonstrate conclusively that Brunelleschi worked out his design empirically, experimenting with ways of laying out the geometry of the dome and testing methods of laying the bricks in conic sections.

11 The standard English translation is: Leon Battista Alberti, *On the Art of Building in Ten Books*, trans. Joseph Rykwert, Neil Leach and Robert Tavernor (Cambridge, MIT Press, 1991).

12 See Robert Williams, *Art, Theory and Culture in Sixteenth Century Italy* (Cambridge, Cambridge Univ. Press: 1997): 1–72.

13 The process is well documented in Walter Isaacson, *Leonardo da Vinci* (New York, Simon & Schuster: 2017): 160–69.

14 Leonardo met Francesco di Giorgio in 1490 when the two collaborated on a design for a "tiburio" for Milan Cathedral, and later traveled with him to Pavia. On that trip they were asked to consult with the masons building Pavia cathedral, by Leonardo's patron, Ludovico Sforza. See Isaacson, 140–52.

15 For a discussion of the emergence of architecture as a "profession" in Italy see Elizabeth Merrill, "The *Professione di Architetto* in Renaissance Italy," *JSAH* 70: 1 (March 2017): 13–35. Her discussion focuses on literature and the formation of educational institutions, such as the artistic academies, in the 15th and 16th centuries.

16 The clearest short exposition of the complex design of the cathedral is in Wolfgang Lotz and Deborah Howard, *Architecture in Italy 1500–1600* (New Haven, Yale Univ. Press: 1995): 1–98.

17 See for instance, Christof Luitpold Frommel, "St. Peters: The Early History," in Millon, Ed. *The Renaissance*, 399–423.

18 Henry A. Millon, "Models in Renaissance Architecture," in Millon, Ed. op. cit., 18–74; and C.F. Frommel, "Reflections on the Early Architectural Drawings," ibid., 101–22

19 See Michael P. Farrell, *Collaborative Circles: Friendship Dynamics and Creative Work* (Chicago, Univ. of Chicago Press: 2001).

20 The most comprehensive current history of the basilica is H. Brandenburg, A. Ballardini & C. Thoenes, *Saint Peter's: The Story of a Monument* (Vatican City, Editoriale Jaca Book SpA, Milano: 2015).

21 Meg Light, "I Ragionamenti - Visualizing St. Peter's," *JSAH* 44:2 (May 1985): 111–28.

22 See Ann C. Huppert, *Becoming and Architect in Renaissance Italy: Art, Science and the Career of Baldassare Peruzzi* (New Haven, Yale Univ. Press: 2015): 105–18.

23 Huppert cannot date it precisely but admits that it must refer to Bramante's schemes of 1507. She suggests that it may be a working drawing, 115.

24 Ann C. Huppert, "Envisioning St. Peter's: Perspectival Drawings and the Process of Design," *JSAH* 68:2 (June 2009): 158–77.

25 For a complete history, see Christof Thoenes, "The New St. Peter's," in Brandenburg, et al., *Saint Peter's: The Story of a Monument* (Milan, Editoriale Jaca Book SpA: 2015): 165–303.

26 Cammy Brothers, *Michelangelo, Drawing, and the Invention of Architecture* (New Haven, Yale Univ. Press: 2008).

27 Frommel, in Millon, op. cit., 120.

28 See, for instance, "A neurobiological appraisal of Vermeer and Michelangelo," in Zeki, *Inner Vision*, 22–36.

29 As Mauro Mussolin has noted, "the long-established tradition was transformed in Michelangelo's hands, attaining a material, technological, and typological flexibility associated with architectural drawing." "Michelangelo and the Experience of Space," in Bambach, cited below, 273.

30 Much of this is documented in Carmen Bambach, *Michelangelo: Divine Draftsman and Designer* (New York: Metropolitan Museum of Art: 2017).

31 Paul T. Frankl, *Principles of Architectural History* Trans. O'Gorman (Cambridge, MIT Press 1969).

32 James S. Ackerman, *The Architecture of Michelangelo* (Harmondsworth and Baltimore, Penguin: 1971): 44.

33 Rudolf Wittkower, "The Biblioteca Laurentiana," in *Wittkower, Idea and Image: Studies in Renaissance Architecture*. 10–72.

34 This is noted and analyzed in Caroline Elam, "Prose and Poetry in Michelangelo's Architectural Drawings," in Bambach, *Michelangelo*, 268–72.

35 Chen Liu, "Between Perception and Expression: The Codex Coner and the Genre of Architectural Sketchbooks," Unpublished Ph.D. Dissertation, Princeton University, 2011.

36 R. Wittkower, "The Menicantonio Sketchbook," in *Idea and Image*, 90–107.

37 On Pacioli's influential book, see John Onians, *Bearers of Meaning: The Classical Orders in Antiquity, the Middle Ages, and the Renaissance* (Princeton, Princeton Univ. Press: 1988): 216–24.

38 See Cristina Arcidini Luchinat, *Taddeo e Federico Zuccari: Fratelli pittori del Cinquecento* (Milan, Archivi del arte antica): 1,998–999, 2 vols.

39 Bambach, *Michelangelo*, 256–58.

40 The best synopsis of his treatise is found in Williams, *Art, Theory and Culture*: 135–50. He suggests that the "aim of Zuccaro's treatise is to show that *disegno* is the fundamental principle, indeed the *cause* of all human thought." 136.

41 See John S. Hendrix, "Humanism and Disegno: Neoplatonism at the Accademia di San Lucca in Rome," (2007) *School of Architecture, Art and Historic Preservation Publications*, Paper 1; accessed online at docs.rwu.edu/saahp_fp/1.

42 Notably Panofsky, *Idea: A Concept in Art Theory* (Columbia, Univ. of South Carolina Press: 1968); and *Gothic Architecture and Scholasticism* (New York, New American Library: 1976).

43 Kathryn Blair Moore, "Ficino's idea of architecture: the 'mind's-eye' view in Quattrocento architectural drawings, *Renaissance Studies* 24:2 (June 2010): 332–52. Moore underlines Ficino's belief in the importance of drawing in the Neo-Platonic theory of *disegno*.

44 On the Baroque see Pevsner, *Academies of Art*, 67–82.

45 Noe, *Action in Perception*, 60.

[6]

SCENOGRAPHY AND CRAFT IN THE BAROQUE

In many ways the apotheosis of embodied artistic production occurred during the Baroque era, both in Europe and in its far-flung colonies. From 1600 until 1750 a confluence of historical and cultural forces propelled the creation of new forms such as opera, interior decoration, modern scenography, plaster casting and gilding, marquetry in woodworking, soft paste porcelain in Europe, expert silversmithing, Venetian glassblowing, and many other trades that catered to a newly wealthy bourgeoisie in France, Spain, Italy, and Holland. As Peter Thornton has observed, France created not only a modern commercial economy during the mid-seventeenth century, but also established standards of aristocratic taste that rapidly spread across Europe and into Russia.[1] Even Spain's wealthy colonies in Central America, and Holland's in the East Indies, created elaborate decorative art for a world market. Moreover, trade with China and India brought rich materials to artisans in Europe and the Americas.

The most far-reaching cultural force of the seventeenth century was undoubtedly the art of modern scenography, which served not only the opera stage but also the royal courts of France, England, and Spain. Masques, ballets, musical festivals, and formal balls called for the production of elaborate machinery, sets, costumes, and even new musical instruments.[2] The composers who invented opera as a dramatic medium—Monteverdi, A. Scarlatti, Pergolesi, Rameau, Gluck, Handel—also advanced the language of music immeasurably. The great librettist Pietro Metastasio (1698–1782) virtually created the template for *opera seria*.[3] His friend Farinelli (1705–1782) revolutionized the art of singing with his striking castrato voice.

The new scientific ethos of the era expanded the reach of representation into new realms. Particularly far-reaching was the art of medical illustration, which influenced painting and architecture as it had in the time of Leonardo and Durer. Etching and lithography gave draftsmen new, more precise, means of depicting forms and space. Scenographers and architects, often the same person, used these new media to their advantage on Italian, English, French, and Viennese stages. Advances in printing and bookmaking increased the dissemination of images and ideas. Literally hundreds of illustrated books appeared on art, architecture, and symbolic decoration.[4]

Nothing, however, matched the inventiveness displayed in theaters and their productions. Mastery of illusionistic stage design passed from Italy to France to England, and reached its height in the "Teatrum Sacrum" of rococo stucco decoration in Bavarian and Middle European churches.

With a new market, traditional trades and crafts expanded to produce many of the most elaborate masterpieces among the decorative arts. French baroque ornament morphed into the rocaille style that eventually flourished in the German principalities and Vienna. Families of craftsman, such as the brothers Asam, came to dominate the production of these wood, gilt, plaster, and metal interior scenes, eventually for churches and abbeys. The wild, fantastic, anthropomorphic surfaces of baroque decoration enclosed some of the most complex, emotionally moving spaces ever created. Sometimes it appeared that artisans had molded "space into light" in their superb creations.

In this chapter we shall encounter a few of the artists who created these new forms and examine their connection to creative faculties in the human brain. Some remarkable exograms helped designers to fashion and make these artifacts.

64. Andrea Pozzo, Teatrum Sacrum, from his treatise on quadrattura painting and perspective. The scene was actually realized in a church. (Author collection)

Memory, Theater, and Fantasy

Every drawing or representational type used by designers and builders, from Egypt until the present day, served one cognitive function above all others: a mnemonic device for testing visual and conceptual aspects of the object to be designed, as they developed their ideas. This fact contradicts a view commonly held today: that representation of the world serves mainly as a means of communicating meaning to others. Though such a function is clearly a role of language, non-verbal cognition does not always share this characteristic. More significant is the fact that in many eras the theater provided the venue for making such "fantastic" and inventive images manifest in the brains of designers.[5] We saw that in ancient Greece and Rome architecture and the theater were intimately linked. As human voices and bodies imitated the social and moral actions of their world, Greek and Roman designers also represented the physical and spiritual cosmos with painted scenery.

FIG. LXXI.

It is important to recognize the interconnected, and often ambiguous, relationship between drawn "capriccios" or fantasy sketches, stagecraft, interior design, and architecture during the Baroque period.[6] One of the first artists to explore this relationship was Padre Andrea Pozzo (1642–1709), a Roman priest and designer associated with the Jesuits during the Counter-Reformation. Like Baldassare Peruzzi and his pupil Sebastiano Serlio, Pozzo designed temporary settings for masses, celebrations, rituals, and musical performances in churches throughout Rome. Like most artists of his time he was fascinated by the geometry behind perspective construction, but was more interested in extending the illusion of depth that could be achieved on curved and vaulted surfaces. His painted ceiling at the church of Sant Ignazio is one of the marvels of trompe l'oeil painting in Europe.[7]

Pozzo was but one of a large group of Italian artists who helped define the art of scene design during the seventeenth and eighteenth centuries. While some specialized in church or villa murals and decorative painting, many in Rome, Venice, and Naples turned to opera as a vehicle for their imaginations. Early music dramas were staged by high-church officials such as Cardinal Pietro Ottoboni, beginning with sacred oratorios and chamber operas. The first opera, Monteverdi's *Orfeo* (1607), was called a "favola in musica," and had relatively modest sets when staged in Mantua. The center for composers and singers gradually shifted to Naples, where members of the large Bibiena family produced operas, built theaters, and established a new entertainment industry. As Joseph Kerman has shown, the monumental power of the first operas to create a multi-sensory dramatic catharsis was a cultural force in and of itself.[8] Every art form emulated operatic fantasy and musical emotionalism from the moment of their first stagings in Italy. It is easy to see why now that we understand the fusion of sensory and intellectual apprehension in the brain—here was the first fully synesthetic art form in Europe.

One of the most far-reaching and influential of Pozzo's designs was his "Teatrum Sacrum" for a 1685 marriage celebration in the church of the Gésu in Rome, home of the counter-reformation's "soldiers" of propaganda: the Jesuits. [64] In this design, engraved for presentation in his treatise, Pozzo depicted a multi-layered altar in the form of classical gates, frames, and towers that receded into the picture plane in a one-point perspective construction. Unlike the false fronts of Serlio's Renaissance scenes, the entire illusion remained on a single plane, a canvas fresco. Pozzo bragged that his set was so realistic because he had actually designed it in plan and elevation before creating the perspective: "all was performed with such exactness that the work seemed to consist of solid stones rather than wrought by the painter's hand."[9]

Pozzo's intent in creating this particular setting was to suggest a set of buildings receding into space using one-point perspective, with an upward-tilted

65. Giuseppe Galli Bibiena, *scena per angolo* setting, plate 8, from *Architectural and Perspective Designs*, Dover reprint.

point of view. He was necessarily limited in how many layers, and the degree of spatial depth, that he could portray. Though impressive, the single plane illusion could not represent a range of spaces or buildings at more than a small distance from the picture plane. This problem was recognized, and solved, by the next group of painter/designers who worked in the Neopolitan and Austro-Hungarian courts following Pozzo's successes in Rome.

As in many guilds throughout Europe, the artisans in Bologna emerged in groups, many from the same family, over generations. The most successful and influential of these were the Galli-Bibienas, or da Bibienas, who flourished throughout the late Renaissance and Baroque eras. So prolific were its members that some 80% of extant eighteenth-century stage designs have been attributed to them.[10] They served as architects, designing horseshoe shaped theaters in towns throughout Europe (including Bayreuth and Vienna), as well as producers and scene designers in their theaters. Most significant, they introduced a style and representational type into the canon of design: the *scena per angolo* or angled view.

Historians attribute the popularization of this two-point perspective scene to Fernando Bibiena, who published a collection of scene designs in 1703 and explained the technique "per angolo" in his *l'Architettura Civile* in 1711. [65] In contrast to the one-point perspective scene first popularized in Serlio's *Five Books* in the 1530s, the *scena per angolo* was constructed from a diagonal station point using two-point perspective. The angle view offers the possibility of presenting complex architectural spaces by hinting at the ancillary areas that lay behind the main playing area of the stage. The one-point perspective scenes necessarily limited the complexity of architectural fantasies that the scenographer sought to portray as evidence of his inventive virtuosity. The angle view allowed the depiction of curvilinear stairs, oblique wall masses, domes behind domes, and many different levels all within the same ensemble. Often the station point was artificially lowered in order to impart a Lilliputian scale to the stage players and aggrandize the scene.

Juvarra and Baroque Illumination

Though many scene painters and architects followed Pozzo in extending pictorial illusionism in Baroque buildings, none had the talent or influence of Filippo Juvarra (1678–1736).[11] Though relatively unknown today outside of Italy, Juvarra achieved many of his breakthroughs as a result of his facility with the *scena per angolo*. He was the inheritor of a rich tradition of church design that continued through the Renaissance and blossomed during the Baroque period throughout Europe.[12] More important, he was the first architect to pursue fully open structure in his vaulting, and to introduce hidden light sources into his churches and royal residences.

The son of a silversmith in Messina (Sicily), Juvarra was exposed to both Roman architecture and Neopolitan theater design during his education. His father, Pietro, and four brothers, were accomplished silversmiths. Filippo acquired firsthand knowledge of this craft during his youth, but also began to study architecture at the age of 12. He was talented enough at this age to make copies of drawings in various architectural treatises—Alberti, Palladio, Serlio, and Vignola's books were available in his father's workshop. He is said to have been proficient at plan, section, and elevation drawings during his late teens. He also studied the perspective treatise of Padre Pozzo during his youth, while at the seminary preparing for the priesthood.[13]

66. Filippo Juvarra. "Camera," scene design. (MO22334, Courtesy, Museo Civico, Turin.)

67. Juvarra, scene design, (MO22339, Museo Civico, Turin).

He began his architectural training in the Roman studio of Carlo Fontana in 1704, where he immediately distinguished himself as a quick study and accomplished draftsman. One of his biographers relates the story of young Filippo's first design problem in Fontana's studio: a Corinthian capital. It seems that the maestro assigned this difficult problem to his apprentices on the very first day of Juvarra's tenure. Before his compatriots could even begin their designs the young man had "in a very short time designed at large scale the Corinthian capital with great exactitude," afterwards putting the sketch on a table and leaving, while the other apprentices marveled at the "polish and propriety" of the drawing and the speed of its execution.[14] Though probably not as versatile as Leonardo, Juvarra showed powers of imagination and drawing skill that were rare among architects of his time and which remain unsurpassed to this day. Fontana became an advocate for the young man, eventually hiring him to teach perspective at his school.[15]

He made a trip to Naples, the operatic and theatrical center of Italy, in 1706, and during the months of January through April produced his first studies for scene designs, undoubtedly influenced by what he saw there—sketches of rooms in palaces intended to demonstrate different lighting in morning evening and afternoon. [66, 67] Both use the recently developed "scena per angolo," or angle view technique. He quickly mastered the technique, and utilized it with extraordinary facility when he returned to Rome.[16]

In July of 1709, Juvarra entered the service of Cardinal Pietro Ottoboni. There he perfected his already accomplished draftsmanship, while pouring out an extraordinary number of scene designs for such operatic and dramatic works as Constantino Pio (1710), Teodosio il Giovane, and Guinio Bruto. These designs challenged the supremacy of the Bibienas—Juvarra was not merely a follower but an innovator. He departed from the strict 45-degree-angle view of the Neopolitans by presenting circular, semicircular, octagonal, or longitudinal plan spaces from indeterminate angles, affecting a feeling of infinitude to the space. The architectural complexity of the spaces show him to be more than simply a painter, while the rendering and line quality of the drawings place him on par with the finest artists of the time (at their best recalling the Tiepolos later in the century).The media in the drawings—sepia line and gray or brown wash—were common for painters and scenographers of the day. Wash drawing first became popular in the early 16th century and remained a relatively quick medium for studies in which no subtlety of modeling was required. Juvarra's wash technique was unparalleled for an architect of his time—even Bernini, almost 100 years before, could not match his virtuosity in depicting complex architectural volumes. One feature in the drawings of the Piedmontese master underscores his genius: his understanding of light and his ability to render its complex effects in interior spaces.[17]

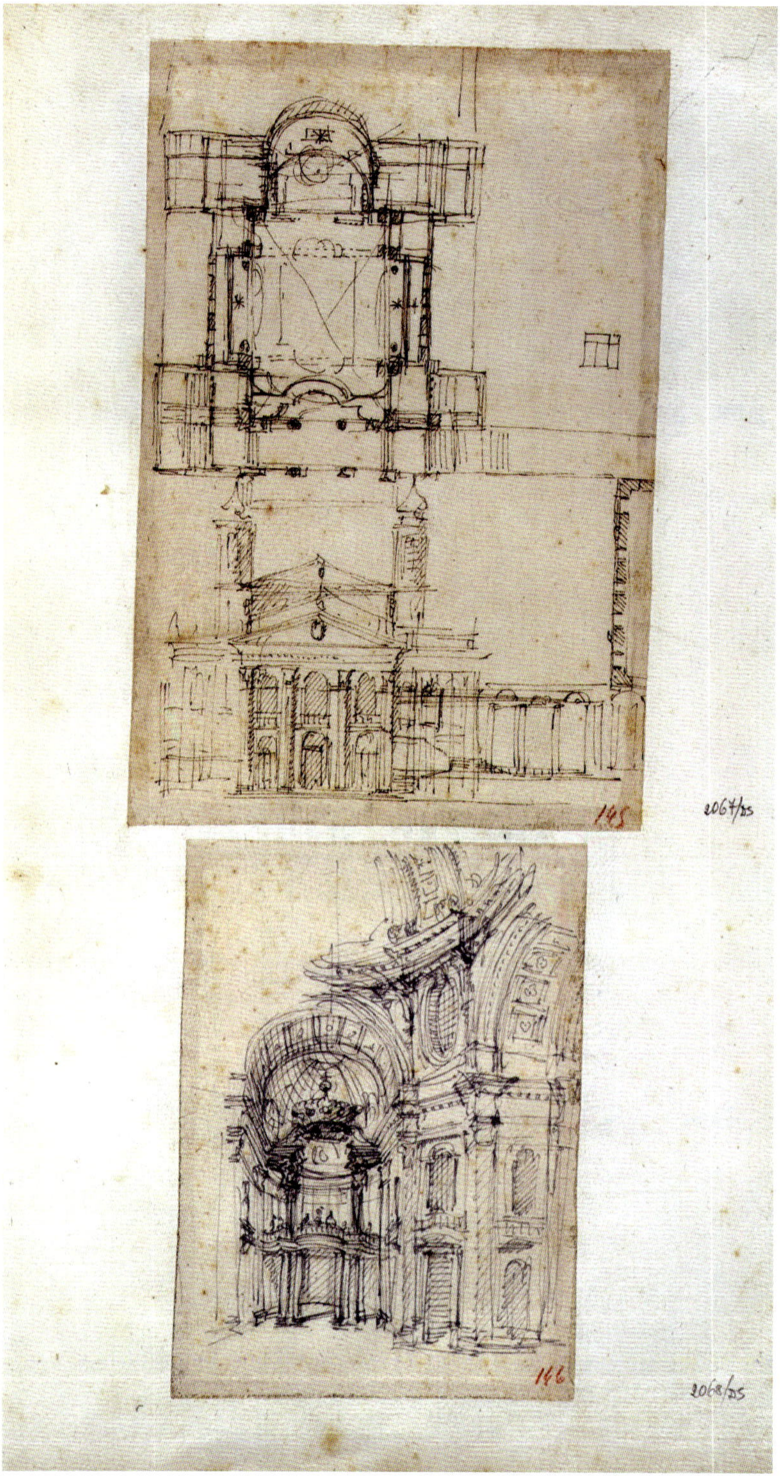

68. Juvarra, S. Uberto, Venaria Reale, plan, elevation and perspective sketches (MO24564, Museo Civico, Turin).

69. S. Uberto, Venaria Reale, interior of nave. (Turin museums website).

70. Juvarra, S. Uberto, Venaria Reale, dome. (Turin Museums website).

Light, of course, is one of the primary elements in theatrical fantasy. Juvarra's sketches for scene designs show better even than the engraved designs the strong contrasts in light and shade which he could suggest with only a few quick washes, creating an intense and dramatic illusionistic impression of space. As Richard Pommer has pointed out, these lighting effects were to reappear in his architecture later on.[18] Rather than focusing on the objects within the primary space or playing area, Juvarra often threw these areas into shade, instead lighting areas beyond the foreground. This produces an aura of mystery in scenographic space because the boundaries appear to be ambiguous. As Susan Munshower has pointed out, his scene designs were often conceived as "real" architectural spaces, with accompanying plans or sections, indicating that he did not distinguish between scenographic space and architecture, even within the theater itself.[19]

Juvarra executed nothing of consequence architecturally until he became "First Architect" to Vittorio Amedeo II of Savoy in Turin near the end of the year 1714. During the next 22 years until his death in 1736 he built prolifically in Turin and its environs, earning a distinguished international reputation as an architect. Five churches, four royal residences, four large city palaces, and the entire quarters of two of Turin's major thoroughfares may be credited to him during those years. Although Juvarra produced a large number of studies and projects for centralized buildings during his lifetime—including his daring Concorso Clementino project, studies for the Duomo Nuovo in Turin, for S. Raffaele and S. Andrea in Chieri—he built only two churches in that mode.[20]

The first, the chapel of S. Uberto (b. 1716) at the royal palace at Venaria Reale, was his earliest building commission for the court of Savoy. [68, 69, 70] The plan is in many respects conservative, recalling the early Renaissance schemes for St. Peter's by Bramante and his circle. It consists of an equal-armed Greek cross nave, an apsidal choir with a columnar screen, recalling Palladio's II Redentore, and four circular star-shaped chapels located between the arms of the cross. Features in the handling of the façade are reminiscent of Pietro da Cortona at SS. Martina e Luca in Rome. Though the details of the nave trabeation and articulation are in many ways novel, the bi-axial organization of the plan itself is not. What is unorthodox is the treatment of the chapels, which seem to burst like stars from the poché of the wall. Each pier is perforated by apertures on two levels, allowing a view into the mysterious realm of the chapels. Each chapel is fitted with dome, lantern, and a window to the exterior. Light filters through the apertures in the piers from the chapels to the nave. The chapels are very like the ones that Leonardo sketched in the late 1400s.

Characteristically, Juvarra studied the effect of this perforated pier in an early perspective sketch, in fact showing a strange oval window in the pendentive itself, a feature that would have been structurally daring. His sketch provided

71. Juvarra, *capriccio* of Torino and Superga church. (Biblioteca Nationale, Torino, Res. 59.1).

a powerful means of studying the problem of representing a three-dimensional object, the pier, in the context of the nave space. [68] This could not have been solved in a simple section drawing. Even a number of sections could not have told the designer what this single perspective study reveals. Furthermore, the method of construction—angle view, two-point perspective—was one with which Juvarra was intimately acquainted. The conventions of perspective in scenography were part of his memory as well as habitual methods of drawing.

Juvarra had the immediate opportunity to surpass himself with the commission for one of the largest and most important churches in the Piedmont, the Reale Chiesa di Superga (1717–31). In its plan he continued to explore the ideas of open structure begun in S. Uberto. Here, however, he was able to do away with the pier completely by substituting eight engaged columns, and to integrate the ancillary chapels into the nave space, preserving the primacy of the vertical drum in plan. In many ways this design was the culmination of ideas first explored by Leonardo over two centuries before.

The church was unusual in a number of respects, not least for the events that led to its construction. Duke Amedeo II of Savoy, having recently also taken over the Kingdom of Sicily, from which Juvarra came, was at war with the French over territory to the north and west of Torino. In 1716 he ascended the highest hill within sight of the city, and remarked to his associate that, should he defeat the French in the next critical battle he would erect a church on the very mountain on which they stood. God granted his wish and he fulfilled his promise.

Nino Carboneri has suggested, on solid grounds, that Juvarra had a model for the Superga in his portfolio even before he came north to work in the Kingdom of Savoy: a *cappriccio* for a church with two elongated towers from his Roman period[21] He was also enamored of Borromini's recently completed Sant Agnese in Agone in the Piazza Navona, a two-towered composition with a Greek cross plan. He could not, however, have imagined the vertiginous site that

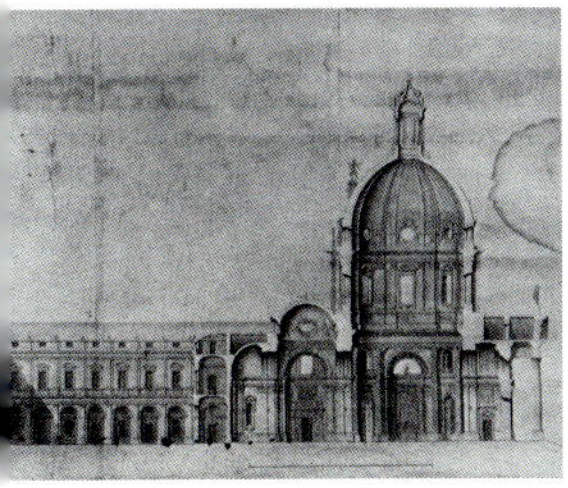

72. Juvarra studio, section through church and cloister. (Archivio di Stato, Turin).

73. Juvarra, first "esquisse" of Superga church, c. 1716. (MOO 20672, Museo Civico, Turin).

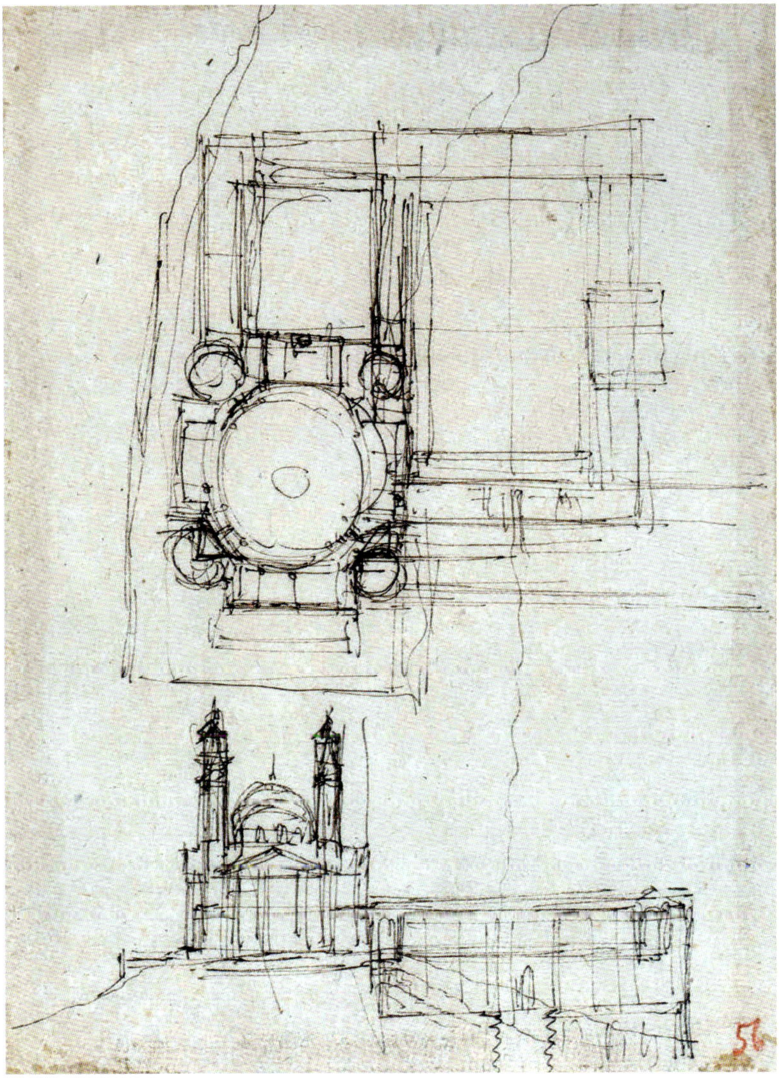

he first sketched in about 1716. His church would be visible from a vast basin stretching south from Torino. [71]

A small sketch of the initial design is preserved in the *Biblioteca Nationale*, dating from the middle of 1716. Evoking Leonardo, the architect created a tiny plan of the proposed basilica, and immediately sketched an elevation/section below it. Most of the key features of the final design were already in his mind, including a dome on an attenuated drum, and two very slender towers. He had not, however, come to grips with the difficulty of attaching the church to the convent associated with it, and placed its cloister off the main axis of the basilica, an unworkable scheme for the hilltop site. [73]

74. Juvarra, sketch of pendentives, dome interior and exterior. (Museo Civico, Turin).

75. Reale chiesa di Superga, Turin, constructed 1717-31, aerial view. (Author collection)

Letters reveal that the architect and his workshop struggled for more than a year to design a plinth on which the church would stand, allowing it to project over the brow of the mountain toward Torino. Recognizing that a low dome could not easily be seen from the city, he created several sketches for an attenuated drum and dome, and also gave up the idea of a standard Pantheon-type porch and pediment. All were perspectives constructed from an angle downhill from the mountain's crest. Without perspectives Juvarra could not anticipate the height needed to keep the dome visible as one ascended the hillside. Nor could he test his hypothesis that an unusual porch design would be required if it were to be visible to pilgrims on their way to the summit. [72]

Indeed, these conditions forced him to extend the porch several bays farther forward than conventional, and also to stretch the columns to their visual limit for the Corinthian Order. In elevations, done by the workshop in 1717, the porch appears not only too long but also rather diminutive in width. Only when seen from a downhill angle does it fit the overall composition. Like a *capriccio* done for a scene in an opera, the tiny temple front takes its place in the view of spectators in a great outdoor theater.

Even more daring than the exterior, Juvarra's interiors for the Superga broke new ground in the technics of natural lighting for central plan buildings. With such a tall drum, the nave of the church became a light filled lantern, elegant and soaring. The chapels, however, were buried behind a tall attic story, thus requiring hidden sources of natural illumination.[22] [72]

Just as he did at Venaria Reale, Juvarra chose to light the chapels from hidden skylights in the roof, concealed by parapets. His studies for the Superga were done with standard section cuts, as he knew what he was trying to achieve from a perceptual standpoint: the illusion of light emerging from above with no visible aperture. The chapels also provided "borrowed light" to the nave itself, through windows in the second story of each pier. In the strong light of the summer, beams stream downward as if shone from heaven itself. The tall dome and drum in the nave are also more dramatic than those of contemporary baroque churches by virtue of their lighting and relatively simple color scheme. There is a conventional lantern at the top of the dome, but the drum boasts some highly unusual, Guarini-esque windows that are distorted to increase the feeling of height as one looks upward. Light from both the lantern and the windows bathes the dome and accentuates the burnished gold color of the ornament in the coffers. [74, 75]

The Superga is an unusually eclectic church, but it is one that influenced countless German and Eastern European baroque churches during the eighteenth century. Moreover, it concluded a series of experiments in central plan designs over several centuries. Juvarra's conceptual mode was key to its ultimate resolution, and his drawings ultimately made such a mode possible. Without the creative connections between concepts in his brain and drawing conventions on paper, he would not have made innovations such as the perforated pier or cross-axial chapels.[23] And yet, unless craftsmen were available to bring his designs to life, no church or theatrical fantasy would have come to light

The Give and Take of Wood

For much of my life I have been fascinated by things made of wood. Growing up in the Pacific Northwest, I knew the art of the totem pole carvers and boat builders of the Native American tribes. When I was 15, an art teacher introduced me to a 90-year-old Anglo carver who had made poles during the early twentieth century. Perhaps because of this, I wrote a book on the architecture of Gustav Stickley, and studied log cabins as folkways.[24] When I reached my late 40s I began to make wooden furniture, using mainly hand tools. I later learned to carve wooden spoons and relief panels. I have grown to understand wood as only a craftsman can, through embodied practice.

There is little question that the greatest achievements in western wood carving came during the seventeenth and eighteenth centuries, both in Europe and the Americas. The complex leaf, flower, and drapery carving of German, Spanish, Portuguese, and French artisans has seldom been matched. Likewise, extraordinary furniture forms emerged during their period, such as the cabriole leg, bombé chest, Chippendale chair, marquetry table, and anthropomorphic chimney piece. The wizardry of design in a reredos from a Spanish, New World,

76. David Esterly, portion of overmantel carved in the style of Grinling Gibbons. (Courtesy of the artist)

or German baroque altarpiece still impresses humans with iPhones, computers, and video monitors that fill stadiums.

Wood is a material that demands hand skills, even if the craftsman is operating a machine. This is because it is a living thing that maintains its organic properties when under the manipulation of the artisan. Moisture, grain direction, hardness, and other properties direct craftsmen as they remove material or bend the wood. Moreover, wood must be formed by a process of removal of "waste" material. Any errors, especially large ones, can spoil an entire work piece, requiring a restart. The expense of hardwood makes mistakes very costly. Patience, a virtue in all art, is at a premium.

Japanese woodworkers, who have maintained folk practices for centuries with little modernization, are impressive to watch. Since we cannot view videos of Spanish carvers, French gilders, or Thomas Chippendale, modern woodworkers often marvel at the skills of master craftsmen from Japan. All traditional furniture is made with simple hand tools. The tools are in turn fabricated by masters, using age-old techniques. Perhaps most interesting is the position of the body when the artisan is in the "flow" of creative work.

Japanese carpenters sit on a mat while working, and often bring the workpiece close to their bodies. No area of the body is idle during the working process—legs, arms, feet, hands, and the head are moving. The tool and the wooden bench are a part of the action, as the body grips and manipulates them. The artisan uses the entire environment in fashioning his art. In this way he exemplifies the concept of embodied cognition—humans using brain, body and

77. David Esterly, musical trophy, 2004. (courtesy of the artist).

surroundings in their everyday lives. The simple tools used in these processes—chisels, planes, measuring sticks, saws, and squares—have not changed in centuries. Some master craftsmen will not use a machine to fashion their wood pieces; others combine them with hand tools. Unfortunately for westerners, most of the virtuosity in Japanese woodcraft is concealed behind smooth oiled surfaces, as joints are meant to be hidden.

Both in Japan and in Europe artisans were intimately involved with buildings until the last century. Without them no ornamental enrichment would be possible. Wood carvers often presaged the invention of particular kinds of ornament in furniture or wood paneling before similar motifs appeared in plaster, metal, or stone. Their materials were strong and pliable enough to withstand undercutting, so that thin leaflike details could be carved with relative ease. Limewood in particular proved to be a versatile and fine-grained medium for intricate detail—English and German carvers became renowned for their work in it.

Grinling Gibbons (1648–1721) was undoubtedly the most renowned of European wood carvers by the time of his death. Like many of the best artisans in wood he was trained in Holland before coming to England to work. Born to English parents in Rotterdam, he entered the Draper's Guild but quickly took to wood carving as a youth. By 1680 he had become known as "the King's carver" and worked for Charles II and James II, and at St. Paul's cathedral under Sir Christopher Wren. Horace Walpole said that no one before Gibbons "gave wood the loose and airy lightness of flowers." [76]

No wood carver since Gibbons time has studied or understood his working methods as well as the contemporary American master, David Esterly (1944-2019), who was hired to replicate and restore the famous Gibbons panels at Hampton Court Palace following the disastrous fire on March 31, 1986 that took a toll on the finest rooms designed by Wren. A Californian, Esterly had studied carving following a Ph.D. In English literature that had brought him to the United Kingdom. Once exposed to the miraculous lime woodwork of Gibbons, Esterley endeavored to find a way to make similar masterpieces, and has succeeded brilliantly since then.[25] Today there is no one who can approach his genius. [77]

The most difficult property of wood for a carver to assess is its grain—both direction and density. Growth rings in wood are composed of straw like groups of cells that run vertically up the tree in continuous strips. The grain of the wood follows the direction of these cellular straws. When milled, the wood generally follows the grain, but, since there are branches and knots in the trunk, grain shifts direction. In addition, the milling process affects the consistency of the grain. A woodcarver can be tormented by inconsistent grain, which is why lime wood is so versatile. Most devilish is the fact that grain is sometimes hidden—even end grain does not tell the carver the whole story. This is why artisanal knowledge is required for mastery of woodcarving—the carver must "feel" the grain while cutting into the wood. Esterly recounts an instance when, during the carving of a string-like stalk in a Gibbons restoration, he lost his sense of the grain the stalk snapped under his chisel, a carver's nightmare. Luckily there was a modern glue that could correct his mistake.

Esterly and Gibbons created carvings that are so intricate, and deeply undercut, that it is impossible for the layman to understand the complexity of their making. Even if a drawing is composed to guide the carver in his or her work, an exigency such as a knot or grain change can require a revision in the design. In this way the artisan works in real time to adjust to the "give and take" that his or her material provides. Wood is by no means the only material to require this flexibility, but is one of the most delicate.

Any master carver needs a work place and bench that can accommodate many different positions—in the workpiece and in the worker. If such a space is not available the task is more difficult, though not impossible. Fatigue sets in if the workpiece is not in the right position or close at hand. In this way the interaction between the maker and the immediate physical environment is manifest—seldom do tech workers reach the same intimacy with their objects of study. And it is not merely sufficient for the maker to have an accurate drawing or even model of the piece during this process. Artisans must continuously assess their progress and conditions of their pieces while making them, something that no alternate representation can do.

As Esterly wrote in his book on Hampton Court: "You take up where you left off the night before. You come to the workbench every morning and ask yourself "Where was I?" Inspiration is for amateurs, somebody once said. The creativity

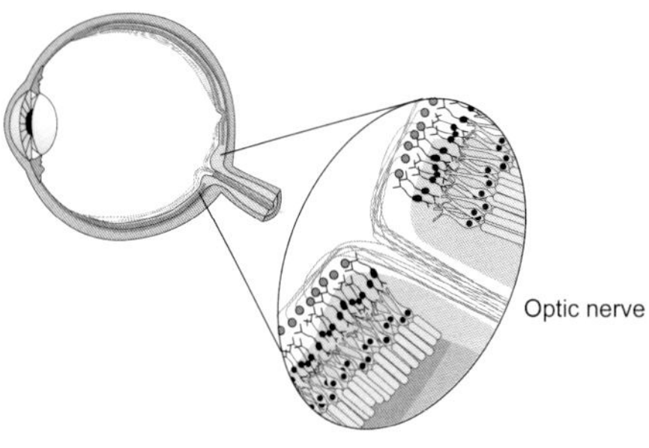

78. Close up of optic nerve and retina, from Groh, *Making Space*. The cells nearest the retina wall are small, foveal ganglion cells which specialize in sorting dark and light. Further from the optic nerve are larger ones that process motion, space, position and depth. (Courtesy Jennifer Groh).

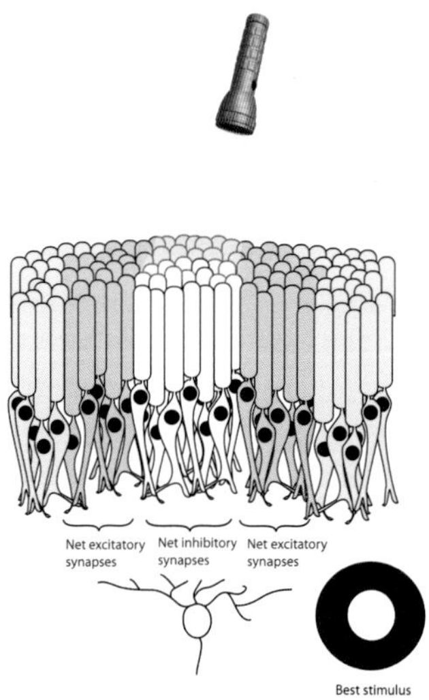

79. Center surround ganglions in retina, from Groh, *Making Space*. These tiny bipolar (on/off) cells are essential to identifying edges and contours. (Courtesy Jennifer Groh)

is incremental, not divorced from the making. You invent while you make. You work in the churn of the moment, and the forms seem to determine their own shape. You think with your hands. The carving thinks with your hands."[26]

When the environment, material, shape, and character of an artifact are in perfect harmony with the artisan, masterworks are possible. Without this harmony, humans can make repetitive, mass-produced products but certainly not ones of ineffable beauty like the panels at Petworth House or Hampton Court.

Corporeal versus Pictorial Illusionism

All of the extraordinary invention, craftsmanship, and illusionistic magic of the high Baroque achieved even greater brightness during the Rococo period, especially in the south German and Austro-Hungarian city-states. Karsten Harries suggests, with others, that one of the distinctions between the two eras was the tension between pictorial illusionism—that practiced by Pozzo and the Bibienas on planar surfaces—and the fully sculptural decoration produced by Bernini, Juvarra, and other baroque architects in their mature works.[27]

The human brain distinguishes which objects are in front of or behind others using *occlusion*, one of the most important perceptual faculties used in moving through the environment. If one object blocks the view of another, it has been *occluded*. It seems that the human visual system evolved ways of picking out objects in the visual field by deciding whether they are "figures" that may be in front of a "ground" of other patterns of light and dark. To do this, the brain must first use a kind of bull's eye that is known as a "center-surround" configuration of neurons. [78, 79]

The cells that participate in this trick are ganglions located immediately in front of photoreceptors (rods and cones) in the retina. These long cells are arranged in ring-like groups that have a high contrast (light sensitive) circle surrounded by a low-contrast band. When strongly contrasting groups of photons strike the retina, these center-surround groups go into action, generating electrochemical spikes that alert the visual cortex to objects or edges that separate one shape or form from another. They are almost like the "radar" blips on a ship's screen or the glowing creatures in an old Atari game like Space Invaders.[28] Some neurons respond to dark, and are "off," while others respond to light and are "on."

Just as the brain needs to assess direction, movement, color, and shapes in order to form abstract concepts of a scene or object, so too the visual system is sensitive to contrasting areas of light and dark. We use this sensitivity to ascertain the position—front or back—of objects in relation to the spaces they inhabit. Much like the "move to front" command in Photoshop, the visual system can assign a position to a "figure" using various visual cues about contrast and shading. This isn't just a matter of sorting black from white.[29] One of the

80. Gray value illusion. The upper and lower gray values are the same, but cannot be "seen" until the shaded center line of the illustration is obscured. (Author collection)

most compelling illusions used to demonstrate this is Fig. 80, a picture showing two adjacent squares of gray separated by bands. The lower middle band contains graded white to black wash suggesting a shadow cast from a light source above the frame.

It is impossible for the uninitiated viewer to see the upper and lower squares as identical in color and value, even though this is the case. Only when the "edge" of the shadow band is removed can the brain ascertain the true equivalence of the colored or neutral gray squares. The experiment may be done using an entirely monochrome image with the same result. What is happening here is a kind of figure-ground illusion, but the figure is ambiguous as to its position. The brain "places" one square in the background and assumes that the foreground square is receiving incident lighting; hence it must be lighter than its neighbor. Such a complex reading of the visual scene would normally be attributed to cortical thinking; in fact, the perceptual construct occurs at the pre-cognitive level, making it immune to reversal even when "explained" by reasoning.

As in so many developments in the history of art and culture, great achievements are often made possible by the chance births or inspired education of key protagonists. The flowering of rococo ornament and decorative design in the German city-states was no coincidence. Without a network of artisans, patrons, and theological narratives none of the churches and monasteries in Austria, Bavaria, or Bohemia would have been constructed. And, whereas the old guilds in Italy became outmoded in the sixteenth century, they were still flourishing in Central Europe two centuries later.[30]

Like the Bibiena family of Bologna, the Asams of Bavaria emerged as a market for their particular talents grew, with patrons who demanded work in the Italian manner. Cosmas Damien Asam (1686–1739) followed Filippo Juvarra in his studies at the Academia di San Luca in Rome, but trained as a painter under Carlo Maratta. Like many Bavarian artists he opened a studio that could produce many kinds of decorative sculpture, painting, and allied arts. With his brother, Egid Quirin Asam (1692–1750) he created masterpieces throughout Bavaria.

The Zimmermans were trained both as visual artists and as stone masons; they too participated in the creation of theatrical settings for Roman Catholic worship and pilgrimage. Johann Baptist Zimmerman (1680–1758) produced the extraordinary frescoes for Die Wies, a pilgrimage church built from 1753 to 1754; his brother Dominikus (1685–1766) designed the building and some of the framing ornament. Though small by comparison to palaces and cathedrals, there is more dramatic illusionism in this building than in virtually any contemporaneous sanctuary.

Scholars have noted two critical ideas behind the development of stucco and fresco decoration in German rococo churches: the importance of framed panels surrounding illusionistic painted frescoes, and the increasingly ambiguous character of the stucco cartouches, moldings, and other framing devices employed by master artisans in church decoration. The Asam brothers contributed significantly to the former, and the Zimmermans to the latter.

Since the frame helps to define the position of objects in the visual field, frames have been essential in architecture since the beginning of western classical art. John Summerson pointed out in a famous essay that *aediculae*, Roman temple motifs, appear in most Gothic buildings as well as in subsequent classical buildings because they define house-like frames for icons and sculpture.[51] In Baroque architecture these *aediculae* become alive with twisting, turning, rotated, and amorphous moldings, most often related to natural forms such as shells, stems, flowers, and *groteschi* (fantastic beasts and human faces). The enrichments carved into, or painted onto, the moldings give even more anthropomorphic energy to the compositions. More important, any shadows cast by the moldings, even if illusory or painted on, enhance the brain's apprehension of the character of surfaces and their relationships in space. Painters, stucco artisans, and architects played with the illusion of depth in their rococo decoration, particularly during the second quarter of the eighteenth century. They understood implicitly that the brain could be fooled by painted shadows and edges, and even more so by a combination of three-dimensional (corporeal) and two-dimensional (pictorial) illusionism.

In Bavaria, a school of stucco artisans—led by the Zimmermans—worked on monastic, parish, and royal church commissions throughout the middle of the century. Centered at the Wessobrunn Abbey in Bavaria, the south German *stuccatori* trained in both bricklaying and plasterwork emerged as the most sophisticated artisans in Europe, far surpassing their Italian mentors. Like Gibbons and his Anglo-Flemish wood-carving brethren, these artisans pushed each other to new heights of invention and technical mastery with each commission from wealthy monasteries and courts. Their creations remain unsurpassed to this day.[32]

When embodied knowledge such as that shown in the rococo decoration of the late Baroque period is in full flower, art advances. But it does so mainly through groups of collaborative artists, not simply with individuals. The Asam brothers were acutely aware of Gianlorenzo Bernini's remarkable *concetti* such as the altar at San Andrea al Quirinale in Rome. At the beginning of their careers they employed similar materials—marble, bronze, painted stucco—to create powerful dramatic scenes from the lives of saints that became centerpieces in religious buildings in Munich and surrounding towns. Their interpretation of the "teatrum sacrum" emphasized integration with the walls of the church, as the sculptural frame surrounding the scene seemed to grow from the fictive body of the building. This occurs in their masterworks for the priory church at Rohr and the abbey church at Weltenburg, both executed between 1721 and 1725.

81. Cosmas Damian and Egid Quirin Asam, Rohr Augustian priory church, 1721-25, choir and altar. From Harries, *Bavarian Roccoco Church*. (Courtesy, Karsten Harries).

82. Cosmas and Egid Asam, Rohr Augustian priory church, 1721–25, detail of high altar. As in a Bernini liturgical concetto, the figures around the basin enact a dramatic mis en scène. From Harries, *Bavarian Roccoco Church* (Courtesy, Karsten Harries).

At Rohr, the subject is the ascension of the virgin, a Marian apotheosis essential to Counter-Reformation theology. The composition is divided into three zones: a lower sarcophagus surrounded by bewildered apostles, a middle zone occupied by the virgin and two angels in mid-air, and an upper one attached to the pediment of the aedicula, depicting the Holy Spirit. A panoply of materials were used in the altarpiece: marble, stucco, wood, steel rods for supporting the virgin, and painted elements. All of the figural sculpture and decoration relates consistently to the altar framework, which in turn supports a one-point perspective view from the nave. The closer one comes to the piece, the more three-dimensional it becomes; from the back of the church it might well be mistaken for a painted fresco.[33] In this regard it is similar, if more sophisticated, version of Pozzo's teatrum sacrum, or of Bernini's St. Theresa in agony. [81, 82]]

At Weltenberg the composition is more dynamic. Hidden light from windows behind the altar pushes the main sculpture—St. George on a charger—to the foreground, while also emphasizing the arched frame of the reredos surrounding the altar. The spatial layers in the composition are complex, as the saint floats in a lighted void framed by no less than three ranks of sculptural *aediculae*. The foreground contains the altar and surrounding frame; the middle ground statues of the patrons of the project, and the intermediate one statues of the dragon and the virgin. An airy fresco in the rear apse is ambiguous because there is no visual reference to its position in any of the frontal zones. With these compositional devices the Asams have begun to dematerialize their sculptures and meld them into the painted frames that have been conventional references for tromp d'oeuil frescos.[34] [83, 84]

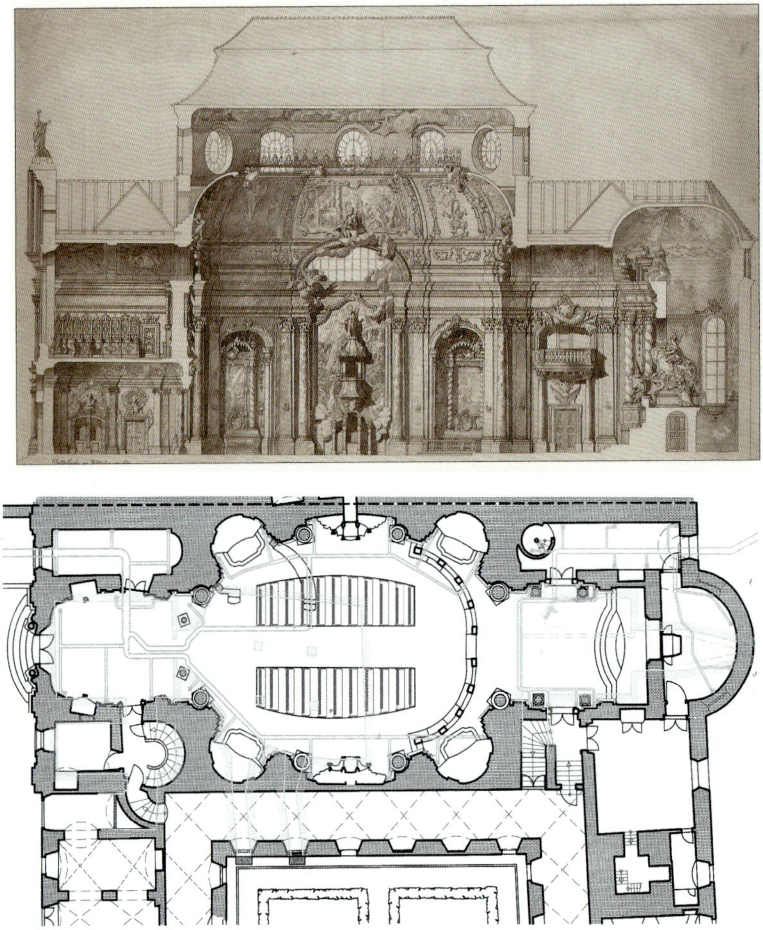

83. Cosmas and Egid Asam, Weltenberg abbey church, 1721-25, plan and section. The statues of St. Michael and the dragon are in the space behind the altar. (Courtesy, Weltenberg Abbey).

Even as the these signal altarpieces were under construction, the Zimmerman brothers—inspired by their father and uncle—undertook a series of commissions that fully integrated the panel paintings and sculptures of the typical Bavarian rococo chapel and altarpiece into the body of the church as a whole. The result was a new conceptual modality as well as an inventive synthesis of previous artisanal traditions among like-minded designers and their patrons.

Dominikus Zimmerman began his first masterwork, the church at Steinhausen, in 1727, completing it in 1733. Assisted by his brother as well as by lesser *stuccatori*, he worked on the church continuously and without interruption, integrating the decorative design, painting, and sculpture as a seamless enterprise. Such a complete *gesamkunstwerke* demonstrates the power of embodied practice: all the contributing artisans had a part in the success of the whole. However, the unified design concept could not have come to fruition without a single genius at the center.

84. Cosmas and Egid Asam, Weltenberg abbey church, 1721-25, choir and altar, showing statue of St. Michael in deep space. Compare the statue's location to the plan and section. In the photo it appears to hover in a bright void without reference to the aedicular frames surrounding it. (Courtesy, Weltenberg Abbey)

Zimmerman invented the aisle-type rococo church at Steinhausen and perfected it at Die Wies. Such churches were built to cater to pilgrims and devout followers of Mary the Virgin. As Harries point out, visitors to either church are segregated in their experience of the theological and symbolic theater that the artists created. Those traversing the aisles are aware of the intricate stucco decoration in the votive chapels, centering on sinuous vaults and windows in each bay. Those seated in the sanctuary are more aware of the ceiling fresco and its relationship to each pier/column order. In either case, the white stucco

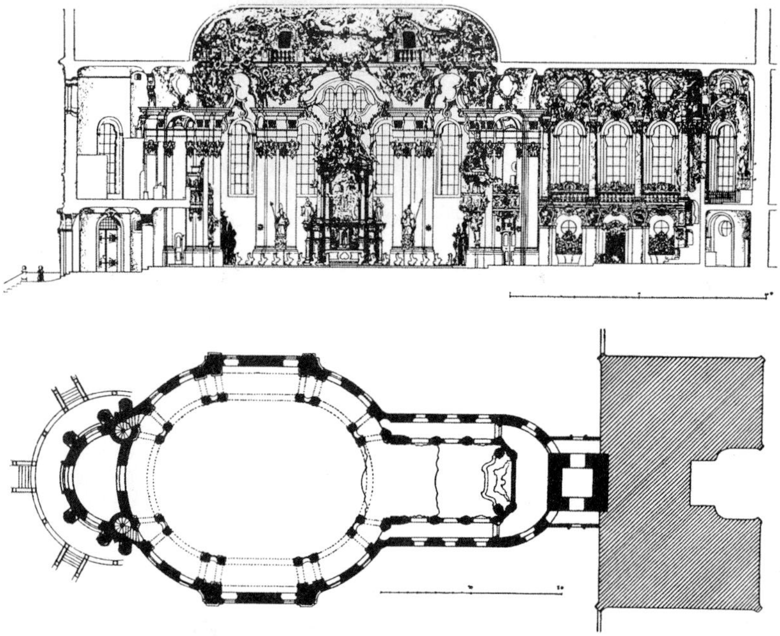

85. Dominikus Zimmerman, Die Wies, 1730s, plan and section. The ambulatory, defined by columns, is for pilgrims, while the nave and altar are for those attending a mass. From Harries, *Bavarian Roccoco Church* (Courtesy, Karsten Harries).

surfaces provide more than a ground; they form a seamless corporeal setting for the intense, intoxicating drama of miracles, saints, and resurrection. The pictorial conceits of early Baroque art give way to more ambiguous, and more cathartic, corporeal illusions.[35]

If Zimmerman and his studio worked together to build and decorate these churches, were they designed entirely "on paper" before construction began? Doubtful. Though plans were necessary, and rudimentary sections, it is unlikely that most of the ornament was determined prior to its execution in three dimensions by stucco workers. If the maestro, Dominikus, was present during the majority of the construction process, he was capable of judging and critiquing the work of his colleagues on a daily or weekly basis. The body of the building rose, and was molded, by artisanal collaboration. It could not have been otherwise. In this respect, south German studios, organized by guild, were not entirely directed from a distance by artist/architects, as was increasingly the case in Italy.

When Zimmerman was asked to undertake the design and construction of the small country church of "die Wies," (in the field or meadow), he had angered his patrons at Steinhausen due to the exorbitant costs of construction. However, his family was directly connected to those responsible for the funding of the new pilgrimage church, one that commemorated a miracle. [85] In the 1730s the abbot at nearby Steingassen began using a stuffed effigy of Christ in Lenten and Easter processions. When it wore out or proved inadequate it was acquired

86, Die Wies, nave looking toward altar. The carved aediculae around the altar appear in the fresco zone above, as if being lifted toward the heavens. (Wikimedia Commons)

by a local innkeeper, who in turn gave it to a peasant cousin who lived in a rural area about an hour from the monastery. The woman, Maria Lori, discovered the statue with real tears on its face, and a miracle was proclaimed. Pilgrims came to the woman's house, were cured, and the holy site acquired legendary status. In a matter of a few years hundreds were making the journey to what started as a small chapel, producing funds for a larger and grander church. More importantly, Dominikus's son, Franz, married Maria Lori and built a house near the site of the new church. His father, expelled from Steinhausen, also made his dwelling there as he began the design of one of the most unique religious buildings in Europe.[36]

The subject of the religious drama at Die Wies is Christ's Resurrection, and everything in the decorative program supports this theme. The billowing oval vault at its center shows Christ in Judgment; his tiny figure atop a rainbow appears so far from the ground it is almost lost to the eye amidst crowds of biblical figures. The architectural frame around the scene is clearly a part of the church, but its corporeal reality is indeterminate. Where does the gilded stucco end and the fresco begin? The occluding edges move sinuously around the oval in a riot of pediments, shells, and *aediculae*. Only from certain angles does the perspective reveal a difference between pictorial and corporeal ornament. Moreover, when standing at the entrance or moving around the ambulatory the visitor cannot easily determine which elements are in the foreground and which are behind. The entire screen of double columns and their upper, arched openings blend into one another. Only the more intense colors of the altar apse bring the eye to rest. [86, 87]

87. Die Wies, ceiling fresco. The corporeal ornament in the piers fuses with the painted ornament above in a seamless illusion, leading the eye upward toward the apotheosis of Christ and the Resurrection. (Wikimedia Commons)

Harries trenchantly observes that all of the theatrical play and sensuous illusionism of the Bavarian rococo church serves a theological and social purpose; the perspective that controlled the composition of Italian baroque art has given way to a multi-sensory, multi-viewer conception of corporeal form. Noting the popularity of anamorphic images in frescoes, he suggests that art has become emblematic of "the illusoriness and unreliability of the world." He writes, "The baroque theater is itself an emblem of the finitude of human existence, caught in changing perspectives. Far from simply placing the power of perspective at the service of dramatic presentations, the baroque theater uses a multiplicity of perspectives to put the rule of perspective into question."[37] By the 1740s the representational mode that captured the imagination of Renaissance painters and architects had broken down amidst doubts about the efficacy of illusion to portray the human condition.

Indeed, by the late eighteenth century art schools in France and Italy had begun to stigmatize perspective as a representational tool and replace it with more conventional drawing types. In the next chapter we will explore the pedagogy behind these academic institutions, and discuss their implications on the history of design thinking.

Notes

1 See Peter Thornton, *Seventeenth Century Interior Decoration in England, France and Holland* (New Haven: Yale University Press: 1978). "By 1660 there was no longer any doubt that it was the French who set the pattern of civilised life indoors. The lay out and arrangement of their buildings was widely studied; the details of their interior decoration was [sic] imitated; and the pattern of life led in such surroundings was copied." 25.

2 See John Summerson, *The Architecture of the Eighteenth Century* (London, Thames and Hudson: 1969).

3 See Ulrich Weisstein, Ed., *The Essence of Opera* (New York, W.W. Norton: 1964): 99–104.

4 On the influence of printing in architecture see Mario Carpo, *Architecture in the age of printing: orality, writing, typography and printed images in the history of architectural theory* (Cambridge, MIT Press: 2001).

5 Even in primitive, Neolithic societies a kind of ritualized "theater" was preserved in the layout and forms of settlements. See Trevor Watkins, "Architecture and 'theater of memory' in Neolithic Southwest Asia."

6 For an examination of this confluence, see William Alexander McClung, "A Place for a Time: The Architecture of Festivals and Theaters," in Eve Blau and Edward N. Kaufman, Eds. *Architecture and Its Image* (Montreal, Canadian Center for Architecture: 1989): 86–108.

7 Pozzo collected his theory and works in a two volume treatise, *De Perspectiva Pictorum et Architectorum* (Rome 1693, Vienna, 1700). On his work for the theater see Donald Oenslager, *Stage Design: Four Centuries of Scenic Invention* (New York, Viking: 1975): 54–57.

8 Joseph Kerman, *Opera As Drama* (New York, Vintage Books: 1952): 25–38 on Monteverdi.

9 Quoted in Oenslager, *Stage Design*, 56.

10 See Janos Stolz, Ed., *Baroque and Romantic Stage Design* (New York, E.P. Dutton: 1962): Plates 22–48 for examples of their work.

11 For a concise biography, see Cristina Ruggero, *Filippo Juvarra* (Venaria Reale, Celid: 2014). The author provides an up to date summary of the life and work of this prolific architect, using Italian scholarship from recent decades.

12 Mark Hewitt, "Juvarra's Drawings for Central Plan Buildings." *Dichotomy*, The University of Detroit School of Architecture, 1983, 17–29. My early essay was one of the first attempts to relate scenography to architecture in Juvarra's conceptual methodology. It contains numerous errors and was superseded by Millon and Munshower, cited below.

13 See Susan Scott Munshower, "Filippo Juvarra's Spatial Concepts and Italian Stage Design: The Consummation of a Renaissance Discovery," Ph.D. dissertation, Pennsylvania State University, 1995, 230. This study is the best available source on Juvarra's unique abilities as both architect and scene designer.

14 M. Viale Ferrero, *Juvarra Scenografo*, 8, quoting Anonimo (Francesco Juvarra), Vita, Peruga (A. Rossi) 1874. This biography was written by the architect's brother, and is one of two that chronicle his life from first-hand accounts.

15 Munshower, 221. See also Henry A. Millon "Filippo Juvarra and the Accademia di San Luca in the Early Eighteenth Century" in Hager & Munshower, Ed. *Projects and Monuments in the Period of the Early Roman Baroque*, 14–15.

16 Munshower presents a detailed analysis of the precedents for these new scenographic techniques, 345–49. She discusses, on 368 apartments of Flavia in Giunio Brutto 1711 shows his ability to hint at spaces beyond the foreground in his per angolo designs. Page 371 discusses Victoria and Albert album and Theodosio il Giovane and talks about a ground plan in the margin of the scene design.

17 The most comprehensive study of all the Roman drawings is, Henry A. Millon *Filippo Juvarra: Drawings from the Roman Period, 1704–1714*. Part I, Corpus Juvarrianum. Rome, 1984.

18 See Richard Pommer *Eighteenth-Century Architecture in Piedmont. The Open Structures of Juvarra, Alfieri, and Vittone* (New York and London, Zwemmer: 1967). This brilliant book was the earliest to relate the scenographic "capriccios" of Juvarra to his later church and palace designs. Pommer noted that it was often impossible to distinguish a design study for a building from one of his many fantastic drawings for the theater. See also Munshower, diss., 388 "Our analysis of Juvarra's scenographic design methods has revealed what appears to be conclusive evidence about Juvarra's thinking as a designer."

19 Munshower, 340 "Juvarra's ability to conceive this kind of visual architectural reality within the confined dimensions in which he worked as a stage designer is a clue to the means by which his conceptual process functioned."

20 Boscarino, Salvatore. "Filippo Juvarra tra Scienza e Technica." *Restauro*, 7, 1978. 3–90; and Boscarino, Salvatore. *Juvarra architetto*. Rome, 1973.

21 Nino Carboneri, *La Reale Chiesa di Superga* (Torino: 1978).

22 See for instance, Gian Franco Gritella. "La cupola della chiesa juvarriana di Soperga a Torino. Analisi della Struttura e indagini sulla fase costruttiva." *Palladio*, n.s., 4,1991, H. 7, 91–106.

23 This conclusion is well supported by Munshower: 388 "Our analysis of Juvarra's scenographic design methods has revealed what appears to be conclusive evidence about Juvarra's thinking as a designer."

24 Mark Alan Hewitt, *Gustav Stickley's Craftsman Farms: The Quest for an Arts and Crafts Utopia* (Syracuse, Syracuse University Press: 2001).

25 See David Esterly, *The Lost Carving: A journey to the heart o making* (New York: Viking Penguin: 2012).

26 Ibid., 206.

27 Karsten Harries, *The Bavarian Rococo Church: Between Faith and Aestheticism* (New Haven, Yale Univ. Press: 1983).

28 Groh, op. cit., 75–80. The patterns are sensitive to light and dark such that they form what are called "center surround" responses based upon inhibitory or excitatory synapses.

29 See Gibson, *Ecological Approach*, op. cit.

30 Christian Norberg-Schultz, *Late Baroque and Rococo Architecture* (New York, Electa/Rizzoli: 1980): 53–89.

31 John Summerson, *Heavenly Mansions and Other Essays* (New York, W.W. Norton: 1963).

32 Johannes Goldner, et al., *Wessobrunner Stukkatorenschule* (Freilassing: Pannonia. 1992).

33 Harries, op. cit., 128–30.

34 Ibid., 130–35.

35 Ibid., on die Wies, 138–45.

36 Ibid., 145–46. Following the death of his wife, Zimmerman had requested a room in the monastery of Schussenried, where another son was a monk, but was refused.

37 Ibid., 149.

[7]

TEACHING DESIGN IN THE MODERN ERA

We saw in the previous chapters that the Renaissance initiated the pedagogy of using *disegno* to inculcate judgment in the minds of artists—cultivating an ability to assess the quality and beauty of their designs. Students were given drawing exercises that would prime their brains with the best historical precedents. We noted that memories of great buildings and places filled their minds with appropriate images that could be synthesized while drawing. As the great educator Paul Philippe Cret observed in the 1940s, this system had clear weaknesses, but like the case law system in legal education, proved to be a close approximation of the kind of situations that most architects faced when designing for institutional clients. This system has both similarities and key differences with those of the Italian art academies and their French counterparts of the seventeenth century. In this chapter we will explore the dominant forms of pedagogy used to train architects over the past 300 years. This is the period in which designers focused on the depiction of buildings, not entirely renouncing craft-based practices, but overlaying them with forms of drawing, and developing systems that allowed the entire building to be rendered prior to construction. It was also a period in which modern ideas of scientific, rational thinking began to invade the craft-based world of architecture.

Le système des Beaux Arts

Though the Ecole des Beaux Arts in Paris taught architecture only from 1819 until 1968, it remains the most influential educational institute on of its kind to this day. The late Jean Paul Carlhian (1920–2012), a French-American architect who attended the Ecole, maintained that "le système" was a paragon of clarity and logic reflective of everything French.[1] Understood as a strict, cohesive program for teaching architecture, with little room for "creative freedom," the school has been pilloried as much as praised. In fact, it served as the bête noir of Modernists, who still view its methods as the antithesis of their own enlightened educational precepts.[2] Many elements of "academic" pedagogy remain in contemporary university programs, but crucial pieces of the system were gradually weeded out during the last century, much to the detriment of young architects.[3] [88]

88. Salle des prix, Ecole des Beaux Arts, Paris, 1929. The school was closed, with great fanfare, after students rioted in Paris in 1968. (Author collection)

Several aspects of the Beaux Arts system bear examination in relation to questions of how to inculcate visual judgment in architects, and, more fundamentally how the neuroscience of perception was borne out in early architectural pedagogy. The first connects to memories and mimesis as they build visual networks in the human brain. The second is related to habits of mind followed by architects after they leave school. The third underlines the effectiveness of Beaux Arts programs in imparting a clear sense of visual judgment through the *concours* system. The last, and most important, involved the use of socially sanctioned precedents as inspiration for new designs.[4]

We have already hinted at the benefits of learning drawing by literally building natural forms out of *geons* or blocks that seem to reside in areas of the visual cortex. When painters, sculptors, and architects entered the French academy, they had already been exposed to the practice of drawing from abstracted models, usually plaster casts, before attempting live or al fresco drawing. Richard Morris Hunt, the first American to attend the architectural program in Paris, studied with a Beaux Arts trained teacher for several years in Europe before applying to the Ecole. In this he was similar to French "aspirants" seeking entrance into the competitive program in Paris.[5]

89. Atelier Pascal, Ecole des Beaux Arts c. 1890s. Students benefitted from learning together with older, more experienced mentors, and from professional architects in a functioning office. (Author collection)

Once at the school, young architects were immersed in the full experience of design from the start, but only at the most basic level. Their daily work occurred in the *atelier* of a Beaux Arts trained architect (*patron*), who provided design training for a fee or stipend. These masters were not compensated directly by the Ecole, but by each individual "elève" or pupil. There they were encouraged to sketch buildings in the city, hone their drafting skills under a *massier* or head tutor, and do short design exercises. The social bonding was also important, for students pursuing similar crafts teach each other quite effectively.[6] [89]

Every art, sculpture, and architecture student at the Ecole shared one studio drawing experience that proved critical to his or her success: *drawing from the cast*. Because both precision in outline drawing and a keen understanding of shades and shadows was needed in all three disciplines, students spent hours in the school's large halls, each filled with casts of historic sculpture, architecture, and decorative pieces. Félix Duban, the architect of the Ecole, arranged these

90. Alvar Aalto, esquisse for church at Lahti, 1952. Aalto received a Beaux Arts education as a student in Helsinki, and used such sketches throughout his working career. (Alvar Aalto Museo, Jyvaskyla)

91. Felix Duban, plan of the Ecole des Beaux Arts, Paris. At the lower left is the Palais des études, in which several halls displayed casts of antique and modern ornament, sculpture, and pieces of buildings. (Author collection)

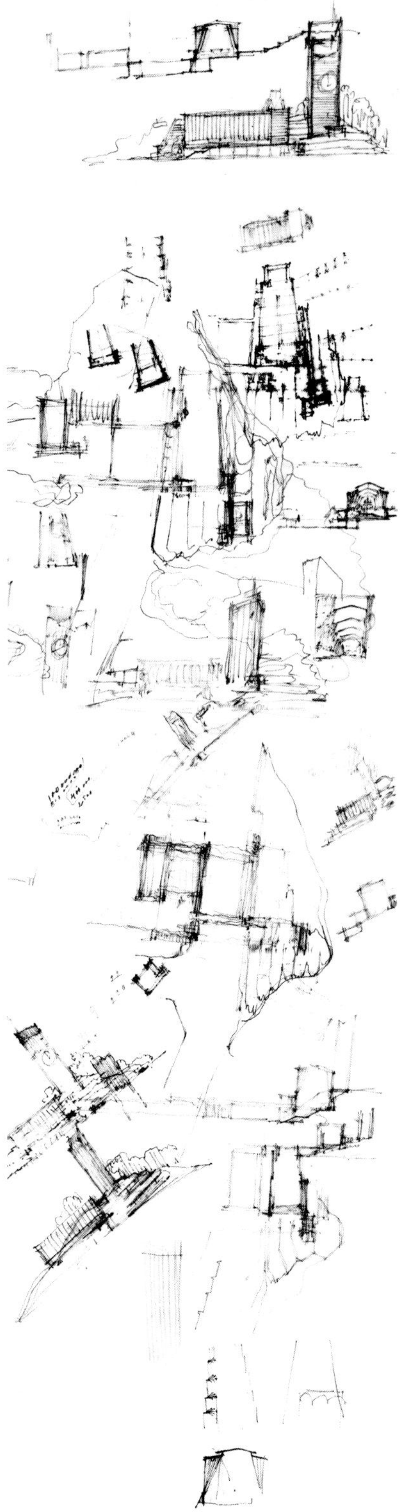

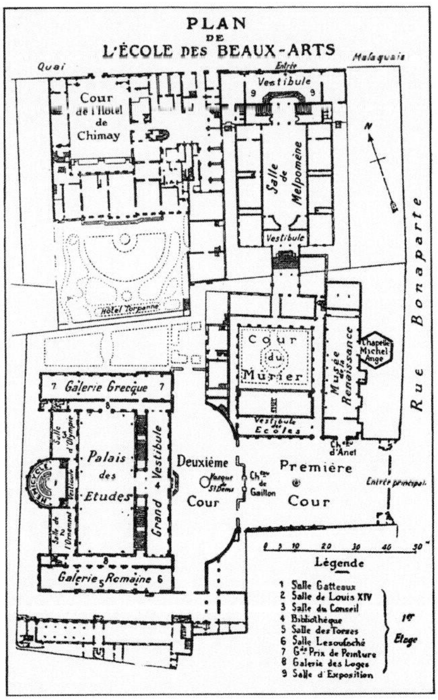

92. Ecole des Beaux Arts, Salle des Etudes, in a recent photo, without the many examples of cast ornament that served as exemplars for students. The glass roof provided excellent diffused light. (Wikimedia Commons)

rooms in sequences that allowed the élèves to move chronologically from one key period to the next. [92, 93] As they filled their albums with drawings, they also memorized key examples of say, an Ionic capital or frieze. Just as a piano student might absorb a Chopin étude following hours of practice, so these students filled their brains with the necessary models for making buildings.

More importantly, drawing from the cast reinforced perceptual schemata that linked the hand and the eye so that the basic visual thinking involved in design became habitual, hard-wired in current parlance. These relatively rote exercises benefitted students in myriad ways; yet there were similar complaints about the time spent to those heard in music conservatories throughout the world. Alvar Aalto often said that school was a waste of time, but he absorbed the Beaux Arts system while there and benefitted from it cognitively in his later work. [90]

The basic drawing pedagogy followed perception by first isolating the object in the studio, in monochrome, with light from a single source, usually above and at an oblique angle. [93] Removing color, movement, and environment as distractions allowed the student to first concentrate on the outline of the object, its form and mass, before looking at its topography. Initially, students drew only in pencil or charcoal lines, without shading. According to psychologist David Marr, this precisely tracks the first stage in visual recognition: contour or outline.[7] (This is the way the brain sorts visual data, from V1 to V4.) The initial outlines were drawn lightly, completing the figure. Standing away from the drawing, the student was taught to apply "la correction" to the initial line drawing in successive, darker lines, until a satisfactory mimetic image was achieved. In this way the memory could be used to hone in on a more perfect figuration of the cast in the visual field. After mastering the occluded edges—all objects

93. Head of St. Jerome, student drawing from a cast, New York Academy of Art, 2000s. Only recently has the practice been revived in art schools. (Courtesy, The Classicist)

in space are in front of something else such as a wall—the student gradually learned about how light reflects, refracts, and illuminates surfaces. The visual cortex must rely on both shadows and occlusion in order to define and place any object. Marr suggests that this is the second crucial assessment made in the visual brain. Once trained to see according to this schema, students could create abstract shades and shadows on invented masses and forms. Ernst Gombrich brilliantly analyzed this increasing skill in *Art and Illusion*. Architecture students then moved on to a course on the theory of shades and shadows, usually taught by a master with some scientific acumen.[8] Color perception, the final stage in recognition, was introduced in drawing only after line and shadow were fully understood in monochrome.

94. Louis Kahn, student drawing, drawn from the cast, University of Pennsylvania, 1920s. Kahn was a student of Paul Cret, the noted Beaux Arts exponent (John Harbeson, *The Study of Architectural Design*).

Critics of the academy during the early twentieth century often portrayed these courses as vacuous and uncreative, when in fact the skills taught were powerful and the ideas quite sophisticated. Advocates for "Pure Design" in the U.S., such as Arthur Dow and Denman Ross, developed teaching methods that purported to teach more abstract concepts than could be imbued by drawing ornament, sculptures, or casts from nature. It is difficult, in the absence of comparative studies, to discern whether such "modern" methods were more effective than traditional ones, but historians often note that all of the "Modern Masters"—Le Corbusier, Mies van der Rohe, Alvar Aalto, and Frank Lloyd Wright—were trained to draw from the cast, and often to design using Beaux Arts methods.[9] Bauhaus training, using some Gestalt principles, was also based

95. Michael Graves, San Giovanni in Laterano, Rome Prize sketch, July 1961. (Courtesy, Michael Graves Architecture & Design)

upon tabula rasa requirements, but no Bauhaus trained architect attained the distinction of these "form-givers."

The next clear benefit of Ecole training was what Cret cited as case studies, or learning from exemplary precedents. As noted above, any French trained student could draw on a large storehouse of historic buildings that he/she memorized during his or her time in Paris. Law students also learn effectively from case studies, as these precedents form the basis for tort law. In addition, most Ecole students were encouraged to take a "Grand Tour" following school, with sketchbook in hand, to record even more key monuments *in situ*. Winners of the coveted Prix de Rome had the richest experiences while spending several years in the Eternal City. As a student at Penn during the 1970s I benefitted from such a program when I earned a special scholarship to make a European sketching trip after my first year. The experience changed my entire attitude toward architecture and still charges my creative batteries, after more than 30 years. As psychological experiments have proven, one thing that quickly distinguishes an expert from a novice is his or her knowledge of outstanding precedents. [94]

There are countless examples of architects who have begun or enriched their careers with sketching trips or residencies at the American Academy in Rome, an offshoot of U.S. Beaux Arts programs founded by Charles McKim. The late Michael Graves published a superb book of sketches and photos ten years ago; [95] Antoine Predock and Errol Barron also created printed versions of their remarkable sketchbooks. These artistic compendiums are reflective of what talented architects do to build neuronal and memory storehouses for later work. They are in many ways a continuation of the Renaissance sketchbooks that we examined in Chapter 5. The legacy of the grand tour is still very much alive in professional practice today.[10] Unfortunately, it has given way to learning through photographs projected on digital screens in most academic settings.

Most first-year architecture students have difficulty with time management and decision making as they undertake early design projects in the studio. One of the reasons for this is that they have no intellectual foundation for visual judgment, and every problem requires selection of the best options from a range of possible solutions—a hard mental problem. Another reason is that our society emphasizes freedom of expression over mental discipline when making "art." Studies of novice architects and students indicate that they use the pre-frontal cortex almost exclusively while designing, while trained architects rely on experience, using mainly their parietal cortex. The former is associated with choices and decision making, while that latter is associated with integration of sensory and cognitive signals, a kind of synthesizing role in the brain.[11]

Young architect-élèves at the Ecole had no such difficulty after completing their first or second "esquisse-rendu" problem. They were given the "program" or requirements for the competitive exercise in a meeting with the professor who "set" the *concours* (essentially a race or circumscribed period in which to complete required drawings). In short design problems such as a small park pavilion the student was required to enter a "loge" or alcove at the school, with all other competitors, and spend up to 8 hours, sometimes 12, on a first draft or "esquisse" of their proposed solution. These "sketches" (the English word does not fully define the term) had highly specific layouts and drawing conventions. Students were not to leave the premises during the period "en loge" and could only submit one esquisse at the end of the session. No one would argue that such a compressed period was ideal if the best or most creative solution was desired, but that was not the point. Many students no doubt complained about time, but to no avail. Each designer was compelled to chose a specific "parti" or diagrammatic solution within the period en loge and create a drawing that showed all of its salient features, using only a single plan, single section, and single elevation. Usually the plan was placed in the center of a vertical sheet, with a section below and an elevation above it, drawn to the same scale.

Following the submission of the esquisses, the professor judged each student's submission and marked some of the efforts "hors de concours" or not fit to continue. In early exercises the percentage "making the cut" was small, teaching students to attend closely to the professor's instructions. Eventually every student acquired the ability to not only create a successful esquisse, but also enough visual judgment to determine its chance of success. Indeed, the esquisse was a powerful tool for teaching the most basic precepts, such as composition, proportion, scale, and character, to a young architect.

The advantage of such a highly codified method was that a student had no option but to decide among plausible alternative designs after a relatively short but intense period of analysis and study. Of course, no contact was offered with a teacher during this time, and such a dry run would leave contemporary students bewildered and dismayed. However, Beaux Arts students were permitted to discuss their work with their peers, and older students could help less experienced designers. Peer learning is essential when studying design because humans interact by imitating their siblings and friends.

The brain benefits from such a regimen of choosing between *available* alternatives, and leaving all unavailable ones aside. In current pedagogy the student is never compelled to abandon the full range of solutions to any given problem, merely to present his or her current favorite choice to critics, who will state the obvious weaknesses and perhaps commend the strong points. Visual judgment accrues from a repetition of such critical sessions, and eventually the best students surmise how to create solutions that meet a "professional" standard. The down side of this system is that weak students may never achieve such awareness, and that strong students may spend unnecessary months guessing how to please critics, when they might have achieved such mastery in a matter of weeks in a Beaux Arts based system—cruel though the experience may be of spending long hours in a closet with only a drawing board and some pencils.

A finely tuned, truly advanced sense of how to design beautiful things—whether paintings, graphics, buildings, or gardens—takes many years of field experience. However, getting the brain to square one—visual judgment about good and bad options—is not difficult if one follows the path that perceptual scientists, and post-Renaissance masters, have laid out. Research on the brain suggests that this system has proven benefits, if one looks at its pedagogy objectively, stripping away biases that have accumulated since the 1920s.

Disegno and Déssin Géometrale

The word *disegno* has at least two meanings in Italian: drawing, and design. During the Renaissance it acquired even more layers of significance, as painter architects read Latin treatises and explored the ruins of Roman buildings in search of Vitruvian secrets. Giovanni Paolo Lomazzo could state by 1584 that those who begin their education in painting and sculpture "and then, well armed with *disegno* turn to architecture … miraculously make into buildings what comes into their minds." *Disegno* was the capacity to judge and create beauty on aesthetic grounds.[12] It was also a system of thinking about form and proportion that required years of training.

"Good design" as defined by this Italian term was always a visual and formal concept, not a linguistic one. Though many Vitruvian and Albertian terms had etymological roots in Latin rhetoric, this powerful word was always connected with the "trained eye" of an artist.[13] In many respects the Italians still adhere to this faith in the "bella figura" as judged by the eyes—their fashion, furniture, sports cars, and decorative arts industries depend upon it.

When French architects returned from their tours of Italy in the late seventeenth century they brought with them a reverence for Rome and for the towering achievements of the Renaissance. So, when royal art academies formed under Colbert and Louis XIV, a central feature of their training was *déssin*, the closest linguistic term to the Italian. However, neoclassical architects began to refine and even change the concept of visual judgment, especially under Quatremere de Quincy. By the early nineteenth century the Ecole des Beaux Arts taught students under a complex theoretical schema that was passed down to such masters as Charles Garnier, Henri Labrouste, and Paul Cret. As Richard Moore described it during the 1970s:

> In contrast to the Italian concept of *disegno*, which stressed equally contour and internal details, French *dessin* emphasized contour, called *dessin au trait*. By means of *dessin* theory, the artist was taught to see planimetrically in terms of the single wholeness of the *contour exterieur*, rather than in terms of a multiplicity of details, the *contours interieurs*. This quality of *dessin* was traditionally known as l*a correction*.

In employing *la correction*, the artist was expected to avoid perspective effects. The anti-perspective doctrine created a strong distinction between *dessin perspectif* and *dessin geometral*. The purpose of *dessin geometral* was to see objectively in undistorted sections and elevations based on a grid of horizontals and verticals.[14]

An illustration of the process of "la correction" from *déssin au trait* shows how closely the pedagogy reflected the brain's increasing apprehension of detail, following the identification of shape and direction. The occluding edges of the object must also be distinguished from the background through contour recognition, which has been studied extensively. [See Chapter 6] Once shades and shadows are filled in the object can be located in space and in the light. [96, 97]

We know that the brain's architecture contains grid-like patterns of neurons as well as specific neurons for assessing direction and even positions in space. Because the human body has a strongly cross-axial disposition and the limbic system keeps track of our balance about the torso, we tend to experience space in axial terms. Our spatial "umwelt" is based upon a 180-degree visual field. Déssin géometrale was an extreme extension of the cardinal axes, as all Beaux Arts plans were to have multiple cross-axial visual and corporeal relationships. It is interesting to note that many experienced architects use hand gestures and even bodily poses to describe aspects of design to their peers and students, a kind of embodied critique of others' designs.

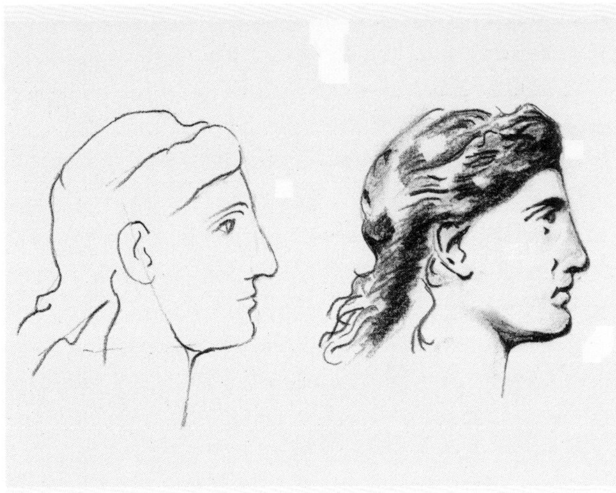

96. Déssin au trait, *développment*, Ecole des Beaux Arts, showing geon-like forms. Students were taught to understand geometric solids before applying them to building form. (Author)

97. Déssin au trait, use of *la correction*, author, after A. Yon, 1867. The outline, at left, is drawn and refined before the shaded drawing, at right, can be made. The student uses acute, foveal ganglions in the first exercise, only later engaging the larger depth-defining ganglions of the surrounding retina.

By the late nineteenth century students understood that creating a plan that could be assessed "tout en coup et tout suite" was the highest attainment in their training. Professors looked for simple relationships between the parts of a program, and a "disposition" of elements that were balanced on major and minor axes, always bi-axially symmetrical on rectangular sites. Paul Cret criticized the tendency to over emphasize planning over other aspects of design the search for the "beau plan" ideal drove many architects to create overwrought fantasies with little functional or social merit. The academy earned its scorn, and the professors were called "pompiers" as a nod to Roman classicism.[15]

Nevertheless, the basic pedagogy behind disegno/dessin represented a long-standing and time-tested method for cultivating visual judgment, and one that would yield readily to analysis by neuroscientists in the coming century. One reason for this is that research has now made it possible to connect systems of external representation, through memory, with internal representational patterns. Merlin Donald calls these things viseo-symbolic paths or codes. He writes: "Science and art are both dependent upon external memory devices and thus upon the vast numbers or viseo-symbolic codes and conventions stored in the modern brain."[16] The Beaux Arts and Disegno systems are examples of such codes. They began to be supplanted by mathematical and logical forms of thinking during the post-revolutionary rise of nation states in Europe and the Americas. It is no accident that military institutions hastened their demise.

Battles and Building Academies

Sébastien le Prestre de Vauban (1633–1707) is considered the greatest military engineer in history by virtue of his extraordinary fortifications throughout France. Before Vauban, military fortifications were constructed by masters of the "works" of kings and other aristocrats, usually masons, who had no formal training in physics, statics, or materials.[17] Michelangelo and Leonardo, who designed sophisticated fortifications, were trained as artists, but had some knowledge of geometry, physics, and mathematics. After training as a soldier, and successfully mounting several sieges, Vauban became "ingeniere du roi" for Louis XIV, and thus invented the profession of military engineering. His legacy produced the vaunted system of technical and construction training that propelled France to the forefront of both architecture and engineering innovation during the modern period.[18]

Karl Friedrich Schinkel (1781–1841), probably the greatest architect of the nineteenth century, was trained in a Beaux Arts system, but created an educational program that was in some ways its major competitor during the succeeding two centuries: the Bauakademie or building school. Not only did he help to develop its curriculum, he created an embodiment of it in the building that was to be its home. Like all of the architects in this book, he learned architectural design by directly experiencing great buildings in their settings, and by drawing them to understand their form and meaning. Schinkel's genius was far broader than that of any other architect of his time—he excelled in painting, lithography, perspective, decorative arts, and architecture. His stature in Germany is today similar to that of Vauban in France: not simply a designer or builder, but a national hero.[19]

Under Colbert and Louis XIV France flourished both economically and strategically by creating efficient bureaucratic structures. Men with organizational genius rose to lead manufacturing enterprises, the church, decorative arts, landscape design, and, most importantly, the military. Though Vauban was not individually responsible for devising educational institutions, his leadership fostered the development of three kinds of building schools: the Academie Nationale des Beaux Arts, the Ecole des Ponts et Chaussées, and the Ecole du Génie. The latter two were charged with infrastructure construction and military fortifications, respectively. These institutions were run on the model of a military "corps" or unit, with highly autocratic, pyramidal leadership structures, much like the larger Ecole Militaire in Paris.

It is significant that the etymology of the word "engineer" comes through Latin and French terms for genius, ingenuity, and capability. Colbert valued men with the ability to get things done, preferably quickly and cheaply. Jean-Rudophe Perronet, who ran the Ponts et Chaussés from the 1750s onward, developed canals, bridges, and roads throughout the country. His counterparts under the king's works, Germaine Boffrand and Ange-Jacques Gabriel, were trained as architects but supervised construction projects large and small. By stressing efficient organizational skills, the French system advanced the architect-engineer to a dominant role in the building professions.[20]

In Prussia, the largest among many German city-states during the eighteenth century, artisan-based building trades were breaking down under the weight of industrialization. Under Friedrich Wilhelm II (the Great) it became a wealthy military power, only to succumb to French occupation at the turn of the eighteenth century. Following the defeat of Napoleon and the Vienna Congress of 1815, the state reorganized under the enlightened guidance of intellectuals such as Wilhelm von Humboldt, and bureaucrats such as Heinrich von und zum Stein. Using an enlightened, and French influenced, system of municipal administration, education, and manufacturing, Stein created a modern state rivaling that of Europe's dominant powers in a matter of a few decades.[21] In matters of technology and science, Prussia looked to England.

Schinkel became the dominant artist, architect, and designer for the Prussian state during the height of its post-revolutionary building boom. He could not have been better prepared for this role. The son of a local building inspector, he became the protege of David Gilly, the leading architect and educator in the German states at the turn of the century. Though nominally educated at the nascent Bauakademie, the homegrown prodigy was virtually a family member in the Gilly household. More importantly, young Karl Friedrich befriended Friedrich Gilly, the talented son of the French-trained David. As Barry Bergdoll has pointed out, the collaborative studio at Gilly's office provided Schinkel with not only ideas but early commissions, since Friedrich was to tragically pass away at the age of 28 in 1800.[22]

Schinkel's education was not only comprehensive from an academic standpoint (he attended lectures on painting, philosophy, and art history); it was immersive and embodied. The young architect used his increasingly sophisticated skills in drawing, especially perspective composition, to study European architecture at a key moment in its development. After a brief period in practice, Schinkel set out on a two-year sketching tour in 1803, first visiting Mediterranean countries and then going to England, where he visited the architect John Nash and saw the latest industrial technologies around London. His direct experience with buildings in their settings was to inform all of his subsequent work. He left more than 400 drawings from that trip, now in Berlin, among an extensive archive.

One of the finest ink drawings in the collection is a view from Schinkel's quarters on the Monte Pincio in Rome, where he could see virtually every important building in the city (1803). [98] Though a portion of the drawing (a balcony at the left) is awkwardly foreshortened, the topographical details are astoundingly accurate. The artist even annotated the sketch with the names of key monuments, including the Vatican and the Villa Pamphili. Another superb sketch, made at the top of Mount Etna in Sicily, depicts the young architect being blown off his feet in a freak storm. Accompanied by letters and acute descriptions, Schinkel created a "memory palace" from the grand tour that he could use in the future, when fashioning his own designs. Venturing more freely to experience countless examples of both ancient and modern architecture, and landscape, Schinkel had even more opportunities to draw key monuments than his later counterparts at the Ecole des Beaux Arts.

When the young painter-architect returned to Berlin in 1805 he was confronted with a dilemma: how to use his new skills during the French occupation, when building was at a standstill. Providence seemed always to work in his favor, as he found employment as a painter of dioramas, or 180-degree panoramic pictures, of tourist sites throughout the continent. He was also active as a stage designer during this fallow period in architectural practice. Both jobs offered the chance to test drawing skills and work out concepts that would inform his later buildings. Though none of his 40 paintings in the round survive, he did create a circular drawing of one—"Panorama of Palermo"—that gives us an idea of their ambition and optical effects. Like Peruzzi and Juvarra before him, Schinkel excelled at creating persuasive fictive illusions on flat surfaces, often combining real architecture with invented landscapes.

His extraordinary imaginative facility was also evident in a number of stage designs. Whereas Baroque opera sets were often composed of numerous flats layered from front to back, Schinkel and his German colleagues sought to simplify sets by using only single large pictorial backdrop. In his famous sets for Mozart's *Die Zauberflote* (*The Magic Flute*), he created some of the most mystical and evocative scenes in operatic history. His sets for stage plays were no less awe inspiring. Though both panoramic paintings and stage sets used optical effects to create illusions, Schinkel was rehearsing embodied concepts that could be constructed in real urban and rural settings.[23]

98. Karl Friedrich Schinkel, "View from the window of my lodging in Rome, St. Peter's in the distance," 1803. Linecut from published engraving, in the public domain. Schinkel's sketchbooks contain hundreds of highly accomplished drawings, made al fresco, throughout Europe.

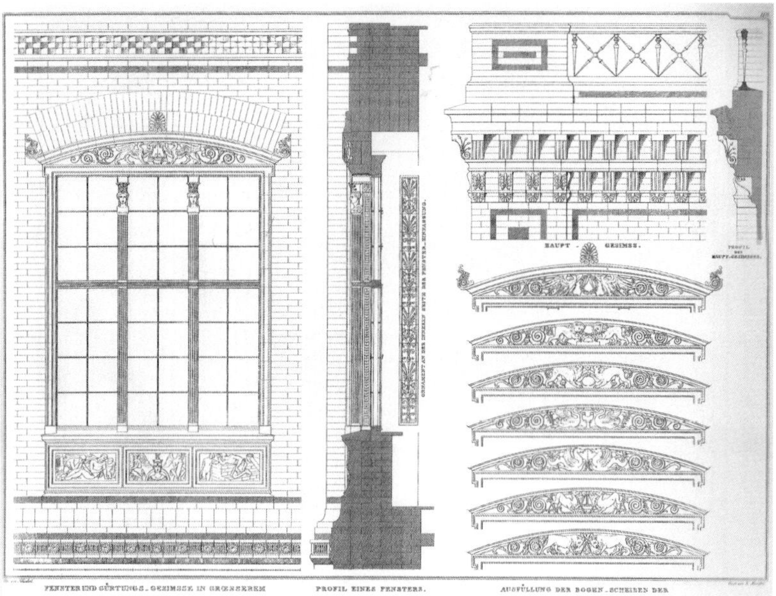

99. K.F. Schinkel, Window elevation, section, and details of Bauakademie, Berlin. From *Sammlung Architektonischer Entwurfe*. The building contained hundreds of didactic carvings and decorative friezes, all for the edification of its students. (Author collection)

He was able to put this knowledge to good use throughout his years as director of public works in Berlin, but it was at the end of his career that the memories and mental maps had their greatest impact. Schinkel's friend and colleague, Peter Beuer, invited him to become professor at the Bauakademie and design its new building in the center of the new Berlin that both had helped to create. His complex and unusual design, using brick-vaulted construction with iron, became a testament to all his ideas about building and art. [99]

The program of study at the Berlin building academy coalesced under Beuer into one of the most pragmatic and powerful means of imparting professional expertise and design judgment on young students. Each year the young designers spent the winter months inside, attending lectures, learning to draw, and completing design exercises. The basic content was similar to that at the French academy. However, during the spring and summer they ventured out to take on internships with practicing architects and builders, visiting works under construction or helping with drawings in preparation for bidding. Such direct experience, following fundamental instruction, linked theory with practice, and made memory archives through embodiment. Schinkel began to develop a textbook, his *Architectonische Lehrbuch*, which would offer students a compendium of different constructional and formal types. In it, he hoped to distill the fundamental principles of building and aesthetics from both modern and historical precedents. Never finished, only leaves exist in the form of sketch drawings.

This view of architecture, and of education, became the theme of Schinkel's building. Using his favorite materials, brick and terra cotta, the architect aligned

100. K. F. Schinkel, Bauakademie, Berlin. The architect not only designed the school building, but also planned its curriculum (Berlin, Institut fur Denkmalpflege).

vaulted bays in rows to form a cellular floor system, which he then stacked to form a four-story block, surmounted by an iron parapet. Each vault appeared in the outer revetment as a bay infilled with a terra cotta and iron aedicula. Within each bay was a narrative, using sculpture and classical decorative motifs, that explained the programs and activities taking place in the academy. As was his wont, Schinkel designed all of the decorative elements himself, later publishing them in his massive "collection" of architectural designs, or *Architectonishe Entwurfe*. As he wrote in his notes for the *Lehrbuch*, "The real task is to make every part of the construction beautiful. In that word—beautiful—lies the whole nature, the whole feeling for conditions" that materials give to form.[24] Densely concentrated, the Bauakademie building epitomized this principle in part and whole. Schinkel's masterpiece (demolished by the East Germans in 1961) has never been equaled as a setting for design education. [100]

Mechanical Drawing and Draftsmanship

Another remarkable aspect of architectural training during the modern era was the way in which basic drafting skills were developed, perfected, and taught. Already, in the seventeenth century European architects had a wide array of precision tools with which to ply their skills. These included dividers, compasses, ruling pens, chalk, charcoal, India ink, and watercolors. Another key tool, the pencil, was not widely available until the early seventeenth century. Using graphite rather than ink, architects could produce very fine, erasable lines with ease.[25] [101]

In the engineering professions, especially in Britain, France, and Germany, techniques developed rapidly during the eighteenth and nineteenth centuries, so that there could be specialization in the field, freeing the designers (thinkers) from the producers of designs (draftsmen).[26] As in typesetting, carpentry, or metalwork, repetitive tasks were delegated to skilled tradesmen, who were

101. Baluster, drawn with a pencil and shaded with softer lead, by the author. Line drawing always preceded shaded renderings.

never in charge of the workshop. Gaspard Monge, who taught at the Ponts et Chaussées, and James Naysmith, who invented by drawing, were the first innovators to use draftsmanship as a tool for machine-making. As the latter wrote, "Mechanical drawing is the alphabet of the engineer. Without this, the workman is merely a 'hand.' With it, he indicates the possession of a 'head.'"[27] Echoing him, many scholars consider the splitting of design from craft a key cause of the dehumanization of work during the industrial revolution. We'll discuss more about this in Chapter 9.

The net result of this bifurcation was not entirely negative, as ranks of the industrial labor force swelled with new specialists and tradesmen. In architecture the benefit was the emergence of a class of artisans whose task was to bring buildings to life, before, during, and after their conception in the mind of a master.[28] Buildings needed to be sold, to be precisely quantified, and finally to be constructed. Moreover, members of a studio were required to quite literally put a building together on paper before construction could begin. The draftsman emerged as a separate laborer in architectural offices.[29] During the 1850s Richard Upjohn employed seven men in this position, running one of the most efficient offices in New York.

In the early nineteenth century the Mechanics Institute in New York began offering courses in technical drawing, something that had already occurred at the Carpenter's Company in Philadelphia a century before.[30] One of the first comprehensive books on drafting was William Minifie's *A Text Book of Geometrical Drawing for the use of Mechanics and Schools* published in Baltimore in 1852.[31] In it the author gave examples of isometric projections, perspective, and various kinds of mechanical illustrations useful in the trades. Later mechanical drawing was distinguished from drafting, especially in courses taught by Louis Rouillion (1866–1949) at the Mechanic's Institute.[32] Today, the General Society building in New York is home to the Institute for Classical Architecture, one of the few organizations still teaching Beaux Arts design in the U.S. During the early twentieth century numerous courses emerged to teach drafting to those wishing for a career in engineering or architecture. Manuals for achieving competence in architectural drafting chronicle some of their pedagogical methods in detail.[33]

One of the first means of teaching the ability to see a building in its parts as well as its entirety, was the principle of the "coupe" or section cut. The French word has a slightly stronger connotation than anything in English, more like a blow with a butcher's knife than slicing an orange in half. In both the Polytechnique and the Academy the student was required to dissect many different solids using sections before undertaking the design of something so complex as a building. If two sections were cut through the middle of the building professors stressed that there would be no hidden features that the student might ignore in creating his or her esquisse. Often sections were presented at

102. Victor Blavette, Analytique, Temple of Hercules, Cori. Ecole des Beaux Arts, Paris. Like most drawings of its type, it features a plan, section, and elevation of the building at different scales (Harbeson, *The Study of Architectural Design*)

right angles to each other, demonstrating the way in which a draftsman would create the drawing via projecting lines.

Nowhere was this technique more important than in the so-called "analytique" exercise. Though this fundamental design problem probably showed up early in the second class at the Ecole in Paris, and was not a focus of training there, it became the cornerstone of the American system developed by members of the Beaux Arts Institute of Design. John Harbeson, a student of Paul Cret, outlined the methods of this influential program in his book, *The Study of Architectural Design*. This widely used textbook was a record of a curriculum that provided competitions and exercises for students outside of university systems.[34]

As the term suggests, the *analytique* problem requires a student to deconstruct a small building or portion of a building, surmise its salient features, and then represent the building on a single sheet of paper in such a way that a trained observer may understand its interior, exterior, structure, and ornamentation. As you can imagine, such a task was downright frightening to beginning students. Where even to begin such a thing?

Of course, some of the foregoing discussion allows us to understand the way in which drawing unveiled the process of making the building as well as perceiving it in space. A section through the piece to be analyzed, both as an exercise and as a representation of its body, was the key. Once drawn, the brain could apprehend much about the object (turn it around, look inside, etc.). Like Leonardo da Vinci, who seems to have been able to cut sections in

103. Office John Russell Pope, detail, Temple of the Scottish Rite. This working, shop drawing was produced to be used on the job site. Line weights are used to define edges and contours in all such drafted details. (Clute, *Drafting Room Practice*)

his mind, a trained architect, or draftsman, can readily create a section after standing inside, and walking around, a small building. Moreover, sections can be generated while a building is in design, to develop relationships between its interior and exterior. [102]

A good *analytique* would impress studio masters if it were well composed on the sheet, relatively comprehensive in its representations, and clever at showing the best aspects of the building in question. It is likely that many students enrolled in Institute programs, which were held throughout the U.S. for several decades, did not go on to become architects, but rather used the training to become staff in various offices requiring "draftsmen." My grandfather worked in this capacity during his early years as a mechanical engineer, and learned his skills through a popular correspondence course.

He worked in the United States after emigrating from Slovakia, joining tens of thousands of highly skilled people who created the drawings required to build most of the infrastructure we use today. In crucial respects, the first half of the twentieth century might be called the Golden Age of hand draftsmanship, since so many individuals perfected their drawings skills during that era.[35] Indeed, no one who examines a drawing on linen or paper produced during that time will fail to notice its high degree of accuracy and polish. Standards were necessarily high when the precision of such drawings could make or break a construction project like the Empire State Building (1929) or Manhattan Bridge (1918).[36] [103]

All working drawings are made using precise, ruled lines. Initially these line drawings are created to scale as a part of design development, as preparatory tracings for a final document. After the introduction of tissue tracing paper, draftsmen could re-draw details many times in pencil for review by supervisors before committing anything to ink. That did not mean that sloppiness was condoned, for a section sketch made at ½" to the foot might require accuracy to the half or quarter inch (a bit thicker than a hairline). Such drawing acumen demanded a fully mature visual cortex and deep memory banks of previous hand-eye patterns.[37] Simply drawing a straight line using a pencil required hours of intense practice in what some called the "bulb" technique.

I learned this technique at Penn from upper-level students who were quick to stress the importance of getting it down before undertaking any graded projects. It was probably developed about two centuries earlier in, you guessed it, Paris, or perhaps even earlier, in Rome. In order to draw a line that appears to be precise and perfectly level, a draftsman must actually twirl his pencil while varying the pressure of the stroke. Moreover, each line, no matter what its length, must begin with a dark "bulb" or thickened end, stretch out like taffy, and end with the same thick stroke. Pencil lines must always cross at their corners, or the eye/brain will not perceive them as fully joined.

This technique became the cornerstone for draftsmanship all over the world, and appears on every drawing produced during the Golden Age. Of course, it is only one of dozens of drawing "tricks" or techniques that were known by competent draftsmen/women. Some of these appeared in books such as Philip Knobloch's *Good Practice in Construction*, or in the magazine *Pencil Points*. Many may be lost when the last of the trained draftsmen or "green visors" dies. Some, however, remain in publications such as Eugene Clute's valuable 1928 book, *Drafting Room Practice*.

In it, Clute provides detailed rubrics for various kinds of drawing modes and model making. Particularly valuable is a long chapter on "Working Drawings," which sets out the methods by which design ideas were refined from the earliest sketches to finished "contract" documents that were provided to builders. Edgerton Swartwout, a valued associate at McKim, Mead & White before opening his own office, provides the most detailed account of his own methods. He stresses that the basis of all good construction drawings was the initial sketch, or *parti*, roughly drawn in a soft pencil or charcoal pencil. Like Alvar Aalto decades

104. William Hough, student esquisse, University of Pennsylvania, c. 1920s. A similar process of refinement from a first sketch was used in most U.S. offices from 1900 until the 1950s. Hough founded one of the most successful of all Philadelphia firms with John Harbeson, his classmate (Architectural Archives, University of Pennsylvania).

105. Rendu of same design, with all the elements from the sketch.

later, his draftsmen used small-scale, roughly drawn sketches to develop larger scale, more detailed drawings as the building took shape. This method followed what we now know to be the steps that the brain uses in processing visual information, from the most abstract outlines to more refined understanding of color, light, and position. We may also note that Swartwout's Beaux Arts education informed his later professional practice.

The media and scale of each drawing reflected its increasing development of detail, the kind of understanding relevant to builders and tradesman executing the building contract. The process begins with a drawing at 1/8" scale, on tracing paper, that will become a reference for all future details. Most architects know that such drawings are excellent for understanding the whole building. A larger scale drawing, perhaps at ¾" to the foot, would then be drawn to "lay out the bones of it with the greatest care, to be sure it scales to a fraction of an inch … We study over this in freehand and render it in pencil or charcoal and sometimes rough-out some of it in full size to be sure of the scale … When we're finished we know that the [building] as detailed would, reduced, look exactly like the eighth scale, or, rather, like the design."[38] [104, 105]

Such a progression from relative abstraction or fuzziness to relative realism or focus is a fully embodied one; it also reproduces the iterative process of most imaginative work on a film, a painting, or a musical composition. When ready for translation into working drawings for a builder, scale pencil drawings could be used as definitive references for a team of draftsmen, as a means of facilitating their production of large scale details. One of the drawbacks of digital computer modeling is that the schematic line model demands more detail in its early stages than can reasonably be provided by any architect, let alone by a beginning student who knows almost nothing about building technics. Slow, but sure, the old-fashioned draftsman went only as far as the master architect could imagine at each step in the design process. Every paper representation advanced the details.

One of the unforeseen drawbacks of precise mechanical drawing was its implicit substitution of technical mastery for artisanal knowledge of materials. If any material could be abstractly presented as lines on paper, its full characteristics could be ignored by the brain of the engineer. Each abstracted part in the assembly could be treated as an identical unit, irrespective of its material properties or limitations. This is why most shop workers in small industries mistrusted such drawings throughout the nineteenth century, preferring instead to rely on full-scale models. In many respects the gradual substitution of technical, mechanistic knowledge for mimetic, embodied knowledge began during the late eighteenth century in France and later spread throughout industrialized Europe and the United States. It did not reach Asia until the twentieth century. The next chapter addresses this issue in greater detail.

Notes

1 See Jean Paul Carlhian, *Americans in Paris: foundations of America's architectural gilded age: architecture students at the Ecole des Beaux Arts, 1846-1946* (New York: Rizzoli: 2014).

2 On the pedagogy, see David G. Van Zanten, "Le système des Beaux-Arts," *Architectural Design* 48: 11–12 (1978): profile 78, 15.

3 See Richard Oliver, Ed., *The Making of an Architect: 1881–1981* (New York, Rizzoli: 1981); Joan Ockman, Ed., *Architecture School: three centuries of educating architects in North America* (Cambridge, MIT Press: 2012).

4 For an excellent summary of the curriculum see, Isabelle Gournay and Marie-Laure Crosnier Leconte, "American Architecture Students in Belle Epoque Paris: Scholastic Strategies and Achievements at the Ecole des Beaux-Arts," in *The Journal of the Gilded Age and Progressive Era*, Vol. 12, No. 2 (April 2013), 154–98.

5 See Paul R. Baker, *Richard Morris Hunt* (Cambridge: MIT Press); Susan R. Stein, Ed., *The Architecture of Richard Morris Hunt* (Chicago, Univ. of Chicago Press: 1986), especially "Hunt in Paris," by Richard Chafee.

6 The most comprehensive analysis in English is in Arthur Drexler, Ed. *The Architecture of the Ecole des Beaux Arts* (Cambridge, MIT Press: 1977).

7 David Marr, *Vision: A computational investigation into human representation and processing of visual information* (San Francisco: W.H. Freeman: 1982). Marr's theories have been revised by more recent research on the passage of signals from the visual cortex to other areas of the brain. The basic sorting remains similar to his assertions.

8 A popular science program on TV, *Brain Games*, uses a shade-shadow illusion to demonstrate the power of this perceptual cue—two square objects in a medium gray are shown separated by a white highlight. Not one observer is able to determine that the objects have the same "value" or darkness because the highlight suggests a lighter value below.

9 An interesting study of Dow, Ross, and Emile Lorch is Marie Frank, "The Theory of Pure Design and American Architectural Education in the Early Twentieth Century," *JSAH*, v. 67, n. 2 (June 2008): 249–73. The author goes to great pains to trace the influence of Dow and Ross's ideas in architecture programs, but cannot show any significant penetration into the leading programs (Columbia, Penn, Cornell, Harvard, MIT, Illinois) during the first half of the century, citing Lorch's impact at Michigan, which was never a front rank architecture school.

10 See for instance Brian M. Ambroziak, *Michael Graves: Images of a Grand Tour* (Princeton, Princeton Architectural Press: 2005).

11 See Mehmet H. Goker, "Effects of experience during design problem solving," *Design Studies* 18 (1997): 405–26. The experiments used neuro-imaging to track cognitive patterns in both students and experienced architects, with clear results. Since students struggle to make choices among solutions, they must employ the pre-frontal cortex extensively. Also Manolya Kavakli, "Sketching as mental imagery processing," *Design Studies* 22 (2001): 347–64.

12 Quoted in Alina Payne, *The Architectural Treatise in the Italian Renaissance* (Cambridge, Cambridge Univ. Press: 1999): 68.

13 Branko Mitrović, "Aesthetic Formalism in Renaissance Architectural Theory," *Zeitschrift für Kunstgeschichte* 66. Bd., H. 3 (2003): 321–39.

14 Richard A. Moore, "Academic Dessin Theory in France After the Reorganization of 1863," *JSAH* 37: 3 (1977) 145–74.

15 On classical principles see Alexander Tzonis and Liane Lefevre, *Classical Architecture: The Poetics of Order* (Cambridge, MIT Press: 1982).

16 Donald, *Modern Mind*, 307.

17 See for instance Howard Colvin, *A History of the King's Works*, 2 vols. (London, Her Majesty's Printing Office: 1963).

18 Andrew Saint, *Architect and Engineer: A Study in Sibling Rivalry* (New Haven and London, Yale Univ. Press: 2007) presents a good summary of Vauban's career and influence, 9–33.

19 For a substantive introduction to the man and his work, see Michael Snodin, Ed., *Karl Friedrich Schinkel: A Universal Man* (New Haven and London, Yale Univ. Press: 1991).

20 Marvin J. Anderson, "The Architectural Education of Nineteenth-Century American Engineers: Dennis Hart Mahan at West Point," *Journal of the Society of Architectural Historians*, 67:2 (June 2008): 222–47. Mahan was trained in France and was influenced by Monge, Rondelet, and Durand. His students, trained between 1840 and 1871, formed the first cadre of professional structural engineers in the U.S. All were advocates for rationalism and structural determinism.

21 See David Watkin and Tilman Mellinghof, *German Architecture and the Neoclassical Ideal* (Cambridge, MIT Press: 1987): 8–15.

22 Barry Bergdoll, *Karl Friedrich Schinkel: An architecture for Prussia* (New York, Rizzoli: 1994): 12–16. The younger Gilly set up a small circle of architects who met to discuss theory, work on projects, and advance the profession in Berlin. Such clubs were popular during the nineteenth and early twentieth centuries in many cities.

23 See Bergdoll, *Schinkel*: 24–33.

24 Quoted in Bergdoll, 216.

25 See Elisabeth Kieven, "Mostrar l'inventione: The Role of Roman Architects in the Baroque Period," in Henry Millon, Ed. *The Triumph of the Baroque: Architecture in Europe: 1600–1750*, (New York: Rizzoli: 1999): 173–206.

26 For a summary of this development, see Saint, *Architect and Engineer*: 131–70.

27 Naysmith, quoted in Johnson, *The Birth of the Modern*, 583.

28 A good example of the drawings made by such tradesman is: Elwood, Franklin George. *Architectural drawing plates* (Peoria, Ill., Manual arts Press, 1919). The style is unmistakeable.

29 The most comprehensive treatment of this era to date is: George Barnett Johnston, *Drafting Culture: A Social History of Architectural Graphic Standards* (Cambridge: MIT Press: 2008), particularly 58–91. His point of view is indebted to the work of Pierre Bourdieu on professional standing and cultural capital.

30 Jeff Cohen, in *Drawing Toward Building: Philadelphia Architectural Graphics 1732–1986* (Philadelphia, Penna. Academy of Fine Arts: 1986).

31 On the development of courses in architecture see Dell Upton, "Defining the Profession," in Ockman Ed., *Architectural School*, 36–65.

32 On Rouillion see Johnston, *Drafting Culture*, 25–37. Johnston fails to provide context for the emergence of manual trades education before the 1890s, placing too much emphasis on twentieth-century standards for mechanical drawing.

33 For one such course, see Edminster, C. Franklin, *Architectural drawing*, 3d ed., enl. (Brooklyn, The Author: 1905).

34 John Harbeson, *The Study of Architectural Design* (New York, Pencil Points Press: 1926).

35 Svensen, Carl L. *Architectural drafting* (New York: D. Van Nostrand: 1929) shows the results of this development by the eve of the Depression.

36 For example, see Mark A. Hewitt, "Beaux Arts Representation and the Golden Age of American Draughtsmanship," *The Classicist*, 3, 27–34.

37 For an excellent analysis of how these techniques were used in a single building, see Alexander Ortenberg, "Joy in the Act of Drawing: Maybeck's Palace of Fine Arts," *JSAH* 70: March 1, 2011: 38–63.

38 Eugene Clute, *Drafting Room Practice* (New York, Pencil Points Press: 1928): 177.

[8]

ENGINEERING, SCIENCE, AND THE MACHINE

Throughout the late Baroque era in Europe and the Americas most architects practiced their craft according to haptic and visual modes of conception such as those used in disegno-based programs. Artisans and artists were brethren. In the German principalities a strong tradition of craft-based design allowed architects to work closely with plaster craftsmen, painters, masons, metalsmiths, and carpenters in the production of churches, palaces, and public buildings. Likewise, in Spain, Portugal, and around the Mediterranean, artisans produced buildings under the direction of artists. Even at the court of Louis XIV the workshops started by Colbert maintained an artisanal tenor. Once empirical "science" became the dominant epistemology during the late seventeenth century, all of this changed dramatically. One of the results of this change was an increasing emphasis on individual invention over collective artistry in studios and workshops, later to become factories. In many respects, increased production changed the character of design. And with rapidly advancing techniques of mass production, designers began to look at their tasks as problems, maximizing the efficiency of assembling machines and other artifacts that might make life easier for the common person.

Knowledge itself became more complex and assembly based. Pamela H. Smith describes the pre-Enlightenment position of material producers as "artisanal literacy," as distinguished from academic or linguistic literacy (the dominant form in our era). Knowledge gained by working with nature and her materials was embodied or directly experienced. Such knowledge was always acquired through the inventions of the mimetic faculty in the body and brain together. Smith traces a period in which artisanal literacy was linked to both art and science during the sixteenth century, but asks "how do we get at this knowledge at the distance of centuries and over the great divide of the Scientific Revolution, at which time scientists assumed the place of experts on nature."[1] Indeed, the epistemological divide between our modern view of hand work and that of European artisans in 1500 is vast.

As we saw in Chapter 5, the profession of architecture emerged during the Renaissance, when designers of buildings ceased to work among the guilds of masons, carpenters, and other members of the building trades. After designers began directing the building process from a position of distanced authority, they ceased to acquire knowledge at the level of the artisans under their direction. More importantly, humanist scholars and writers began to devalue artisans, as we saw in the case of Leonardo. As John Fitchen has pointed out, "the men adept at words—those who had been formally educated in philosophy or other polite studies, the 'gentlemen' of each era—have traditionally shown their superiority by denigrating the artisan and relegating him to an inferior status in society."[2] This included the first gentlemen scientists who looked at the natural world.

The famous account of Francesco Borromini's challenge to the sculptor, Gianlorenzo Bernini, when he condemned the latter's design for the foundations of towers on either side of Carlo Maderno's St. Peter's façade, is a case in point. Borromini, trained in Milan as a lowly stone mason, understood that the soil would not support Bernini's tall stone towers. Though he pointed this out, his authority could not match that of the artist, and his advice went unheeded. The towers were built, and taken down, when he was proved correct. He was nonetheless not offered the position of Vatican architect—an artisan had none of the cachet of a painter. As we consider the cognitive value of "artisanal literacy" we must understand how, and when, such knowledge waned in Western society, and why it was supplanted by the "inventive" capacity of engineers.[3]

The Death of Mimesis

Modern definitions of the "fine arts" differ significantly from pre-eighteenth century concepts of knowledge and professional labor. These differences are manifested in such publications as the *Encyclopédie* (1751–1772) of Diderot and L'Alembert, published in France but disseminated throughout Europe.[4] Whereas the Greek concept of seven Liberal Arts did not distinguish between art and science, post-Enlightenment thinkers created divisions in knowledge disciplines that now seem natural and inevitable. Even the term "beaux arts" was an invention of the late eighteenth century that hastened the demise of mimetic practices among artisans.

As Larry Shiner has made clear, the common view of an artisan-artist as a maker of things that might be used, perceived, and enjoyed by everyone fell apart as a result of new economic and social forces during the post-revolutionary period in Europe. An empirical view of knowledge, divorcing science from the arts and humanities, emerged only after the Enlightenment. Further eroding the mimetic view of artistic production was a new class-based social and economic structure following the toppling of the French aristocracy in 1789. In philosophy, thinkers such as Kant and Schiller pressed for a distinction between aesthetic experience, available only to the educated and elite classes, and popular appreciation of low art made by artisanal labor.[5] The net result of these trends was separation of the higher arts—poetry, painting, sculpture, music, and architecture—from every other kind of artistic production, including fabrics, ceramics, and metalwork. In the older guild systems all of these areas were united.[6]

Architecture suffered most from such a division, because it was the "art" most dependent on the unification of applied with fine arts. It was also the art most connected with mathematics, physics, and other forms of "science." The medieval dictum of "ars sine scientia nihil est" would suffer its greatest challenge in the face of an industrializing world. In industrialized, consumer societies, the commodification of building encouraged humans to look at assemblies of parts, each of which had an economic value, rather than at the whole.

Pre-enlightenment thinkers would not have considered architecture as scientific. Indeed, one of the most important theoretical arguments to emerge out of late twentieth-century critiques of Modernism was prompted by scholars studying classical architecture during the Enlightenment in France, Italy, and England. Joseph Rykwert, among others, began to distinguish early forms of Neo-Classicism and empirical scientific studies of the classical orders from previous architecture that, superficially, used the same stylistic idioms. Rather than locating "modern" architectural thinking in the nineteenth century, as had been common during the early twentieth century, Rykwert suggested that a major new understanding of building was part of the general Enlightenment turn toward empiricism.[7] He rejected some of the assumptions made by earlier historians in their interpretations of the rise of Modernism.

Work in the history of science by such thinkers as Thomas Kuhn and Carl Popper caused art historians to re-evaluate some widely held views of what constituted modern design methods.[8] Whereas Renaissance architects such as Palladio and Alberti tied all proportional systems to the human body, and described their art as one imitating the laws of nature, Claude Perrault (1619–1683, trained as a physician) argued that such systems were arbitrary, and based only on customs.[9] Empirical study, and a skepticism of antique precepts, caused Perrault to question not only the supposed natural basis of proportion, but also to separate perceptual from a priori understanding of architectural form.

Enlightenment science obliterated centuries of wisdom concerning the orders of the natural world, especially aesthetic "truths."[10]

In France, the dissolution of the classical notion of mimesis and the waning of *disegno* as an artistic ideal revolved around a debate between the "Ancients and the Moderns." Perrault and his fellow scientists at the Academie Royale jousted with architects such as Francois Blondel (1617–1686), who defended Vitruvian standards of proportion and beauty against rationalist critiques. Throughout the West, Neoclassical views of architecture were undergirded by empirical and skeptical positions among leading theorists, who expressed their ideas in printed treatises and "parallels" of the orders that placed ancient precepts under a microscope for dissection. It was immediately clear that no consistent, purely mathematical systems governed the Classical Orders, especially when comparing Attic Greek, Imperial Roman, and Renaissance treatises. As James McQuillan has observed, "After nearly two millennia, the traditional European architectural treatise and its intrinsic connection to the concept of mimesis came to an end."[11]

Renaissance artists such as Leonardo and Albrecht Dürer used drawing to probe the workings of the natural world, thereby opening up a new form of inquiry that presaged experimental methods during the following centuries. We noted in Chapter 5 that the culture of the *botteghe* of the sixteenth century was steeped in habits and methods common to artisans in many trades. Guild practices and apprenticeships developed through the Middle Ages in both Europe and the Ottoman Empire. The so-called Age of Discovery would not have been possible without highly sophisticated crafts and trades such as ship-building and blacksmithing. Historians of science link the use of these "technological" practices with fundamental scientific breakthroughs.

However, the use of the term "technology" implies certain biases and mentalities that were not present in the world of Leonardo, Jan van Eyck, and Paracelsus. These artists practiced in a milieu that valued forms of embodied knowledge that would not today be classified under the rubric of "rational" thinking. As Alberto Perez-Gomez and Louise Pelletier have noted, "While an increasing rationalization of visual perception began to influence the geometry of depth in sixteenth-century painting, artists still relied on their predominantly tactile experience of the world to create perspective images."[12] Drawing itself, and the dependence upon visual thinking in general, seemed to place artists and craftsman in a different realm from early scientists such as Newton and Leibnitz. Even mathematical perspective was used to disengage the observer from the physical world. As Perez-Gomez noted in his critique of modern science, "In order to achieve his objectives, however, Perrault had to reject the traditional symbolic implications of architectural proportion … Architectural proportion lost in Perrault's system its quality of absolute truth."[13] In essence, the visual basis of architecture lost its hold on the cultural leaders of the Enlightenment, paving the way for our modern views of technology and functionalism. Only Giambattista Vico defended the humanistic view of poetry, art, and history from the increasing skepticism of the empirical sciences.[14]

106. Comparison of Egyptian and Greek water vessels. Gottfried Semper, der Stil. Semper did not use "tectonic" in the way that the term is used by many architects today. His analysis was ethnographic and cultural; very different from that of Viollet-le-Duc in France.

The turn away from experiential or embodied forms of knowledge work, both in the sciences and in the professions such as law, medicine, and engineering, was a gradual process that probably did not reach its apogee until the mid-twentieth century, though its beginnings were in seventeenth-century philosophy and mathematics. Pamela Smith argues that the early scientific revolution was actually nurtured by artisans and alchemists, who insisted upon an experiential, sensory appreciation of the natural world. Her recognition of the contributions of Paracelsus, the controversial Swiss physician, is unusual in this regard, though Carl Jung singled him out in his writings as early as the 1930s. Only after the royal academies (in England, France, and even Russia) began to assert their control over "academic" discourse in the sciences did rationalism begin to erode the influence of craft practices and the guilds among those employing "techne" in their disciplines. Most historians of science cite the influence of Francis Bacon on Isaac Newton, Robert Hooke, and Descartes, who increasingly insisted on mathematical and fact-based proofs of scientific ideas.[15] We noted in Chapter 7 that "mechanical" drawing became a recognized skill during this period.

We also noted in Chapter 3 that in the fine arts mimetic practices, especially in the training of visual artists, remained dominant throughout the nineteenth century, eventually spurring the *salons des réfusées* and modern art. Museums were filled with art students, copying Old Masters. Every museum had a hall of casts. During the Arts & Crafts Movement in Europe and the U.S. progressive social thinkers encouraged the resurgence of small handicraft workshops in the face of overwhelming industrial capitalism, but their efforts were relatively brief and ineffectual in keeping mimetic conceptual modes alive. The evidence of industrialized methods replacing handwork can be seen in publications dedicated to traditional trades, such as the works of André Roubot, Denis Diderot, and Joseph Moxon, among others. [115] Woodcuts and engravings of tools, artisanal training, craft production, and material processing can be contrasted with those describing machines and industry, often in the same books. Such "instrumental" forms of representation resulted in a shift in the understanding of how architecture could be conceived and built. Not only did machines interpose themselves between designers and fabricators, but a whole new definition of design emerged, divorced from the concept of *disegno* altogether.

Naturalism, Science, and Tectonic Form

Empirical study of the earth, its flora and fauna flourished in the early nineteenth century, spurred by the efforts of two Prussian brothers: Wilhelm

(1767–1835) and Alexander (1769–1859) von Humboldt. We have already noted the influence of the former on Schinkel and his architecture. The latter was to have a profound impact on the work of another architect, Gottfried Semper (1803–1879). Sometimes called the "second greatest" German architect, this protean figure lived through the turmoil of the 1848 Congress and remained in exile for much of his life because of his militant activities in Dresden. He was drawn to scientific inquiry during his early education, and became the architect most associated with *naturalism* in the modern sense through his extensive teaching and writing.

The profound interest in nature and her process of growth and change that obsessed many nineteenth century thinkers, including Goethe, Schiller, Wordsworth, and other literary figures, was of course pursued first by gentleman scientists. Scientific academies were founded by monarchs throughout Europe in the previous century, providing both status and support for research on many aspects of natural history. Without the academies Newton, Leibnitz, Descartes, and the Perrault brothers would not have been able to discuss and publish ideas about the natural world. Moreover, the academies provided prizes for discoveries that advanced many areas of social and natural science, such as astronomy, mathematics, and optics. As industries grew and prospered, their capitalistic goals often directed science, just as they do today.

In Prussia the pursuit of science was underwritten by enlightened monarchs such as Friedrich Wilhelm II, but with far less insistence on sanctioned subjects and results than in England or France. Wilhelm von Humboldt was a royal advisor who used his scientific approach not only to study languages and societies but also to improve governance. His brother traveled extensively in the Americas to learn about the natural world, practicing what he called "terrestrial physics." Both published research easily and quickly under state patronage, and became internationally renowned, influential beyond the borders of the German states. Alexander's protean abilities as a researcher, writer, and explorer made him particularly notable among Germans of his time.

He spent most of his life exploring the wildest parts of the globe, spending many years in Central and South America. He also made multi-year excursions to North America and Russia. The last and most ambitious of Alexander's projects was *Kosmos*, a multi-volume study of the earth and its inhabitants that presaged what we now called ecological or environmental science.[16] Gottfried Semper was well aware, following trips to Berlin to look at Schinkel's buildings, of this far-reaching scientific treatise. He was to use some of its ideas in his subsequent work. Indeed, no German intellectual in the mid-nineteenth century would have been immune to its influence.

Semper was quite similar to von Humboldt as a prolific writer who also remained active as social servant during a long career. Though his extensive writings have often been interpreted as laying a critical foundation for twentieth-century architecture, particularly as an expression of the logic of construction, his ideas were complex, contradictory, and even antithetical to modern design theories.[17] He was involved in numerous cultural, scientific and educational enterprises, and may have been hampered as much as encouraged by nineteenth-century advances in knowledge. According to Harry Francis Mallgrave, Semper's reputation suffered serious damage during the early twentieth century, and his ideas were often taken out of context, leading to misapprehensions that persist in contemporary theory.[18]

Semper had ample time during his hiatus from Germany to write his first study of architectural history, *Die vier Elemente der Baukunst* (1851).[19] In it he was able to draw on recent archaeological discoveries in the Middle East, particularly related to the Assyrian/Chaldean civilization. His fascination with these builders was not limited to their "technological" achievements—he looked at culture, society, and the natural world in everything he studied. He was very like a modern anthropologist or material culture specialist in his research methods and goals. In general he followed the same empirical, observation-based techniques that had been introduced by von Humboldt, and even used the word "cosmos" to describe nature's sense of order and adornment.[20] He was one of the most learned and widely read architects of the nineteenth century, and his years in London and Zurich gave him access to many leading thinkers. He was also a first-hand observer of the 1851 Exposition in London, later to influence his notions about architecture.

The most often cited of Semper's many ideas about architecture and society is his concept of "tectonics." In *Der Stil* (2 vols., 1863), his massive study of building in many cultures, he uses the term to describe a form of stick construction that can correspond to carpentry, metal frames, or lightweight skeletal structures. This type is contrasted with woven textiles, ceramics, and masonry as one of the five craft-like material processes that history shows to be virtually universal among humans. It is important to note that Semper does not mention science in any of his writing on the "technical arts" or building. Further, he does not insist that tectonic construction is the only form acceptable to modern builders, as do many theorists who use his ideas today. He is not a materialist or functionalist in the contemporary sense. Though biologists had begun to study the morphology of plant and animal forms during the mid-nineteenth century, Semper was only an observer of their work, impressed by taxonomies. [106]

He later used "tectonics" in a more universal sense, as a rubric for discussing formal laws that govern the natural and physical world. "Tectonics is an art that takes nature as a model—not nature's concrete phenomena but the uniformity and the rules by which she exists and creates," he wrote in 1859, echoing the *natura naturans/natura naturata* distinction in Renaissance aesthetics.[21] His three

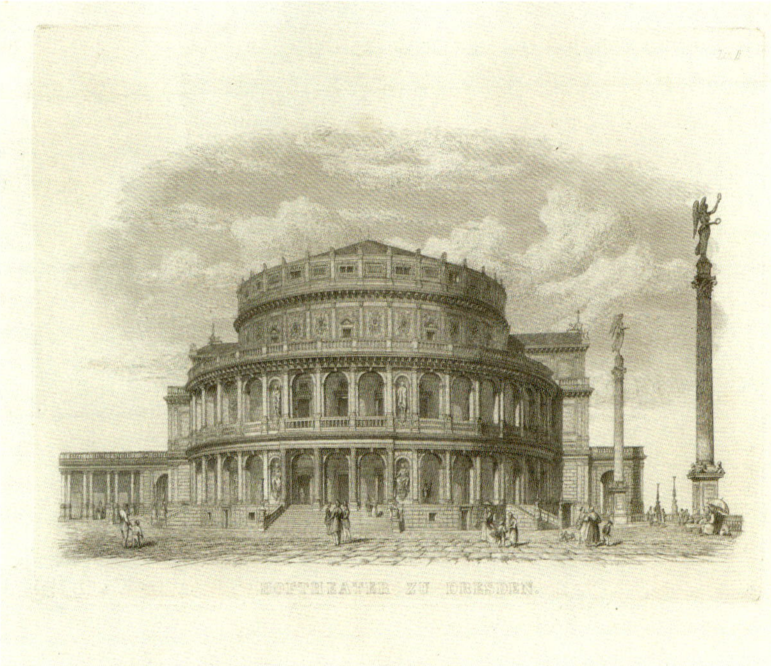

107. Gottried Semper, Hoftheater, Dresden, 1839–41. All of Semper's major buildings are classical, and use elaborate ornament to articulate building elements and create a fictive embodiment of the activities that occur within. (Author collection)

attributes of formal beauty were: "(1) macrocosmic unit or order (symmetry); (2) microcosmic unit or order (proportionality); and (3) unit or order of direction (direction)." These concepts are also entirely consistent with principles in many previous treatises, such as those of Alberti, Palladio, and Vitruvius, which were in turn based on a mimetic understanding of the arts. In essence, this purportedly scientific thinker was pursuing a generally humanistic, and naturalist/primitive (in Rousseau's romantic conception) trajectory in all of his writing. When he attempted to show these aesthetic principles at work in the natural world, he could at once pursue Humboldtian science while also seeing bodies (celestial, human, plant, and animal) as part of a universal *taxis* or ordering system.

In fact, his theory of formal aesthetics was essentially classical in its basic statements, and all of his buildings were based on classical models. He was also, not surprisingly, one of he most innovative theater and concert hall designers among modern architects; with extensive knowledge of music and dance he was able to design both stages and auditoria that met the most exacting requirements of dramaturgy. We can see this in his first Dresden theater, his magnificent opera house there, and his unbuilt Munich theater project for Richard Wagner. Semper would not abandon the figural, classical ideal of embodied artistic activity in all human societies. Those who see him as a precursor of modern notions of tectonics have not understood the tenor of his writings or essence of his thought. As Harry Mallgrave noted in his authoritative study of Semper's life and work,

"Whatever may be said about Semper's reasoning today, it represented one of the first attempts in modern times to interpret architecture and evolution in cultural terms rather than on the basis of quasi-mythical values."[22] [107]

As Dalibor Veseley wisely observes, "we do not as a rule think about architecture as a mimetic art. This disinclination has partly to do with the well-established tradition in which architectural mimesis is reduced to the imitation of reified precedents, such as the primitive hut, the Solomonic Temple, exemplary buildings, and so on, or to such generalized solutions as the 'imitation of nature.' Mimesis is not the same as imitation." He attributes some of the confusion over representation's role in modern architecture to this fundamental misunderstanding of human history and culture.[23] He also points to the crucial role of memory in understanding why architecture engages the human imagination so powerfully and directly. In many respects architecture activates memories of bodily states and movements, sensory pleasures, and even particular feelings about places and things that are charged with emotion. No other art form has the capacity to bring such a wide range of sensory recollections to the surface of consciousness.[24] Why, we might ask, has this fact been forgotten by contemporary designers? The answer can be traced to the emergence of engineering as a discipline. Today we would refer to the engineer as a designer of useful objects, machines, and technological instruments, though no such understanding existed until the early nineteenth century.

The Rise of the Engineer

Engineers emerged as building specialists only after the Industrial Revolution, when fabrication processes became too complex for crafters to understand. During the early modern era, sixteenth- and seventeenth-century engineering was the province of artisans and architects. But when French military officers began to specialize in the design of fortifications, their expertise was gradually collected and taught in special schools, as we learned in Chapter 7. Paul Johnson has observed that the Ponts et Chaussées had a profound affect on warfare, as French armies developed superior communication and logistic systems during the early nineteenth century.[25] Indeed, it was the expansion of railroads as the primary form of transportation that made engineers necessary. Whereas canals could be dredged without complicated machines, and built out of masonry, railways required level tracks, bridges, viaducts, and iron rails. Moreover, engineers could specialize in the process of designing and assembling parts of machines, making certain that those parts worked together in a smooth operating system. They specialized in explaining the *assembly process* to those charged with constructing machines.

The first heroic figure in transportation engineering was Thomas Telford (1757–1834), first president of the Institution of Civil Engineers (1828), and

108. Thomas Telford, Menai Staits Bridge, 1819–24. University Prints, Boston. Telford founded the first engineering society in Britain. (Author collection)

the designer of the Menai Straits Bridge (1826). [108] Like his contemporary Schinkel, Telford was a superb draftsman and master of building materials. Born in Scotland, he learned the mason's trade before coming to England to work as an architect with Sir William Chambers, on the giant Somerset House. Catching the eye of William Pulteney, an M.P. from the west country, he moved to Cheshire to become Surveyor of Public Works at Shrewsbury. He later designed the Ellesmer Canal and the monumental Chirk and Portcysullte aqueducts, built of masonry reinforced with steel. He was then hired to supervise the London-Holyhead road, then the largest public works project in England.

Telford eventually went on to design and supervise construction of an astounding array of roads, canals, bridges, piers, and aqueducts. In the mountainous Highlands of Scotland he transformed a barren region into a marvel of efficient transportation infrastructure. He built no less than 1,117 bridges throughout England, Scotland, and Wales, an achievement not likely to be surpassed. He engineered the world's first modern, all-weather road linking the north and south of Britain, and thereby modernized mail transport and delivery. Learning from French engineers such as Perronet, he created both cast iron bridges and chain suspension bridges such as that at Menai. So profound was his contribution to the industrial revolution that poems were written about him by Robert Southey (1774–1842), England's poet laureate.[26] Despite his status as an inventor, he never renounced architectural aesthetics, or local materials, in his designs.

Telford influenced English education through his professional example and advocacy, but training continued through apprenticeship and trade schools—learning by doing. In France technical training was eventually provided by a non-military academy in Paris, and thus engineering was split from architecture. Both France and the German states created the Ecole Polytechnique or Technische Hochschule to train mechanical and structural engineers, sometimes with and sometimes without architects. Engineers were celebrated in such projects as the Eiffel Tower in Paris, and Firth of Forth Bridge. Their status in France was dramatized by E.E. Viollet-le-Duc's abortive attempt to "reform" the

109. Joseph Paxton, Fox and Henderson. Crystal Palace, London, 1851, the most influential building of its age. Glass would henceforth become the most "modern" of materials. University Prints. (Author collection)

110. Augustus Roebling, Niagara Falls Bridge, 1851. University Prints. Roebling was trained as an architect under a pupil of Schinkel. (Author collection)

Ecole des Beaux Arts in the 1860s. The French Academy rebuffed that celebrated graduate of the Ecole Polytechnique, and retained its *disegno*-based curriculum over one based on mechanistic and rational problem solving.[27]

During the following century the engineer began to assert authority as a master of industrial materials and technologies, outstripping former brethren. New materials such as high-strength steel and reinforced concrete demanded more sophisticated means of manufacturing. Whereas wood could be readily joined with simple male-female, mortise and tenon assemblies, metal parts required threads in rounded sections, and welds in flat pieces, eventually even requiring various kinds of bearings for moving joints. Such assemblies were vastly more intricate and complex than either wood or masonry. Experts in industry, sometimes trained within a corporate environment, attained the status of "engineers" when their products transformed the marketplace, as in the case of Joseph Paxton, John Augustus Roebling, and Rafael Guastavino Sr.

Paxton's Crystal Palace (the second version at Sydenham) amazed Europeans following the London Exposition of 1851, and his fame was assured. Andrew Saint notes that he began his career as a horticulturist, and was helped by the design contributions of architects Owen Jones, Charles Barry, and Matthew Digby Wyatt: "The Crystal Palace was a collaboration: a nursery for engineers, and briefly too an academy for debate between architects, engineers and fabricators of many ages and experiences."[28] Only after Paxton's triumph did British engineers emerge as professionals, at first through trade schools where they could also learn the craft of fabrication. Gottfried Semper was in London during the construction of Paxton's marvelous building, and remained enthralled by it throughout his life as an exemplar of tectonics. [109]

Roebling, the patriarch of one of America's most successful industrial dynasties, came to Pennsylvania from Germany in 1831 following training at the Berlin Bauakademie under one of Schinkel's protégés, Friedrich Stueler. After establishing a reputation in industry for bridges at Niagara Gorge (1851) [110] and in Cincinnati-Covington (1856–67), the Roebling Company began manufacturing steel cable at its plants in Trenton, New Jersey. Their patented technology for twisting wire gave them a virtual monopoly. In America the company dominated the bridge industry for more than a century, supplying cable for the George Washington, Golden Gate, and Verrazano Narrows bridges. Roebling and his sons were not engineers but designer-fabricators, though they were mythologized for their heroic efforts at the Brooklyn Bridge by critics and historians.[29]

Guastavino created one of the most successful construction systems in the world after emigrating from Spain to the United States in the 1890s. Using a centuries-old technique for tile vaulting from his native Catalonia, he patented a method for constructing tile vaults using herringbone patterns and began to secure construction contracts for utilitarian buildings. In 1885 he bid on

111. Rafael Guastavino and Sons, Hamilton Farms Stable, Pottersville, NJ, 1920. Interior of carriage floor showing shallow vaults carrying the load of the stables above (Avery Architectural Archives, Columbia University). Guastavino was also trained as an architect and builder in Spain.

the construction of vaulted spaces in the Boston Public Library, designed by McKim, Mead and White, and saved the city crucial dollars over a conventional stone vault system. His success was assured when Charles McKim praised the Guastavino system, after which many leading architects utilized it in their monumental buildings. Like many engineers, Rafael Guastavino Sr. was trained as an architect and became a contractor; his son grew up in the business and took over from his father during the 1920s. Neither was looking for a new "aesthetic" in their elegant structures, but architects found one when designing tile-vaulted domes, crypts, and naves.[30] The primary market for such vaults was in industrial buildings such as power plants and warehouses. [111]

Le Corbusier praised such achievements, both in *L'Esprit Nouveau* (a magazine) and in his book, *Vers une Architecture* (1923). He himself had tried entrepreneurship after coming to Paris from Switzerland before World War I, establishing a brick kiln and supplying masonry for many small projects, including an abattoir of his own design.[31] Many of his heroes were American: the builders of grain elevators in Buffalo, bridge designers, aircraft makers, and the like. And of course he elevated machines to the status of art in his worship of motorcars and locomotives. He was one of the first architects to extol the *assembly* of buildings, as if they were motorcars, as the essence of modern design.[32]

The polemics of post World War I avant-gardes, trumpeting new technology and design, spun narratives that continue to drive today's architecture, even though engineers made no claims to aesthetic validity, nor did the manufactures of machine-age wonders. In many respects the modern image of the engineer as a master of functional aesthetics is a product of the modern military-industrial complex. Without modern warfare, we might not worship the technological achievements of modern engineering. Both of the World Wars in the twentieth century had profound effects on the practice of architecture, and the technology of buildings, that remain largely unexplored by historians.

In his book, *Architect and Engineer*, Andrew Saint analyzed the fraught relationship between engineers and architects in the modern era. His conclusions do not jibe with conventional views of that relationship, nor does he reveal a logical progression from visual aesthetics to structural determinism as guarantors of legitimacy in architectural design. Indeed, engineering is as much an art, dependent on judgment, as architecture. Its theory and practice are as riven with contradictions as those of any art. But one aspect of engineering distinguishes it from all forms of artisanal production: mathematical and logical thinking, using an assembly of abstract symbols. Another is that most engineering deals with dynamic, not static systems. Architects limit their study of physics to statics and strength of materials.

Representing Machines

Perspective drawing commanded the attention of architects, mathematicians, and artists for several centuries following its codification in the quattrocento. One conundrum confronting those who used this form of representation was its foreshortening of lines in space, resulting in images that were realistic but not useful in laying out buildings or assembling machines. It was necessary for tradesmen to have other means of showing new machines such as gears for milling, artillery mechanisms, and water works. As early as the late fifteenth century architects were employing "military perspective" to answer this need.[33]

Such drawings are difficult to understand for some contemporary observers, but their premise is quite simple. Lines are projected orthogonally to the ground plan of the building or machine, resulting in what is sometimes called a "central" axonometric projection. Though one's impression is of a three-dimensional object, spaces can be measured from the drawing. When laying out fortifications, military engineers often employed such projections, both as design drawings and construction drawings in the field.

One of the first architects to recognize the importance of alternate forms of projection was Francesco di Giorgio Martini, who provided numerous examples of machines in his architectural treatises. Some of his inventions became so famous that later treatise writers copied his drawings in their publications. He was also one of the first architects to advocate "soldier's perspective" as a means of depicting machines with moving parts. As Daniella Lamberini explains it, this form of drawing had clear advantages for hydraulic engineers and mapmakers:

When the inclination of the projectors forms a 45-degree angle with the picture plane, we have military axonometry, which is not very realistic but has the virtue that the unit of measurement of the third scale is equal to that of the first two. Rapidity of measurement was highly appreciated in the tragic urgency of war, and the compass became a fundamental tool for sixteenth-century military engineers, as it was also for hydraulic engineers, who made the most use of military axonometry, and for cartographers.[34]

112. Alberto Sartoris. Ideal villa project, Florence, 1942. Like many Modernist pioneers, Sartoris favored axonometrics for their rational quality and demonstration of geometric purity. (Author collection)

Most such drawings were cast as if objects were seen from above, to maximize the area of view with respect to the machine being designed. When using such drawings in the field it was a simple matter for the builder to measure a wall in plan and transfer it to materials on site. One can imagine the cognitive problems encountered while builders were figuring out how each individual part connected to others—an overall view was essential. It seems that occasionally Francesco and his followers built scale models in wood before preparing their drawings, and sometimes these were placed in a box, which might also appear in the drawing.[35] As we learned above, French military engineers took their early forms of representation from the Italians, but quickly surpassed them in the use of parallel projections and axonometry.

Treatise writers began to use the term "axonometric" consistently only in the early nineteenth century, as Massimo Scolari has pointed out, and then only rarely.[36] In books such as the *Lehrbuch der axonometrischen Projectionslehre* (1885–63), and *Practice of Isometrical Perspective* (1842) authors touted the advantages of measurement and rapid layout, both for architectural and mechanical drawing. These drawings came into wide use at the end of the nineteenth century, both in technical drawing instruction and in the design of machine components. As we saw above, they were not encouraged in the Ecole des Beaux Arts or its offshoots, but were taught in military and technical schools from the late eighteenth century onward. [112]

Following the publication of Gaspard Monge's *Géométrie descriptive* (Paris, 1795) and his foundation of the Ecole Polytechnique, a codified method of drawing using projection spread throughout Europe, and was employed in teaching at the U.S. Military Academy upon its foundation as well. However, a split developed between American and European engineers with regard to the angle of axonometric projection favored by teachers and practitioners alike. So-called "first angle" drawings depicted objects from above and to the right, while "third angle" drawings reverse the order vertically. Americans came to favor the former, Europeans the latter. In either case it was possible to portray machine assemblies persuasively, such that "multiview orthographic projections" became the lingua franca of mechanical engineers by the early twentieth century.[37]

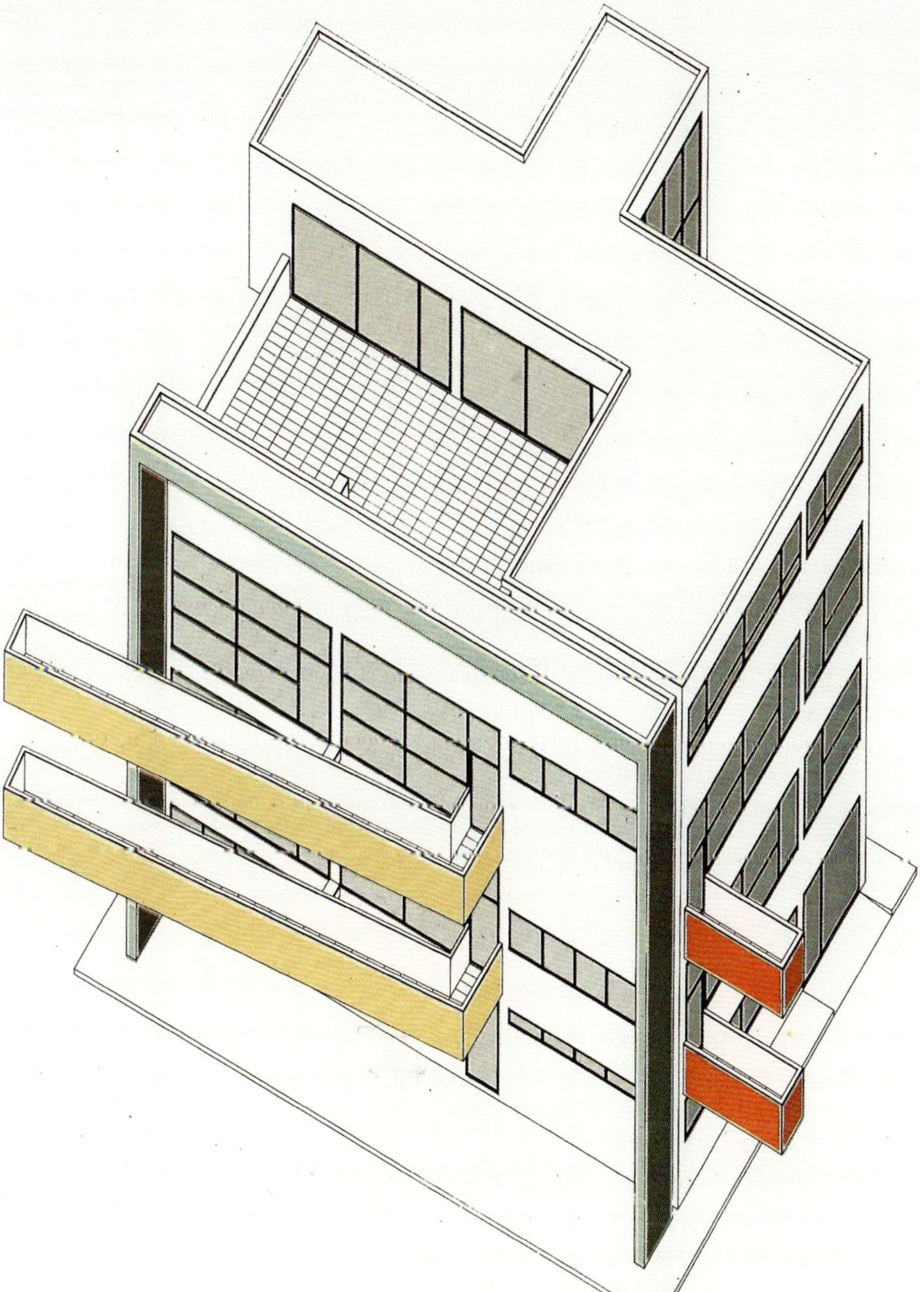

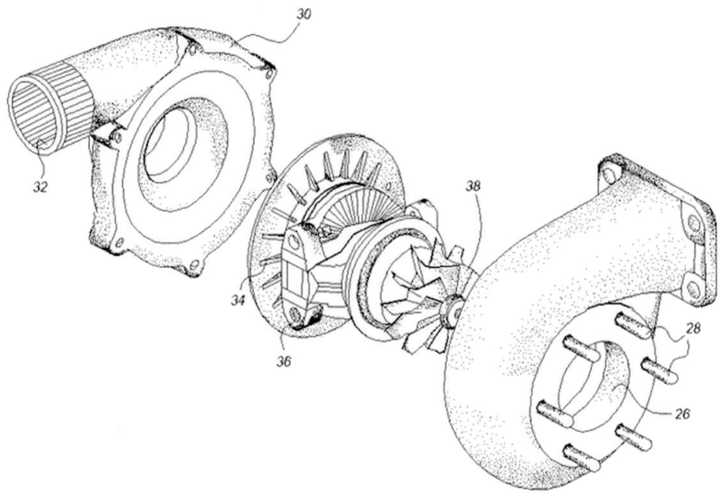

113. Typical patent drawing for a mechanical assembly. The parts are numbered and the machine is exploded to reveal its inner workings. (Wikimedia Commons)

These drawings allowed an overall depiction of the machine to be made while designing, and partial views to be made for production. However, a problem still remained: how to explain the nature and order of assembly required. Think of the instructions included with complex children's toys or boxed furniture and you will understand the issue immediately. Drawings usually present the object as "exploded" to reveal its assembly. [113]

Even in some early Italian machine illustrations the inventors began to pull their contraptions apart, giving readers an opportunity to understand the order of assembly required to produce them. This was clearly necessary in the most complex gear and axle assemblies, even before steam and electricity were used to drive them. In building, tradesmen had conventional means of making walls, roofs, windows, doors, and other components, and needed little more than a section or plan to begin their work. The junctions or connections between non-moving parts such as corner joints or eaves are much simpler than bolted, screwed, welded or bent connections in a typical machine. So it was perhaps inevitable that the deconstructed axonometric view would appear in technical literature.

There is no definitive history of the axonometric, so it is difficult to pick a moment at which exploded views and oblique bird's-eye drawings caught the attention of architects, much less artists.[38] By the mid-twentieth century a number of books on "engineering drawing" were in print, most of which employed various types of axonometric projection. How then, did exploded oblique drawings become *de riguer* among architects?

The answer is that early modern architects, including Le Corbusier, published their designs featuring axonometric projections, and eventually began to pull buildings apart to explain their complex construction joints and assemblies.

To demonstrate the logical/rational process of both building and design it was necessary to prove the concept in a single drawing. The exploded view accomplished this without resorting to the arduous process of laying out perspectives. It wasn't until the advent of Computer Aided Architectural Design that a quicker method became widely available, as we shall see in Chapter 10. However in today's competitive world of starchitects and thought leaders, the most common drawing form found in publications is still the exploded axomometric, the epitome of assembly-based thinking.

Heuristics versus Problem Solving

One of the most important byproducts of engineering during the industrial revolution was the emphasis on "problems" that might be solved using mathematical and rational methods. Every student of mathematics understands that problems are things that can be restated in mathematical terms. Word problems simply create a verbal description of mathematical, often algebraic, formulas. The formulation of such problems emphasizes their finite, rationally deterministic character. Hence, engineers and many scientists deliberately frame their tasks as problems with an array of goals that may be achieved via a logic-based solution.[39] Further, mathematical problem sets are composed of discrete assemblies of related equations that must be treated as parts before they can be understood together.

Psychologists who study creativity are more apt to look at design as a multi-layered mental and physical endeavor that may or may not produce a defined "solution," but which will generally result in an artifact bearing the stamp of the maker.[40] Artists and craft artisans tend to view their work as grounded in habits and steps, sometimes guided only by the unconscious, and often inexplicable.[41] Creative solutions don't always arrive at the end of a linear process that can be categorized in scientific terms. The English furniture designer David Pye writes that design, in all of its manifestations, is concerned with satisfying six requirements, based upon aesthetic principles and practical exigencies:

1) proper arrangement of parts
2) geometric order
3) adequate strength
4) access for the user
5) acceptable cost
6) acceptable appearance

If we look at these criteria in terms of how the brain works, a few things stand out. The sense of order that many researchers have noted in all kinds of experiments is the largest factor in assessing design quality. Part to whole relationships and geometric figuration may be seen as either logical/mathematical or rhetorical/verbal. Fundamentally, order is a property of the brain's wiring and functional organization. In Greek, the term is *taxis*.[42] Classical treatises on many subjects emphasized order as the fundamental principle governing things in the natural world.

Durability and accessibility are functions related to the human body and its limitations. A strong tool is not only useful but also durable. A good tool responds to the hands, arms, torso, legs, etc. Since we "think" with our bodies as well as our heads, these functional properties are not just desirable but necessary in any product we use. Likewise, the economic resources necessary for producing the object constrain its complexity and material value.

The final, purportedly "aesthetic" criterion would suggest that all designed objects be beautiful to look at. Pye uses the word "acceptable" to emphasize the relative importance of attractiveness. Users of buildings prefer rooms and façades that are pleasant to perceive, and brain science is beginning to describe the characteristics of surfaces, forms, colors, and textures that offer pleasure or interest. Experiments will soon show which kinds of façades—the gridded glass curtain wall for instance—produce confusion, boredom, and anxiety in the brain, and which do not.

Designers are responsible for producing useful and beautiful artifacts for humans. Machines are merely tools, and they do not fully dictate forms or methods.[43] Most importantly, learning to design means learning to judge the beauty, utility, and durability of the object to be fabricated, with an understanding of its materials and the economy of its production. Judgment is the critical faculty for all designers. But judgment cannot be achieved without reference to exemplary precedents that have been socially sanctioned. Merely arranging parts in logical order will not result in a satisfactory design.

Heuristic methods, or "rules of thumb" derive from habits learned over many years, and sometimes passed from makers via word of mouth or physical demonstration, as we learned in Chapter 3. There are many heuristic actions that are useful in the design of buildings: lettering, line weights, scales, drafting conventions, and the like. I have argued that these methods are essentially forms of embodied knowledge, and that they differ from rational activities as commonly understood by contemporary engineers and computer scientists. We saw in Chapter 6 that artisans during the Baroque era were entirely beholden to these kinds of methods, and that objects of extraordinary formal complexity and technical virtuosity were produced at that time. It is therefore foolish to argue, as many do today, that such craftsmanship is impossible to achieve using hand skills. One must only learn to think differently.

Most contemporary books on design thinking discuss the distinction between computational reasoning and habitual methods of visual design. Peter

Rowe was among the first architects to remark on the highly heuristic methods used by students and trained architects in their studios. Likewise, Nigel Cross and Bryan Lawson have published books chronicling experiments with designers that indicate an orderly process of assembling program data, working out alternative design concepts, and refining them into suitable final products. As we noted in Chapter 3, modes of conception are almost always heuristic in nature, not mathematical.

It is therefore incumbent on those who teach design, and produce software for designers, to more fully embrace the embodied knowledge that humans employ when making things for their own use. As we will see in Chapter 10, contemporary CADD and other technologies based on algorithmic and logical operations do not necessarily aid designers in their tasks. Nor have the most recent trends in architectural theory led to more effective means of bringing essential, embodied knowledge to architectural students. One of the persistent biases in all pedagogy has been toward "original" solutions to problems, a bias received from engineers and inventors.

Novelty and Artistic Creativity

During the Romantic era many European artists promulgated what has become an unchallenged criterion for creative achievement: the expectation that great art demonstrates novelty to the exclusion of other qualities. If a work was not "new" in reference to all past examples, it was unworthy of praise.[44] The products of engineering were evaluated on their novelty—an invention was something that performed better, using better parts, newer strategies, and more efficient machinery. Art became more like engineering.

Art also became a sacred calling, a kind of secular religion. Romantic artists were valorized in the production of "innovative" or groundbreaking pieces. Auguste Compte, the inventor of positivism, implicitly championed such cultural biases during the middle of the nineteenth century: progress meant dissociation from past ideas, achievements, and even wisdom.[45] In our time these ideas have become entrenched, and often unquestioned.

The historical narrative underpinning what Geoffrey Scott called "the Romantic Fallacy" is well-documented and has been analyzed from a number of points of view, both positive and negative.[46] Early Romanticism was connected to poetic notions of the sublime, "sturm und drang," and emotional extremes. Baudelaire, Coleridge, and Berlioz shocked their public with opium-induced imagery. Artists suffered intensely and made their work a testament to suffering. They distanced themselves from the bourgeoisie by emphasizing the uniqueness of their individual experience of the world. Their antipathy to the status quo, agonistic stance toward society, and radical lifestyles marked them as members of an avant-garde or future-oriented cadre.[47]

With such views, our culture has come to accept the belief that creative work in any domain should transform the world by introducing new paradigms, ideas, forms, or methods into the canon. How is the novelty of these things determined? What distinguishes them from "outmoded" production? What are *new* thoughts? The answers from brain science are startling, and contradict cultural stereotypes.

Scientists studying creativity from a cognitive standpoint use novelty as a criterion[48] while also noting the importance of what is called associative or "content addressable" memory. There is an interesting paradox in the notion that the brain can create new images from old ones, even though associative memory contains only those recorded by experience. As we saw in Chapter 3, the nature of memory storage in the brain may provide the answer to that paradox.

Memories are "coded" in various areas of the cortex, often related to sensory motor modes, but also organized so that information can be retrieved in generalized categories. The information stored in neuron bundles is distributed across a range of individual neurons; though a given neuron may coarsely encode many memories, each memory is encoded in many neurons. This allows the information to be content addressable and available in creative thinking. The human memory system is thus quite different than that of a digital computer, which must store certain types of information at discrete addresses within the storage medium. The human brain is far more flexible, more creative.[49]

Lian Gabora and her colleagues at the University of British Columbia have advanced a theory that "neural cliques" adjacent to "neurds" are active in associative cognition such as design and problem solving. They offer the clever case of how water skiing may have been invented. A neural clique contains memories about related things, such as those that respond to "the flatness of skiis" and the "pointedness of poles." When "mindwandering" occurs (about 15 to 50 percent of our conscious time is spent this way) we play with or recall existing knowledge about unrelated things. When an adjacent area, or neurd, is activated the brain may relate its content to otherwise unrelated knowledge—the neuron responding to the "sleekness of water" or that responding to activity of "being pulled." Hence a combination of unlike characteristics may produce a new concept: skiing on water rather than on snow. Thus, new ideas emerge from existing memories.[50] But is this novelty, or is it merely the way humans think about the world in general? To Gabora and many memory researchers, the brain is always engaged in associative reasoning during creative sessions. [**114**]

The brain has several "associational" areas in the cortex that help to "mix" and direct signals between sensory and motor cortices. These are: a) the *parietal associational cortex*, which provides guidance in motor behavior and spatial awareness; b) *the temporal associational cortex*, which helps in recognition of sensory stimuli and for storage of factual information in memory; c) the *frontal*

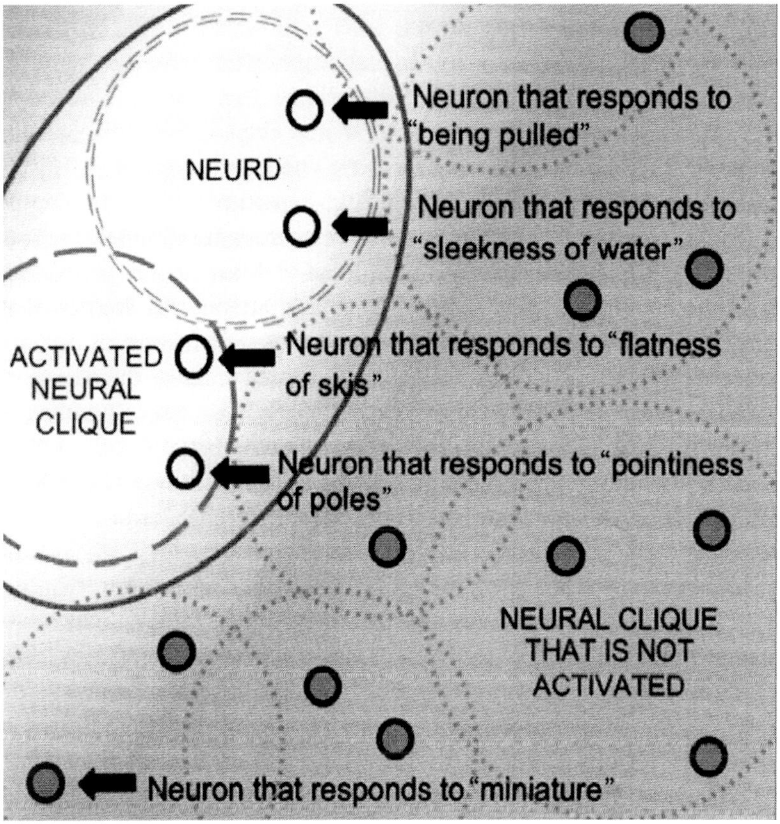

114. Neural cliques, from Lian Gabora, University of British Columbia. Memories formed together or shortly afterward are often stored in the same portions of the cortex.

associational cortex, which plays a key role in organizing behavior and working memory storage, and d) the *limbic associational cortex*, which performs complex functions related to emotions and episodic memory. Signals move between these four areas in either the dorsal (upper) stream or the ventral (lower) stream, according to the kind of information that must be processed in cognition. As might be expected, functional areas that must share critical information are often adjacent to each other. In action-perception loops related to visual processing the parietal cortex serves the critical function of melding and coordinating flows of information necessary to knowing the location of the body in relation to objects in the environment. It is thus important to architectural design and creativity, and is where images can be related to each other when fashioning new artifacts.[51]

To be sure, some humans are better at inventing or creating new or untried things than others. More often than not, as Henry Petroski has suggested, the inventions are combinations of previous useful objects or ideas.[52] However, the unconscious mind is often in control of creative cognition. Though examination of the brains of geniuses, such as Einstein, has hardly been comprehensive, some evidence suggests that development of areas that process spatial and place perception, and integrate these signals with the pre-frontal cortex, is common in geniuses. It has also been suggested that the brains of creative individuals process signals from a large variety of stimuli in a number of different cortical regions. What is certain is that experience in the world, and hard work, are necessary before any invention on the scale of Einstein or Newton can occur. The brain's neural networks, as noted above, are built by stimuli from the world, by experience.[53] Intelligent individuals do not automatically become geniuses. They must develop key neural networks, literally changing their brains throughout their lives, as Eric Kandel has suggested.

In many respects architecture, the most mimetic of the arts, became a prisoner of cultural narratives from both the Enlightenment and Romantic eras. As we noted in our discussion of engineering paradigms for design, and the decline of visual judgment as a criterion for evaluating architecture, these historical trends had far reaching consequences. To a great extent these ideological biases resulted in a decline in genuine creative thinking. By merely solving technical problems that contained finite elements and assemblies, engineers appeared to have a more rational approach than architects, who were unscientific. And, because architects occupied a nebulous realm between fine art and technical disciplines, the field suffered an identity crisis in the twentieth century. In the following chapters we shall review some of the causes for this epistemological shift.

Notes

1 Pamela H. Smith, *Body of the Artisan*, (Chicago, Univ. of Chicago Press: 2004): 8.

2 John Fitchen, *Building Construction Before Mechanization* (Cambridge, MIT Press: 1986): 15.

3 See Robert Mark, Ed. *Architectural Technology Up To The Scientific Revolution* (Cambridge, MIT Press: 1994): 233, for an account of this and other matters related to architects after the Renaissance.

4 This influential compendium of the world's knowledge was intended first to be a mere translation of Ephraim Chambers's *Cyclcopedia* of 1728, an English book.

5 Friedrich Schiller, *On the Aesthetic Education of Man* (New York, Dover Publications: 2004). Schiller did not actually separate the classes in his writings on aesthetics, but wished to see all humans appreciate the combination of reasons and sensual apprehension that intellectual "freedom" could afford.

6 Larry Shiner, *The Invention of Art* (Chicago, Univ. of Chicago Press: 2001).

7 Joseph Rykwert, *The First Moderns: The Architects of the Eighteenth Century* (Cambridge, MIT Press, 1980).

8 Thomas Kuhn, *The Structure of Scientific Revolutions* 4th Ed. (Chicago: Univ. of Chicago Press: 2012); Karl Popper, *The Logic of Scientific Discovery* (London, Routledge: 2002); Karl Popper, *The Self and Its Brain* (New York, Springer, 1977).

9 See Claude Perrault, *Ordonnance of the five kinds of columns after the method of the ancients* (Chicago, Getty Center and Univ. of Chicago Press: 1993).

10 C.f. Indra Kagis McKewen, "Claude Perrault," in Vaughan Hart, Ed., *Paper Palaces: The Rise of the Renaissance Architectural Treatise* (New Haven and London, Yale Univ. Press: 1998): 321–37.

11 James McQuillan, "From Blondel to Blondel: on the decline of the Vitruvian treatise," in Hart, 338–60; quote on 357.

12 Perez-Gomez and Pelletier, *Perspective Hinge*: 33.

13 Perez-Gomez, *Crisis of Modern Science:* 31.

14 Giambattista Vico, *Scienza Nuova* (New York, Penguin, 1999).

15 Smith, *Body of the Artisan*, 215–41 .

16 Alexander von Humboldt, *Kosmos – Entwurf einer physischen Weltbeschreibung* (Berlin, 5 vols, 1845–862). Also Adrea Wulf, *The Invention of Nature: Alexander von Humboldt's world* (New York, Vintage Books: 2016).

17 See Kenneth Frampton, *Studies in Tectonic Culture* (Cambridge, MIT Press: 1995): 4, 5. Frampton relies heavily on outdated studies of Semper's work and influence, as do most other architectural theorists today.

18 Harry Francis Mallgrave, *Gottfried Semper: Architect of the Nineteenth Century* (New Haven, Yale Univ. Press: 1996): 355–81.

19 Semper's famous four elements of architecture: earthwork, hearth, frame/roof, and exterior membrane, incorporated the material and visual aspects of building into a single theory. Wolfgang Hermann, *Gottfried Semper: In Search of Architecture* (Cambridge, MIT Press: 1984): 88–117. In this he follows Aristotle in uniting the material and metaphysical aspects of the world.

20 See Hermann, 219–21 on tectonics as a "cosmic art" with music and dance.

21 Hermann, 219.

22 Mallgrave, *Semper*, 187.

23 Vessely, *Architecture in the Age of Divided Representation*, 366.

24 Vesely, *Divided Representation*, Chapter 2, especially 75–108. The author presents a case for embodied states similar to that proposed here, but insists that all must be understood in terms of "communicative space," a difficult term attached to semiotics, phenomenology, and largely outdated data from psychology. Because he cannot accept a schema based upon measured activities in the brain, his arguments often spiral into circularity.

25 Paul Johnson, *The Birth of the Modern: World Society 1815–1830* (New York, Harper Collins: 1991): 165.

26 Johnson, *Birth of the Modern*, 180–85.

27 Saint, *Sibling Rivalry:* 431–48.

28 Ibid., 134.

29 C.F. Clifford W. Zink, *The Roebling Legacy* (Princeton, Princeton Landmark Publications: 2011).

30 John Ochsendorf, *Guastavino Vaulting: The Art of Structural Tile* (New York: Princeton Architectural Press: 2013).

31 A fascinating account of his early career is found in Nicolas Fox Weber, *Le Corbusier: A Life* (New York, Knopf: 2008).

32 For a new perspective on the early writings of Le Corbusier on this subject see Nina Rosenblatt, "Empathy and Anaesthesia: On the Origins of a French Machine Aesthetic," *Grey Room*, No. 2 (Winter, 2001), 78–97. The author notes that in forming his ideas about machines and objectivity the Swiss architect borrowed much from those studying empathy, subjectivity and the self, a paradox to say the least.

33 See Massimo Scolari, "Elements for a History of Axonometry," in *Oblique Drawing*, pages 1–24. He discusses war in the Renaissance and its requirements for new representational forms.

34 Daniela Lamberini, "Machines in Perspective: Technical Drawings in Unpublished Treatises and Notebooks of the Italian Renaissance" *Studies in the History of Art*, Vol. 59, Symposium Papers XXXVI: The Treatise on Perspective: Published and Unpublished (2003), 212–33, quote on 216.

35 Lamberini, 227–31.

36 Scolari, "Drawing in parallel modo," in *Oblique Drawing*, 325–38.

37 For a comprehensive history and analysis of this practice see Harold Belofsky, "Engineering Drawing—A Universal Language in Two Dialects," *Technology and Culture*, Vol. 32, No. 1 (Jan., 1991), 23–46. The author notes that shop practice in the nineteenth century did not encompass the use of these drawings, as artisans were more comfortable with wooden models as they made their patterns for machine parts.

38 One of the first attempts to address the subject was Ives-Alain Bois, "The Metamorphosis of Axonometry," published in Volume I of *Daidalos*, the short-lived German journal, issued by Bertelsmann on 15 November 1981. At only 124 pages, Bois could not cover a lot of ground in his brief contribution.

39 For an excellent overview of this position see Vinod Goel, "Creative Brains: Designing in the real world," *Frontiers in Human Neuroscience* (Volume 8, April 2014, Article 241): 1–14 (at www.frontiersin.org).

40 C.f. Mihaly Csikszentmihalyi, *Flow: the psychology of optimal experience* (New York: Harper Perennial: 1991).

41 C.f. Hanns Sachs, *The Creative Unconscious: studies in the psychoanalysis of art* (Cambridge, Sci-Art Publishers: 1951). For a more recent treatment by a practicing architect, see Kyna Leski, *The Storm of Creativity* (Cambridge, MIT Press: 2015).

42 See Alexander Tzonis & Lianne Lefrevre, *Classical Architecture: The Poetics of Order* (Cambridge, MIT Press: 1986).

43 David Pye, *The Nature and & Aesthetics of Design* (Bethel, CT, Cambium Press: 1978): 23.

44 One of the clearest treatments of this trend in modern art is in Robert Hughes, *The Shock of the New* (New York, Knopf:1991).

45 As Perez-Gomez asserts, Compte made his theories clear in his *Cours de Philosophie Positive.* "The main thrust of positivistic philosophy was the notion that phenomena were subjected to invariable natural laws." *Modern Science*, 276.

46 His famous analysis is found in Scott, *The Architecture of Humanism: a study in the history of taste* (London, Constable & Company: 1924).

47 For a general history see Tim Blanning, *The Romantic Revolution: A History* (New York, Modern Library: 2012).

48 C.f. Andreas Fink and Mathias Benedek, "The Creative Brain: brain correlates underlying the generation of original ideas," in Vartanian et al., Eds., *Neuroscience of Creativity*: 207–31. They use a common definition: "the ability to produce work that is both novel (unique, original) and useful within a social context."

49 This is noted in a fascinating study by: Lian Gabora and Apara Ranjan, "How Insight Emerges in a Distributed, Content-addressable Memory," in Vartanian, et al., *Neuroscience of Creativity*: 19–43.

50 Ibid., 21–31.

51 Carl R. Olson and Carol L. Colby, "The Organization of Cognition," in Kandel et al., Eds. *Principles of Neural Science*, Fifth Edition (New York, McGraw Hill: 2013): 396–97.

52 C.f. Henry Petroski, *The Evolution of Useful Things* (New York, Vintage: 1994); and Petroski, *Invention by Design: how engineers get from thought to things* (Cambridge, Harvard Univ. Press: 1996).

53 Rex E. Jung and Richard J. Haier, "Creativity and Intelligence: brain networks that link and differentiate the expression of genius," in Vartanian et al., op. cit. 232–54.

[9]

MATERIALS, MODELS, AND MONTAGE

As we saw in Chapters 7 and 8, by the mid-nineteenth century several types of formal education began to shape the practice of architecture in ways that would have far-reaching consequences. On the one hand, Beaux Arts or "academic" art curricula emphasized a largely Neo-Platonic mode of thought that was passed on to architects in the form of esquisse-based drawing. On the other, technological and rational methods of design were entering the building trades via military and engineering schools. Visual judgment and formal aesthetics began to seem out of step with the increasing need for efficient management of production, machine-driven industry, and the speedy solution of technical problems.

As new materials such as high-strength steel, reinforced concrete, metal alloys, electrical conductors, and plate glass entered the building industry architects were forced to confront their impact on types and forms of construction. It wasn't long before engineers, who first invented and mastered this new technology, became competitors, and technology publications proliferated. Gottfried Semper was among the first architects to write about the influence of materials on architectural form in the first half of the nineteenth century. So great was the impact of new technology that a new ideology began to coalesce around the expression of materials and construction methods, which Le Corbusier called the Engineer's Aesthetic.[1]

The tension between those touting the miraculous properties of new materials, and the trade and craft masters who made their living with old

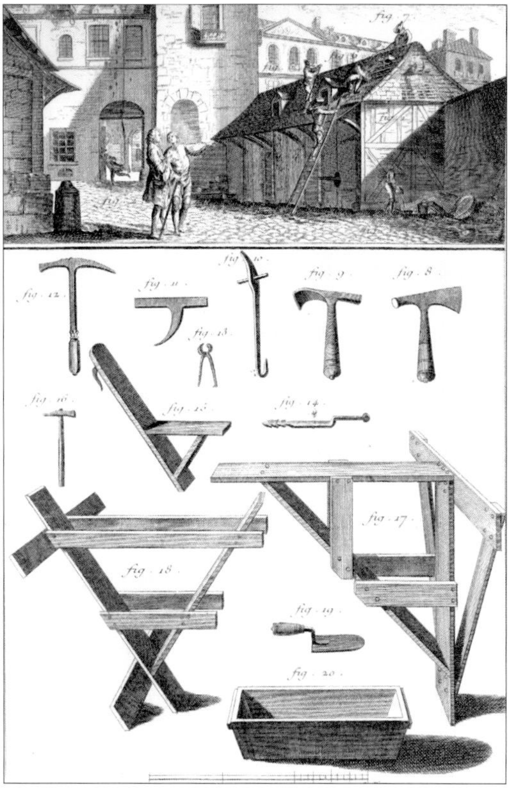

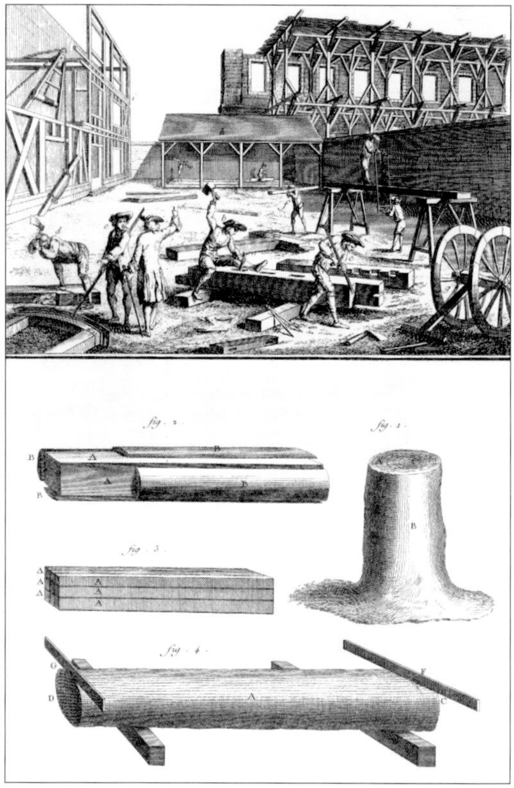

115. Denis Diderot, logging and hewing tools and process, from *Encyclopédie*. Artisans began to lose favor with intellectuals during the eighteenth century when categories of knowledge changed as a result of the scientific revolution, and the rise of the bourgeoisie.

ones, increased as the nineteenth century drew to a close. The Arts and Crafts Movement that emerged in Europe and flourished in the United States after the 1880s was largely a manifestation of this tension, as it exposed the contradictions between industrial production and the arts, between machine manufacturing and human labor, and the nature of capitalism itself.[2] Unfortunately the leaders in the movement, both in the United States and Europe, were often seen as eccentric Luddites or disaffected aristocrats (William Morris and Elbert Hubbard were among the wealthiest men of their time), and thus did not command the respect of heroes such as Thomas Edison or Henry Ford.

As the sheer scale of cities, infrastructure, and building increased dramatically throughout the century so did the scope of representation. Photography, color lithography, high-speed offset printing, and reproduction technologies such as blueprinting increased the speed at which visual information could be transmitted to the public. Mechanical drawing, as noted in the Chapter 7, also benefitted from these new developments. Scholars have noted in particular the influence of photography on art and architecture during the second half of the nineteenth century. The detail that could be shown in the new media was astounding in comparison to earlier methods of drawing, painting, and even sculpture. In many respects the range of devices that could depict the physical world outstripped people's capacity to understand those depictions. [115]

One of the salient characteristics of modern art was this expanded range of media. Another was its expanded point of view, made possible by new scientific optics, color theory, knowledge of electro-magnetism, the physics of light, and the development of modern telescopes and microscopes. Science gave the modern world a new model of the planet, and later the universe. The psychological model of our human environment changed overnight, and with it our embodied sense of ourselves. As Freud discovered at the end of the century, this model was alienating, disturbing, and threatening for many (perhaps most) Europeans.[3] The cataclysm of the Great War only multiplied its effect. This neurotic existential predicament was one of the drivers of modern architecture.

In the following narrative we will explore several of the developments that made the Modern Movement possible, noting some positive and negative consequences for architectural design as a form of human cognition. Indeed, one of the legacies of Modernism was a distancing of design thinking from visual aesthetics and toward problem-based reasoning. Another was the abstraction of design from a process concerned with materials and making to one concerned with concepts and originality. Abstraction allowed designers to pursue pure forms, divorced from any association with archetypes or historic styles.

Materials, Abstraction, and Structural Determinism

Sir Nicolas Pevsner and Siegfried Giedon emerged after World War II as the world's most esteemed European architectural historians. They and their American contemporary, Henry-Russell Hitchcock, largely invented the Modern Movement as a historical construct.[4] Despite criticism during the 1970s and 1980s, their work has stood as doctrine ever since. Pevsner was responsible for the idea of structural determinism as a principle of modern design. Giedion advanced the concept of space-time and open plan structure. Hitchcock trumpeted respect for "nature of materials" as a characteristic tenet of modern architecture and decorative art. Though other salient features of post-war Modernism are part of the historiography, for our purposes these three have proved most problematic.

The autonomy of an artifact as a "structure" composed of discretely articulated materials is largely an intellectual construct derived from nineteenth-century views of history and progress. As technological education emerged in the Ecole Polytechnique, and artisanal literacy waned, all sensory and embodied understanding of artifacts as crafted or made by humans were erased by skeletal, grid-based diagrams; abstractions of type forms rather than attempts to *sense* the nature of the object. Abstract models of "structure" made it possible for engineers to more readily analyze forces, material characteristics, and processes related to industrial production. As a student I remember being struck by the unreality of graphic statical diagrams of trusses when studying engineering. How could such an abstraction capture the vectors and forces in a physical object with such precision? With such models humans began to distance themselves from the things that sustained them on a material basis. This tendency has been noted in recent scholarship, but generally not by architects.

It is not difficult to understand the nature of a small, handmade artifact in relationship to the body; comprehending a 1,000-foot-long steel bridge, 100,000-square-foot warehouse, or twenty-story skyscraper is another matter altogether. New forms of representation were essential to the design of such huge artifacts. City planning as a discipline also required new forms of maps, transit plans, and power grid diagrams. Moreover, testing the strength and characteristics of new materials required new machinery, mathematical algorithms, abstract concepts, and even drawing types such as the exploded axonometric. Merely handling and manipulating terra cotta pieces, steel girders, and cast stone required new machines that were operated at some distance from the mechanic. Tradesmen and women must have felt small in relationship to their new workpieces. They could not "feel" the materials as intimately as in the past. No wonder handicraft began to seem outdated and meaningless.

Paradoxically, theorists began to look at both old and new materials with awe and reverence not seen by artisans. Their distance from the "nature" of these substances required a somewhat spiritual, mystical point of view. Among early adherents to this way of thinking was William R. Lethaby, whose *Architecture, Mysticism and Myth* waxed poetical about primitive and vernacular methods of construction using "native" materials and regional styles of building.[5] Hermann Muthesius, the Anglophile German critic, followed Lethaby in recording the flowering of domestic building, decorative and garden art in the British Isles. Periodicals such as *The Studio, House and Garden,* and *The Builder* were quick to adopt a materials based ethos. In America, Prairie School architects in the circle of Frank Lloyd Wright and Louis Sullivan extolled the "nature of materials" and determinants of form, especially abstract ornament.

The legacy of materialistic views of design is with us today in many forms, not least of which the view of most building materials and methods as technologies and their study as tectonics, terms not even countenanced by late nineteenth-century builders. But aspects of our technocratic ethos can be traced back at least a century, to both European and American concepts of industrial production and positivism. By grouping progress with new inventions and new materials, architects uncritically assumed a new worldview by the early twentieth century, just as the avant-gardes in art accepted the machine aesthetic as their zeitgeist.

Artists involved in Der Stijl, Constructivism, the Wiener Werkstatte, and the Bauhaus were increasingly insistent that the expression of material strength, fabrication, and character should help determine the forms of their creations. But they had already been exposed to a deterministic, instrumentalist point of view in French nineteenth-century architectural history. E.E. Viollet le Duc constructed his massive *Entretiens sur L'architecture* according to his theory of structural rationalism, derived from early studies of Gothic vaulting. So powerful were his arguments about the influence of structural forces on the design of these massive edifices that modernists accepted them with little historical evidence of their veracity. It took more than a century to dislodge them from architectural literature.

As was evident in our discussion of medieval design practices, masons did not conceive their work according to structural force diagrams of any kind. Though they had a working knowledge of arch theory, and a deep respect for the compressive strength of limestone, there is no evidence that these things "determined" the form of ogival arches, flying buttresses, or piers. Robert Mark recently used thermo-plastic analysis and other experimental methods to show that piers, ribs, and buttresses did not fully resolve or transfer loads and forces from the roof to the ground in any major Gothic cathedral.[6] He also developed an alternate hypothesis for how master masons achieved their gravity-defying vault heights. Trial and error, not mathematical analysis, was their secret. So, as humans with embodied intelligence, they quite literally felt their way forward in the design of their massive constructions.

Pevsner in particular, influenced by Ruskinian ideals of "truth" in construction, pressed his colleagues to accept structural rationalism as an antecedent for modernist aesthetics. His *Pioneers of Modern Design*, one of the most influential books of the mid-twentieth century, found protagonists in Britain and Europe among diverse members of the Arts and Crafts Movement, the Werkbund, bridge, and canal engineers, and even furniture designers.[7] All, he argued, were zealous revolutionaries bent on progress, tipping the Hegelian dialectic toward abstract, machine age forms. Recent scholars have generally disputed his narrative, pointing out that many turn of the century designers had other, less radical agendas based on humanistic, not mechanistic, models.[8] William Morris, Herman Muthesius, and C.F.A. Voysey were not interested in machine metaphors or rational design processes, as Pevsner maintained in his highly polemical book. They were intent on reviving the craft of building, and the intimate relationship between the artisan and the work of art.

One telltale of the rational view of design, and the emphasis on assemblies, was the rise of the axonometric drawing as a tool for showing the multiple parts of a machine or building. Any student familiar with mechanical drawing instruction will have made exploded axonometric drawings of bolts, nuts, washers, and other elements required for the construction of metal artifacts. As we saw in the previous chapter, the axonometric began to serve a similar function,

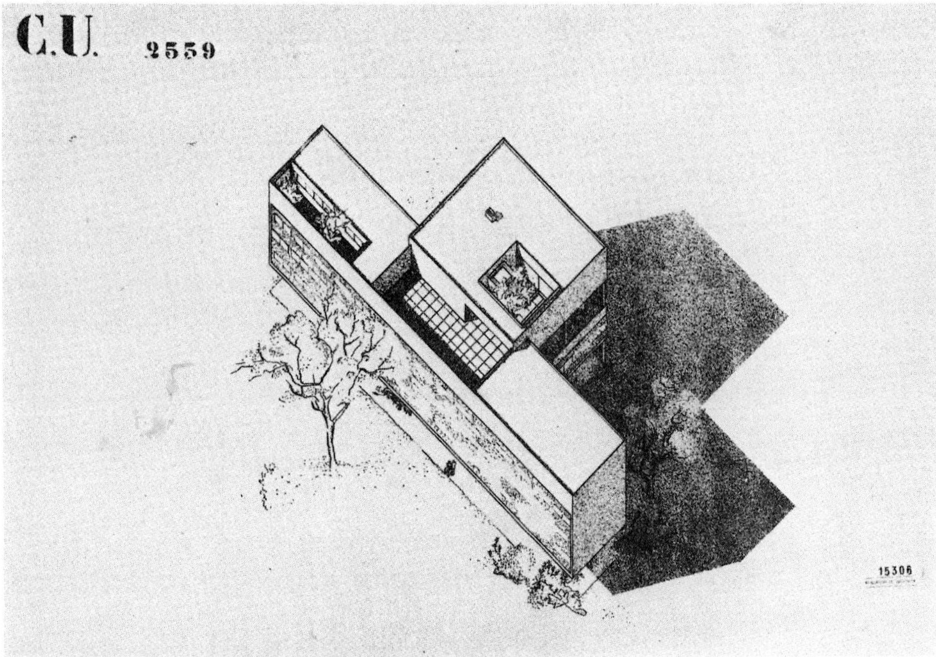

15306 Axonometric

15308 Axonometric

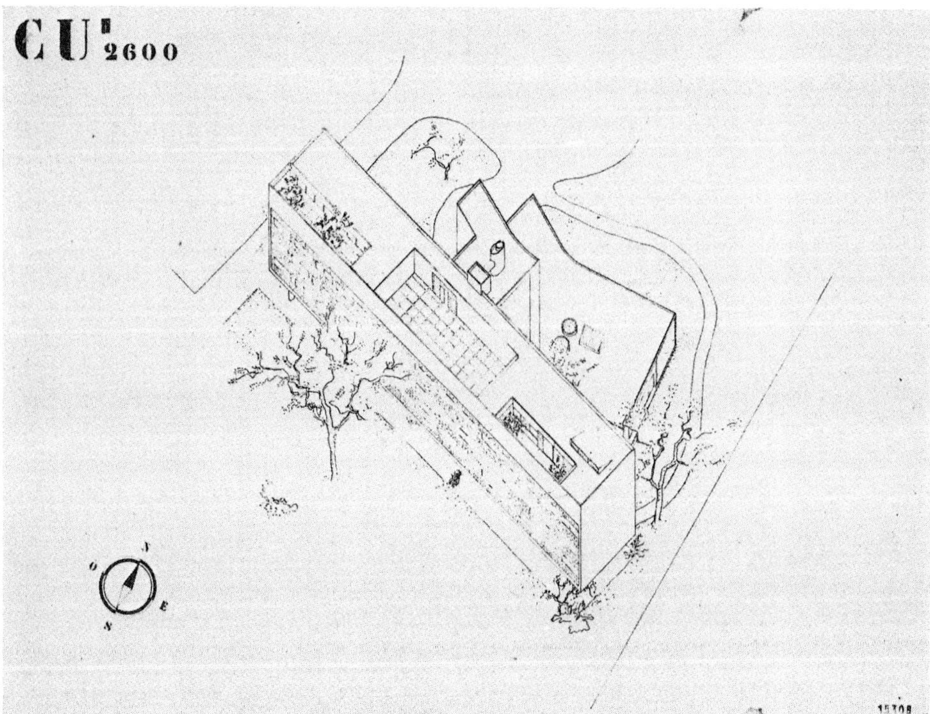

116. Le Corbusier, two axonometrics of the Pavilion Suisse, Cité universitaire, Paris, 1930-32. Le Corbusier was familiar with Dutch axonometric drawings from publications. (Author collection)

allowing designers to disassemble the building in graphic form, either to assess its construction, or to point out its internal logic to clients and the public. One of the most important drivers of the desire to see buildings as assemblies made of abstract parts was the Russian movement aptly called "Constructivism."

Models, Modeling, and Corporeal Form

One of the most persistent legacies of the Modern Movement in architecture is the belief that architecture is a sculptural, corporeal art that can best be represented in three-dimensional models. Indeed, during the late twentieth century, drawings waned in importance in architectural design studios, largely replaced by both physical models and virtual, digitally constructed ones. For some teachers, "designing in model" was the only method acceptable in a studio environment.[9]

Several major art movements influenced architects to use models as studies for built form. Most powerful was the Constructivist Movement in Russia and Eastern Europe during the late teens and twenties. Malevich, Rodchenko, and other early Soviet artists saw their roles as builders of a revolutionary scaffold for reform. Moreover, their intent was to emphasize the physical dynamism of materials arranged in fluid configurations, somewhat in the vein of earlier Italian Futurist compositions. Even two-dimensional art presented three-dimensional information. The picture plane shattered to reveal apparently moving parts. Painters aspired to sculpture, sculptors, to architecture. Architects struggled to transcend the limitations of existing building technology in their daring compositions.[10]

Few movements offered the provocation and excitement of Suprematism and Constructivism. Despite their short-lived creative moment, Russian artists pushed others to explore the relationship between model-like sculptures and other art forms. Ozenfant and Le Corbusier assimilated the maquette into their Purist paintings. [116] Picasso and Braque used collage in many Cubist paintings and found objects in sculptures. Marcel Duchamp created the *Large Glass* as a three-dimensional construction containing several two-dimensional illusions.

The Dutch De Stijl movement offered a major exemplar of how models might be employed in creating more "objective" compositions. Theo van Doesburg began experimenting with interior models and their two-dimensional implications in the early 1920s, following his publication of the magazine *Der Stijl* in 1917 and collaboration with artists such as Piet Mondrian, Bart van der Leck, and El Lissitzky in Germany. Their ideas about plastic art and color had led them to speculate on the influence of the environment on the display of art, and the breakdown of distinctions between sculpture, painting, and architecture. During the early 1920s Lissitzky's *prouns* (room sculptures and wall compositions) were exhibited in Berlin, Amsterdam, and Darmstadt, and it is likely that he knew about van Doesburg's work on color planes and their potential to create abstract architectural compositions.[11]

117. Theo van Doesburg. Design for Amsterdam University Hall, 1923, perspective drawing. In this view the design appears chaotic, whereas plans and elevations were orderly and geometric. De Stil artists soon abandoned perspectives in favor of axonometrics and models. (Author collection)

118. van Doesburg and van Esteren, private house project, model, 1923. Only by creating such physical models could the artists test their ideas about color and space. (Author collection)

In his experimental design for an Amsterdam University Hall (1923) van Doesburg used exploded interior elevations, plans, and quasi-perspectives to study the relationships between color planes and wall/floor mosaics. [117] Van Doesburg also began making axonometric drawings of planar compositions that suggested architecture but did not have functional implications. He was frustrated by the inefficacy of these modes, and began collaborating with the architects Cornelius van Eesteren, J.J.P. Oud, and Gerrit Rietveld on a series of villas for sites near Paris. Their work on the Private House project of 1923 began with models, so as to more faithfully portray the interaction between color planes and openings in the building. [118] Like other artists in the movement he was intent on total abstraction—the complete autonomy of color, form, and geometry in both two and three-dimensional compositions. Their two cardboard models did indeed prove that color could be a material unto itself, not merely an applied medium.[12] None of the projects were constructed, despite some financial backing by a wealthy Parisian patron. However, all were extensively published in "little magazines" in Paris and elsewhere during the 1920s. [119]

119. Theo van Doesburg, axonometric drawing of color planes, 1923. (Wikimedia Commons).

Following his work with Oud and van Doesburg, Rietveld made several models of his Shroder-Muller house in order to determine its final form and color palette. The model was so persuasive that construction of the building followed its form with exactitude, resulting in a building so flimsy it could not withstand the Dutch climate for more that a few decades. It has been restored several times since its original construction in 1924. [120, 121] One of the signal problems with model-based design is this tendency to believe that wood or cardboard elements can be readily translated into walls, windows, roofs, and other constructed assemblies. Paradoxically, for architects purportedly concerned with "truth" in material expression, models were the worst representational tools. As formal experimentation increased during the first half of the twentieth century, models became more common in all architectural studios. In New York, specialist firms emerged that could construct, for very high prices, the most realistic and impressive physical models of proposed skyscrapers and other large buildings. Firms such as Skidmore, Owings and Merrill, Perkins & Will, C.F. Murphy Associates, Minoru Yamasaki, and Eero Saarinen employed such firms while also creating workshops in their offices for the construction of study models. New materials such as plexiglass, cork board, aluminum extrusions, and various kinds of cardboard were employed in making large and impressive models for presentation to clients. Whereas wood and plaster had been common in the past, lighter materials made model construction easier and less costly. Still, it was not uncommon for architects to spend thousands, even tens of thousands, on these presentations.

The office of Eero Saarinen, which was heavily influenced by European Arts & Crafts ideals through his father's practice, used models more expansively than any other leading firm. One of the reasons for this was that the Finnish American master was experimenting with ever more complex, plastic forms for buildings such as Dulles International Airport, the St. Louis Gateway Arch, and Ingalls Hockey Rink at Yale. Saarinen was also interested in the potential of reinforced concrete to create vast, curved spaces for assembly or transportation buildings.[13] [122]

When competing for the Jefferson National Expansion Memorial (Gateway Arch) the office constructed models of wood, cardboard, and plaster (on a metal armature) before finalizing the design that won the competition. This was a relatively simple arch span compared to those at Dulles and Yale. One of the interesting features of these wonderful buildings was their airiness and seamless integration of interior and exterior surfaces. Since concrete is monolithic, Saarinen believed, with many other architects, that virtually any form could be made with it. At Yale he found that steel cables were necessary in order to create a thin shell, concave roof, and that steel was a necessary component in the design. Photos of many Saarinen models show that he conceived the buildings

120. Gerrit Rietveld, Shroder-Muller house, model. Restored along with house. (Wikimedia Commons)

121. Gerrit Rietveld, Shroder-Muller house, Utrecht, 1924. University Prints. (Author collection)

122. Eero Saarinen inspecting one of several models made to study the design of the TWA terminal. Most of America's leading architects had in house model studios during the 1960s and 1970s. (Wikimedia Commons)

123. Eero Saarinen, TWA Airline Terminal, JFK Airport, New York, 1962. (Wikimedia Commons)

without windows sealing interior from exterior, adding glass only after the designs were approved and under design development. As we noted in Chapter 4, models are more helpful when either the interior or the exterior is to be studied.

Large study models such as those produced for the TWA Building at JFK Airport were impressive as works of art in their own right. [123] Like the Shroder house models, they offered a vision for entirely new forms of architecture, but did not specify the materials or construction methods necessary to realize the physical objects depicted. The best architects, such as Saarinen and C.F. Murphy, had engineers and construction specialists who could develop concrete and steel technology capable of realizing their breakthrough designs. This did not come without costly experiments, sometimes failed ones, that could prove the efficacy of the new technologies. Paul Rudolph was eventually driven out of the practice of large-scale concrete building design because his staff failed to understand these problems.[14] It is important to note that, while architects struggled to understand the tectonic and material challenges implicit in their creations, they did not have first-hand knowledge of the materials that would be used in bringing them to life on the building site. They did not have embodied knowledge.

One of the greatest challenges of mid-twentieth-century building was the sheer scale of the efforts required to erect bridges, arenas, stadiums, tall buildings, and roads. In many respects the pace of technological change outstripped the abilities of those charged with managing the design and construction industries. It is remarkable that so much was built during less than a century.

Illusions of Speed, Position, and Parallax

During the first decades of the twentieth century, when Daniel Burnham and American city planners made their bold proposals for Chicago, Cleveland, San Francisco, and Washington, the automobile was just beginning to shape the armature of cities. By the middle of the twentieth century, rail transportation was already giving way to superhighways and high-speed arterials in most U.S. cities. The speed of vehicular traffic presented new challenges to architects and

urbanists hoping to mold the experience of city dwellers in the age of information. As we have seen, perception of space depends upon the accurate location of the body through sensory modalities such as touch, hearing, and vision. When humans are in motion various kinds of compensation must be made in the brain in order to maintain homeostasis and balance.

New media such as film, models, maps, and panoramic illustrations broadened the array of tools available to designers following World War II. Military technology provided some of the means for tracking vehicles and people, but it would take decades for things like GPS to reach the everyday world. During the 1960s, while NASA was launching spacecraft into orbit, transportation planners were expanding highway systems and cars were becoming more and more sophisticated. How would such innovations be incorporated into the fabric of existing cities like New York, Chicago, and Washington, D.C.?

During the late 1950s the Yale School of Architecture established a laboratory for studying some of the new problems posed by superhighways and city traffic controls. Boris Pushkarev and Christopher Tunnard received several grants to create a national study of what would later be called "infrastructure" while engineers rushed to design Eisenhower's Interstate Highway System. The new laboratory at Yale employed mixed media as well as computational methods in search of innovative solutions to the "control" of infrastructure growth. America's cities were growing at astounding rates, and change was everywhere, often resulting in alienation and bewilderment among average citizens working in new environments.

The award-winning book, *Man-Made America: Chaos or Control*, offered intriguing design solutions to infrastructure challenges such as the integration of landscape and highway design, vehicle directional signs, traffic controls, and pedestrian paths along roads.[15] Pushkarev studied perceptual phenomena in order to make generalizations about landscape aesthetics and highway design.[16] They also used film as a means of recording the experience of drivers along highway interchanges. As we have seen, the brain is challenged by increasing speed when perceiving depth and distance. The Yale team was one of the first to seriously consider the increasing scale of highways, their relationship to surrounding environments, and the abilities of humans to perceive these spaces while moving faster than walking speeds. They built many scale models of highway interchanges in order to study these things. They used the technique of montage from cinema in their presentations.

Robert Venturi and Denise Scott Brown came to Yale as guest faculty shortly after the Tunnard and Pushkarev studies were published, and were interested in transportation issues in urban design. Scott Brown had been a student of David Crane and Paul Davidoff at the University of Pennsylvania, following training as an architect in London with her first husband, Robert Scott Brown.

Venturi was emerging as a leading theorist following the publication of his book, *Complexity and Contradiction in Architecture* in 1966, which Vincent Scully cited as the most important work of architectural theory since Le Corbusier's *Vers Une Architecture* in 1923.

In 1967 Venturi and Scott Brown married and were invited to teach a series of four studios at Yale under Paul Rudolph and Charles Moore, successive deans. The first was a study of the New York Subway System, but the second created a stir among American architects that has echoed for decades: "Learning from Las Vegas." One of the many innovative and controversial theses in their subsequent book on the studio was that designers should attend more carefully to the perception of streets and buildings from a driver's point of view. *Learning from Las Vegas* contained numerous images of The Strip taken from moving vehicles, as well as diagrams and street sections designed to point up the significance of "signs and symbols" along the street, apprehended at different speeds and scales. They employed the ideas of Marshall McLuhan and techniques of montage from film in their studies of the street, the city, and the region.[17]

Shortly after the publication of their book, and another studio on Levittown, the two academics decided to quit full-time teaching and devote themselves mainly to a growing architecture and planning practice. The office of Venturi, Rauch, and Scott Brown won several prestigious commissions during the late 1970s that would bring their new ideas into focus, and spur some sharp criticism from the press. One of the largest was the so-called "Western Plaza" project for the Pennsylvania Avenue Development Corporation in Washington, D.C.

The 1901 MacMillan Plan for Washington renewed interest in Pierre Charles L'Enfant's original plan for the city, then a century old. In it the planners suggested that portions of the old plan be reconsidered as a means of bringing order to the expanding federal capital. Pennsylvania Avenue, which connected the Capitol and the White House in L'Enfant's plan, was a messy thoroughfare with a variety of building heights and few breaks along its six-mile length. After the completion of the Federal Triangle, a scheme reminiscent of the L'Enfant design along the avenue, PADC was charged with initiating more projects that would beautify "America's Main Street." During the 1970s Carter Brown, director of the National Gallery, became president of the powerful Art Commission, and pressed PADC to act.

It was Brown who brought the VRSB firm into the discussion of new ideas for Pennsylvania Avenue. The L'Enfant Plan envisioned a small square at the intersection of 13th and 14th Streets and Pennsylvania Avenue, but none was built there, allowing important public buildings to be placed off the main axis of the leading boulevard in the city. One of these was the District Building, or city hall. If a new square were constructed orthogonal to the main street grid, but with visual connections to the diagonal at its corners, the District Building

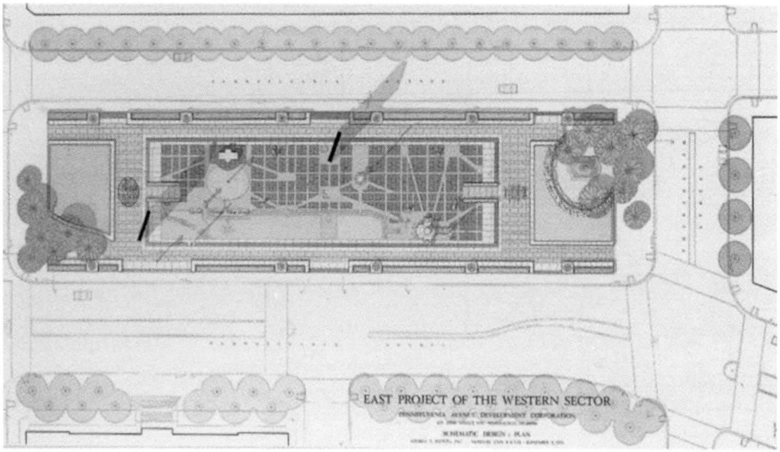

124. Venturi & Rauch, East Project of the Western Sector, PADC, Washington, preliminary plan, 1978. The media used were innovative, but short lived. Initial line drawings were made using Dixon Ticonderoga No. 2 pencils on drafting vellum. These were photo-reproduced on KP-5 opaque mylar sheets. Color was applied using Pantone stick-on film, cut with X-acto knives. The drawings were mounted on large sheets of foam core board. Some were as large as 3' x 8.' (Architectural Archives, University of Pennsylvania)

would front on Pennsylvania Avenue. This was the initial premise under which Venturi and his colleagues were hired.[18]

Venturi and Scott Brown were interested in more than just sprucing up a portion of the city. They understood that L'Enfant's intended visual linkage between the Capitol and the White House had been compromised by the construction of Robert Mills's Treasury Building under Andrew Jackson, and saw an opportunity to reinstate the vista between the two buildings. They also appreciated the perceptual challenges presented by automobile traffic on Pennsylvania Avenue. Using some of their discoveries from the Las Vegas studio, they proposed a daring, multi-scaled design in 1978.[19] [124]

Before even completing the design the VRSB office was marshaled to construct a monumental model of the entire length of Pennsylvania Avenue, with all significant buildings and streets represented. [125] Schemes for the new plaza were constructed and dropped into place so that the partners, including Steve Izenour, could study their efficacy. It quickly became clear that a more three-dimensional design would respond better to the context than a flat, paved green space in the tradition of Parisian squares. Izenour had been a key force in the Las Vegas studio, and suggested that the design incorporate something akin to the "high reader" signs that were so successful at drawing customers to casinos on The Strip.

Another clear premise of the Western Plaza was to commemorate the L'Enfant Plan itself. A significant portion of the plaza would incorporate a large "map" of Washington as envisioned by the French landscape designer in 1799. VRSB had won an AIA Honor Award for their Franklin Court museum (1977) in Philadelphia, which rendered the plan and gardens of Benjamin Franklin's residence in stone "graphics" so that visitors could appreciate something of its spatial characteristics. They used a similar technique to create a two-dimensional

125. Venturi & Rauch, photo of model, showing an auto entering the PADC plaza from Pennsylvania Avenue, taken with a special micro camera by Tom Bernard. (Architectural Archives, University of Pennsylvania)

ground map of L'Enfant's intended design on a portion of the Western Plaza site. As a member of the design team, I recall blowing up woodcuts of the first city plan so that we could accurately trace its outlines in order to determine the pattern of cut stone on the plaza. Andy Warhol's graphic experiments were in everyone's minds as we worked on the drawings. [124]

The design took shape as a collage of several key elements: the map of Washington in 1799, a fountain on the western end, a small park surrounding an existing equestrian statue, and a set of 150-foot-tall "pylons" designed to catch the eye of pedestrians and motorists from any point on the avenue. They would also frame the portico of Mills's Treasury Building at the end of the axis. Photographs of the model taken at the height of a driver's eyes in a moving vehicle tested the visibility of the pylons from various points between the Capitol and the White House.[20] This involved building tiny cameras that could create photos from less than one inch off the surface of the model. The same types of cameras were used at Yale's urban laboratory 10 years before. VRSB staff photographer Tom Bernard labored tirelessly to produce the photos with enough detail and depth of field to represent the driver's point of view accurately. [125]

126. Venturi & Rauch, photo-montage view along Pennsylvania Avenue, toward PADC Western plaza, 1978. The pylons were to act as "shutters" as drivers and pedestrians approached the plaza from the Capitol. The resulting illusion would have been one of the most spectacular in history, anticipating brain science not published for twenty years. (Architectural Archives, University of Pennsylvania)

A central premise behind the placement and form of the pylons was to create illusions about their size and disposition as drivers moved down Pennsylvania Avenue. Our visual brain deals with motion by making assumptions about objects that are approaching or moving away from us. Because of motion parallax we can be fooled into thinking that objects of different size are of the same dimensions, and that objects at different distances are equidistant. Both kinds of illusion could be perceived when viewing the pylons at different speeds and distances.[21]

Working day and night, the office crafted an elaborate presentation for both PADC executives and members of Brown's Commission of Fine Arts. It included gigantic CP5/mylar and Pantone drawings of the plaza at various scales, using standard orthogonal projections as well as large photos of the model arranged in sequence from the Capitol to the White House. Nothing of the kind had ever been seen by Washington officials. The effect was awe inspiring, much as the MacMillan plan presentation had been 75 years before, where a vast scale model of Washington amazed visitors.

The partners returned triumphantly from the first presentation, vindicating themselves before a tired but happy staff.[22] It appeared that the extraordinary effort to show the daring scheme had paid off, though the cost of the presentation was not known at the time. Carter Brown in particular was pleased with the effort, and his support would be key to winning over federal officials and the local District government. However, Brown's sophisticated understanding of visual art made him an ideal client. Other stakeholders would be less receptive to such unusual ideas in urban design and would require more detailed presentation media in order to understand and approve the project.[23]

127. Venturi & Rauch. Bird's-eye view of model for PADC Western Plaza. Robert Mills's Treasury Building is at the top right. Note the use of Pantone paper and stick-on film in the plaza areas. (Architectural Archives, University of Pennsylvania)

Venturi and his partner, John Rauch, were acutely aware of the pitfalls that might submarine a project before federal officials in the Washington bureaucracy. They had gone before such officials during the late 1960s, presenting a far less daring design for a transportation building, and were turned down despite a persuasive narrative and visual presentation. Rauch in particular was concerned that something in the Western Plaza scheme might rankle officials. He was right to be cautious.

Venturi had experimented with monuments that could compress space and distance before. His apple-shaped fountain at the end of Benjamin Franklin Parkway (1966) in Philadelphia was designed as an eye-catching terminus, but was not built. Buildings with "super graphics" like his entry for the football hall of fame were also intended to invoke large-scale advertising such as billboards, and again could not be realized. However, proposing something as daring as two monumental "shutters" for the eyes of drivers on America's Main Street would push the limits of the imagination of any client. The presentation tools needed to win support of such a concept would need to be persuasive on several levels: perceptual, theoretical, practical, and political. [126]

At the level of perception, the designers were experimenting with a concept that had been used by the Ancient Romans, the termination of a vista with a framed arch, leading the eye to the next major monument in the city, even if the axis of the road changed direction. Understanding both the figure-ground relationship between the pylons and the surrounding streetscape, and the principle of occluding edges in the visual field, the VRSB partners were playing a sophisticated visual and conceptual game. Marshall McLuhan and Andy Warhol were doing the same during the 1960s, but their work was in the periphery of American culture. Washington's politicians were not capable of responding intellectually to their design. They had to find another means of making a case for the pylons.

At one point the designers experimented with overtly patriotic symbols, such as the stars and stripes of the American flag, on the pylons, and included scale models of the Capitol and White house as sculptures on the L'Enfant map. However, monuments such as the obelisk commemorating George Washington were white and abstract, not painted like billboards. It was decided to maintain the stark white stone on the pylons, with subtle black accents, rather than create a super graphic collage. This, however, exaggerated their scale from the pedestrian's point of view, and their height was significant, even when reduced about 25 feet to satisfy nervous members of the PADC board. As presentations to various stakeholder groups continued through the year, everything the designers proposed seemed acceptable, except the pylons. [127]

I participated in the creation of a standard eye-level perspective drawing from the corner of the new square, depicting the pylons in front of the District Building. This and other new drawings were intended to assuage the negative

128. Freedom Plaza as built, without pylons but with model of Capitol Building (subsequently removed). Note the optical illusion caused by placing the two domes together on the same visual axis. (Wikimedia Commons)

reaction of the city's mayor and city council. As the only African Americans with a stake in the project, their support would be needed to push toward full approval by the Commission of Fine Arts, giving the green light for construction in 1979, the 200th anniversary of the L'Enfant Plan. The biggest dilemma facing the VRSB team was how create enough graphic and model images to show the project as conceived—at the scale of the city, the street, the driver, the pedestrian, and viewers from surrounding buildings.

The architects understood the necessity for a large-scale element on the plaza better than any of the public officials except Carter Brown. A man of deep learning as well as political standing, Brown would be the pivotal decision maker in the Western Plaza project, and Venturi and his partners believed that he would prevail despite growing opposition from city officials. They also believed, as most architects do, that their vision, or concept, could be shown persuasively using the presentation tools and technology at their disposal.

Venturi, Brown, and the PADC officials filled rooms with new presentation drawings and photographs at each succeeding meeting of the stakeholders, believing that "less" was not only boring, but that more images would make a stronger case. Their strategy ultimately failed. Too much detail in the presentation reduced focus on the main concept—perceptual compression of space along the city's most important monumental boulevard, resulting in a critical linkage between the White House and the Capitol.

In February, 1979 the partners returned from a final, determinative meeting with city and commission officials. Venturi and Izenour were dejected. Rauch was philosophical. Scott Brown was angry. The staff was devastated to learn that the pylons had been cut from the design, and that the partners faced a difficult choice: build the plaza without them or resign the commission.[24]

After soul searching, tense meetings the three partners decided to continue with the project. One deciding factor was the cost of the design phase—Rauch argued that the firm would lose an extraordinary amount of money once the bills came in for staff time, travel, materials, and reproduction costs (many drawings were printed at costs of several hundred dollars apiece). The remaining fee would help the firm recoup its losses. Venturi was so distraught he withdrew from active participation in the project for some months.

The Western Plaza was constructed and remains a historic destination for tourists visiting Washington. The scale models of the White House and Capitol were constructed, but proved difficult to maintain. They are no longer a part of the design. Indeed, drivers passing the plaza, now called "Freedom Plaza," have little awareness of its lively graphic qualities. Who benefits most from looking at the L'Enfant map today? Undoubtedly those high above the street in city and federal government buildings have the best view, and understanding, of what was intended, as if they were looking at a model. [128]

Notes

1 An analysis of the influence of rational thinking can be found in Judi Loach. "Architecture, Science and Purity." *Being Modern: The Cultural Impact of Science in the Early Twentieth Century*, edited by Robert Bud et al., UCL Press, London, 2018, 207–44.

2 See Mark Alan Hewitt, *Gustav Stickley's Craftsman Farms: The Quest for an Arts and Crafts Utopia* (Syracuse, Syracuse Univ. Press: 2001).

3 Sigmund Freud, *Civilization and Its Discontents* (London, Hogarth: 1963).

4 For an in-depth critical study of their influence, see James Stevens Curl, *Making Dystopia: The Strange Rise and Survival of Architectural Barbarism* (Oxford, Oxford University Press: 2018): 49–216.

5 W.R. Lethaby, *Architecture, mysticism and myth* (London, Percival: 1892).

6 Robert Mark, *Light, Wind and Structure: the secrets of the master builders* (Cambridge, MIT Press: 1990).

7 Pevsner, Nikolaus, *Pioneers of Modern Design: from William Morris to Walter Gropius* (Harmondsworth, Penguin: 1975).

8 Curl, 57–79, is particularly skeptical.

9 For a short historical overview, see Massimo Scolari, "The Idea of Model," in *Oblique Drawing*, 137–84.

10 The best general treatment of these movements is in Reyner Banham, *Theory and Design in the First Machine Age* (New York, Praeger: 1960).

11 Nancy J. Troy, *The De Stijl environment* (Cambridge, MIT Press: 1983): 124–29. Her book emphasizes the collaborative nature of the movement and the relationships between key figures.

12 Ibid., 112.

13 Saarinen has not received his full due among modern architects. See Domitilla Dardi, *Eero Saarinen* (Milan, Eero: 2011); Eero Saarinen, *Eero Saarinen: Shaping the Future* (New Haven, Yale University Press: 2006); and Jane Merkel, *Eero Saarinen* (London, Phaidon: 2005).

14 Timothy Rohan, *The Architecture of Paul Rudolph* (New Haven, Yale Univ. Press: 2014).

15 Christopher Tunnard & Boris Pushkarev, *Man Made America: Chaos or Control?* (New Haven, Yale Univ. Press: 1963).

16 Pushkarev, interview with author, October 14, 2019.

17 On montage, see Martino Stierli, *Montage and the Metropolis: architecture, modernity and the representation of space* (New Haven, Yale University Press: 2018).

18 David DeLong & David Brownlee, Eds., *Out of The Ordinary* (New Haven, Yale Univ. Press.: 2001): 73–75.

19 See Paul Goldberger, "Washington Is Planning an Open Plaza to Ease Pennsylvania Avenue Clutter," *New York Times*, July 9, 1978, 40.

20 See Wolf Van Eckardt, "Focus for the Avenue; Sculpture as a Symbolic Solution," *Washington Post*, May 27, 1978, B1.

21 On this phenomenon see Dale Purves, and Lotto, B *Why We See What We Do Redux: A Wholly Empirical Theory of Vision* (Sunderland, MA: Sinauer Associates: 2011). These scientists postulate that the brain has evolved to compensate for distance and motion illusions, though the "real world" disposition of objects does not correspond to our perception of them.

22 On the first presentation see Paul Richards, "New Design for a 'National Square,'" *Washington Post*, May 24, 1978, C13.

23 See Lynn Darling, "To the Gallery Born," *Washington Post*, Oct. 31, 1982, on Carter Brown.

24 DeLong, 75.

[10]

CONCEPTUAL ARCHITECTURE AND THE DIGITAL VOID

Throughout the first half of the twentieth century, during which much of the Modernist canon was codified, psychologists and social scientists were influential in the teaching and practice of architectural design. Following World War I many European architects espousing the *neue sachlichkeit* or "new objectivity" in design were influenced by Gestalt psychology and logical positivism. The fruit of their research came to the United States with Walter Gropius at Harvard and Ludwig Mies van der Rohe at the Illinois Institute of Technology. Science and architecture seemed to be merging, sharing goals, and pursuing research on how to make better environments. This trend continued, especially at technically-oriented architecture schools, for several decades. Indeed, numerous books and articles on architecture referred to vision research up until the mid-1970s.[1] Many studies were first presented at EDRA, the Environmental Design Research Association, where architects and social scientists met yearly to discuss common interests.[2]

For much of the twentieth century the architectural profession debated the question of how to balance art and science in the teaching of design, and proponents of each way of thinking shared goals and demonstrated mutual respect. Few architects would have challenged the view that architecture was fundamentally a visual art like painting and sculpture. And although American architecture schools were influenced more by English and German pedagogy than French Beaux Arts methods, most schools emphasized the development of visual judgment in studio courses. At the beginning of the century it was assumed that buildings were to be created, built, and evaluated in fundamentally visual terms. We saw in the previous two chapters that as an "engineer's aesthetic" began to enter theoretical discourse during the 1920s, non-visual criteria such as functionalism, structural determinism, project management, and technological imperatives began to supplant traditional views. Architecture might not, after all, be an art.

Conceptual Architecture

Throughout this period there was an insidious undercurrent of skepticism regarding the subjectivity of any aesthetic position that used beauty as a criterion; visual laws and principles were inferior to both scientific and semiological ones.[3] Visual art that was seen as beautiful or pleasing by one observer might provoke an opposite reaction in another, especially an uneducated person from the working class. Moreover, science offered numerous paradigms for problem-solving that might streamline the design process. For a time in the 1960s mathematicians used "systems theory" to analyze design, while social scientists researched the relationship between space and the individual "users" of a building. Computer scientists and electrical engineers encouraged architects to frame problems in terms of machine languages and logic problems.

As criticism of International Style modernism arose during the 1960s, many architects lamented the absence of a theory that incorporated meaning in its prescriptions for designing buildings. The cold, impersonal buildings associated with corporate modernism seemed devoid of any symbolic trappings, things that were always a part of previous architectural masterworks. In parallel with these developments, European intellectuals were introducing Structuralism and Marxist criticism as means of studying human culture and behavior. Perhaps most important was the work of semiologists such as Fernand Saussure, Roland Barthes, and Umberto Eco. These thinkers, it seemed, were the new avant garde in literature, linguistics, and philosophy.[4]

In 1969 a book appeared in English that created a wide-ranging debate about architecture as a "language" that might be manipulated and understood by the educated public. Charles Jencks, an American, and George Baird, a Canadian, assembled the leading theorists of the day to write about *Meaning in Architecture*.[5] To many scholars contributing to the book, architecture was poised to take advantage of the most exciting developments in literary criticism and linguistics. Noam Chomsky was introducing his theories of Transformational Grammar and Deep Structure in the search for a universal set of rules governing human language. Could architecture achieve the same thing?

Significantly, American and British universities were then developing design schools, most including architecture, planning, and landscape architecture, that needed faculty with clear agendas for "research." The modern university was a machine for stimulating wide-ranging inquiries into virtually everything, and architecture was a subject with a limited corpus of scholarly literature. Clearly there was a wide swath of territory for researchers to explore.

Post-war architecture departments quickly assembled faculty with backgrounds in the sciences, social sciences, and humanities who could teach a variety of ancillary courses to enrich the design studio experience. What had been a highly specialized, studio or apprentice-based pedagogy became broader and more attuned to mid-century problems of urban development, building technology, mass housing, and transportation. An unexpected consequence of this pluralism was the rise of a cadre of architectural theorists and critics who quickly rose to positions of prominence. Rather than simply building their designs, significant architects were required to write about them using terminology that was increasingly abstruse and borrowed from the humanities.[6] Two of the most influential architects of the 1960s, Aldo Rossi and Robert Venturi, published books with dense, difficult prose and many references to disciplines outside of architecture: *L'Architettura della Città,* and *Complexity and Contradiction in Architecture.* Both were seen as essential contributions to a new theory of architecture.

During the 1970s the field debated the relevance of the Modern Movement, and created a new avant-garde of architectural thinkers who often built little and concentrated on publishing provocative "critical" position papers and utopian projects. A raft of new journals emerged to publish their ideas. Studio education became a vehicle for converting students to one position or another, often amidst rancorous battles within departments among competitive young faculty. Lost in these battles was a clear sense of how to teach students a discipline for practicing in an increasingly complex professional and economic environment. As one historian put it, every student was expected to emulate the image of an architect as "hero and genius," after the example of Howard Roark and Frank Lloyd Wright. Meanwhile, the art world was exploding, questioning its disciplines, adopting radical political postures, and extending its reach into performance, texts, and video. In 1970 *Design Quarterly* published the first exploration of what it called "Conceptual Architecture."[7] In January of 1975 London's Architectural Association sponsored a conference on the same topic, inviting many of Europe's most important architects and theorists to speak and debate the issues.[8]

After a decade of upheaval, it was perhaps inevitable that architects would solidify their philosophical positions using *au currant* schemas from cultural leaders within the academy. In the 1980s the field turned its attention to the philosophy of language, to semiotics, and to critical theory based on linguistic paradigms.[9] The new heroes were Jacques Derrida, Michel Foucault, and Paul de Man. As Branko Mitrovic pointed out, training in visual judgment was supplanted by language-oriented conceptualism in academic architecture programs for the last two decades of the century.[10] Patrick L. Pinnell observed that, between the controversial, ironic gestures of Postmodernism and the arid philosophical probing of Poststructuralism, architecture renounced its position as an arbiter of societal attitudes toward place, civic awareness, and public life:

> Postmodernism's profuse variants all agreed at bottom that architecture's purpose is to tell human beings, through representation, about their place in the world. Poststructuralism, appearing later, also with variations but following a narrower party line, damned representation and said that when looked at hard, architecture tells us instead about everything's fundamental and unavoidable incoherence, difference, and repression.[11]

The impact of this shift was not immediately evident, as most programs continued to value the contributions of practicing architects in all areas of the curriculum. But as in the law, academic journals proliferated, spurring faculty to publish articles rather than building projects. Indeed, by the 1990s many schools began to favor faculty with advanced degrees such as Ph.D.s rather than hiring practicing architects, even for studio education. Further, leadership positions previously given to practicing faculty were more often handed to academic, research-oriented individuals, many of whom were sympathetic to critical theory and written forms of conceptual thinking. Meanwhile, the visual arts were also expanding to include more non-traditional media and performance art. Architects, always concerned that they be placed on a similar platform with "fine" artists, took notice.

As the contemporary art market expanded, galleries and museums began to look at architectural drawings as works of art that could be exhibited and later sold to a clientele that sought new forms and media for their collections. During the 1980s the first New York galleries to exhibit the work of contemporary architects were Leo Castelli and Max Protech, who for a time sold drawings and models at prices comparable to the work of leading artists. Trend-setting architects such as Michael Graves, Peter Eisenman, and Aldo Rossi made elaborate drawings, often of theoretical works, that appeared regularly in New York shows. Some significantly enhanced both their incomes and notoriety by doing so.

Also significant was the impact of museum exhibitions that featured architectural drawings: Ecole des Beaux Arts (Museum of Modern Art, 1977), American architects (Cooper Hewitt Museum, 1977), Antonio Gaudi (The Drawing Center, 1977), Visionary drawings (The Drawing Center, 1979), and Otto Wagner (The Drawing Center, 1987). The Royal Institute of British Architects displayed works from its collection in the U.S. and elsewhere (1983), and several large museums began to actively collect architectural drawings: the Canadian Center for Architecture in Montreal, the Getty Museum in Los Angeles, and Deutsches Arkitecturmuseum in Frankfurt. If architectural drawings could be seen as autonomous art works, what was their relationship to buildings? How were they to be interpreted, displayed, and conserved? Museum directors and historians contemplated such questions well into the 1990s.[12]

129. Rem Koolhaas, Seattle Public Library, completed in 2004. It is already proving difficult to maintain and patrons are losing patience with its bizarre organization and floor structure. Koolhaas seldom draws when designing his buildings, but instead explains his "concepts" to staff. His university training was in screenwriting and film. (Wikimedia Commons)

The net result of these trends, combined with increased information through digital technology, rapidly transformed architectural education, and eventually even practice, from a mimetic discipline to a linguistic one. Since design began with a written program, architects were expected to organize their thoughts, and even create diagrams, based upon a verbal description of the task at hand. During studio reviews students were increasingly encouraged to prepare lengthy verbal presentations to accompany their design drawings in order to please juries interested in exploring concepts, or better, conceits such as those used by poets. My experience as a studio critic in the 1990s indicated a fall-off in the quality of drawings, increased use of schematic models, and even entire studios on non-architectural themes that did not produce building projects. The concurrent rise of computer aided design software only hastened the decline of visual literacy among architecture students throughout the United States.

Benefitting from gallery shows and publishing, many academic architects who made their reputations by writing theoretical papers or designing highly speculative projects began to receive large commissions despite their relative inexperience in building. One of the first of these was Peter Eisenman's Wexner Center at the Ohio State University, a project funded by the founder of The Gap. A major businessman became a patron of a New York architect who made his reputation by designing buildings so complex as to be uninhabitable, taunting his colleagues who used bricks and mortar. Predictably, the Ohio State building frustrated users; but it garnered praise from academic critics, encouraging other patrons to hire theorists. Enter Daniel Liebeskind (winner of the World Trade Center competition), Zaha Hadid (London Olympic Aquatic Center), Bernard Tschumi (Acropolis Museum), Thom Mayne (San Francisco Federal Building), and Rem Koolhaas (Seattle Public Library). [129] All were "paper architects" during the 1980s, but published prolifically in academic journals and monographs. Another unlikely success story was the New York firm of Diller, Scofidio, and Renfro, who were known mainly for their daring conceptual projects during the late 1970s and labored for over a decade with few built projects.[13]

The noted Finnish architect Juhani Pallasmaa recently lamented the absence of sensory training, including hand skills, in architectural education, noting that schools "continue to emphasize and value conceptual, intellectual and verbal knowledge over the tacit and non-conceptual wisdom of our embodied processes."[14] The philosopher and former Boston University president John Silber wrote a book describing his dissatisfaction with contemporary architecture in 2007, suggesting that the worship of "genius" and "novelty" had disfigured "a practical art."[15]

Although architects and architectural theorists have incorporated psychological and philosophical concepts from science sporadically during the past forty years, a curious vacuum has developed in both practice and pedagogy with regard to the advanced research on visual perception discussed above.[16] Ironically, this period produced some of the most revolutionary discoveries ever made on the nature of the "visual brain" and its complex cognitive functions. As Sarah Robinson has noted in *Mind and Architecture*, "while architectural theory was following the other humanities into the vertiginous and barren peaks of semiotics, experts in the various sciences were moving in exactly the opposite direction."[17] So, as architectural educators turned away from this area of science to focus on cultural theory, the field lost one of its essential areas of inquiry. Indeed, many of the premises under which design is taught today are based upon false or outdated concepts of how visual thinking occurs.[18]

Meanwhile, architects sprinted to keep pace with the explosion of information technology during the late twentieth century. With the introduction of computer-aided design into studio education during the 1990s, architectural educators embraced a powerful new tool for conceiving and representing buildings.[19] Amidst some skepticism from older practitioners, school after school took up instruction with CAD integrated into the curriculum during the middle years of professional programs. Some questioned whether the computer was an aid or a hindrance to "design" as commonly practiced.[20] Digital modeling certainly offered powerful imaging capacities, and streamlined the tedious production of construction drawings, but it was unclear what value CAD added to the training of students in visual judgment, aesthetic awareness, or good mental habits of design.[21]

By the second decade of the twenty-first century the architectural profession had jettisoned the core discipline of previous generations of successful architects: a highly developed system of visual design based on centuries of experience. As we have seen in the previous discussion, that system was also grounded in an intuitive grasp of how the brain works, and how humans learn to integrate their sensory awareness of the world with their cognitive abilities. In this way all creative art, indeed any creative endeavor, unites two sides of the human organism, and validates universal aspects of experience. As the critic and historian Russell Sturgis put it more than a century ago: "the purpose of an artist, his mental processes, cannot be explained apart from the movement of his intelligence which partly precedes and partly accompanies the work of his hands."[22]

The renowned American painter, Chuck Close, has condemned the "de-skilling" of artists as a travesty. He lamented the misuse of studios at the American

Academy in Rome, where generations of visual artists made new work: "They [Rome Prize Fellows] sit in these beautiful studios, they put the work on the wall that they used to get the grant [Fellowship] and then they just talk. For a year. It's criminal. You can talk in any room! These rooms should be used to make paintings and sculpture."[23] His thoughts echo those of many critics of digital tools like computers, cell phones, and GPS devices, who suggest that humans are continuously surrendering their creative choices to machines, losing critical thinking, writing, and visual skills.

A Brief History of CAD

Virtually all architects and engineers practicing in developed countries use digital design software and mini-computers in their offices. The software available today is simple to use, powerful, and can generate both design and working drawings quickly when used by a skilled operator. Just as desktop and laptop computers are the fundamental tools of business and commerce, these tools are essential to all sorts of design, from landscapes to small household objects. It isn't hard to see why: computers increase productivity and help generate profits.

Though digital design software was slower to reach the studios of architecture programs in the United States than to professional offices, most architectural schools now rely on computers and large screen displays for teaching design, no matter what the level. Students are often expected to learn the necessary software on their own before coming to a studio. And, given the plethora of visualization, graphics, and mechanical design software on the market, their task is often challenging.[24] Scale models are also commonly required in studio presentations, and even these objects now benefit from 3-D printing and new materials that can be digitally manipulated. Having grown up in a digital world, young architects are seen to be quick learners when it comes to any kind of digital platform. Though they pick up the skills required to generate forms in virtual space, this does not guarantee competence in visual design or judgment.

It would not be fair to say that hand drawing has vanished from architectural education, for many schools offer special instruction in sketching and al fresco drawing. Likewise, many studio instructors ask students to use hand drawings when designing, especially in the early years of their education. There is, however, a large gulf between the kind of visual/haptic thinking discussed in this book and the studio experiences of most architectural students, whatever their level or skill. As we have seen, linguistic and algorithmic thinking predominate in current architectural pedagogy.

More importantly, the adoption of machine languages and information theory has redefined design as a problem-solving, logic-based endeavor rather than one dependent on practical and visual experience. According to the computer scientist Yehuda Kalay, by mid-century contemporary architects

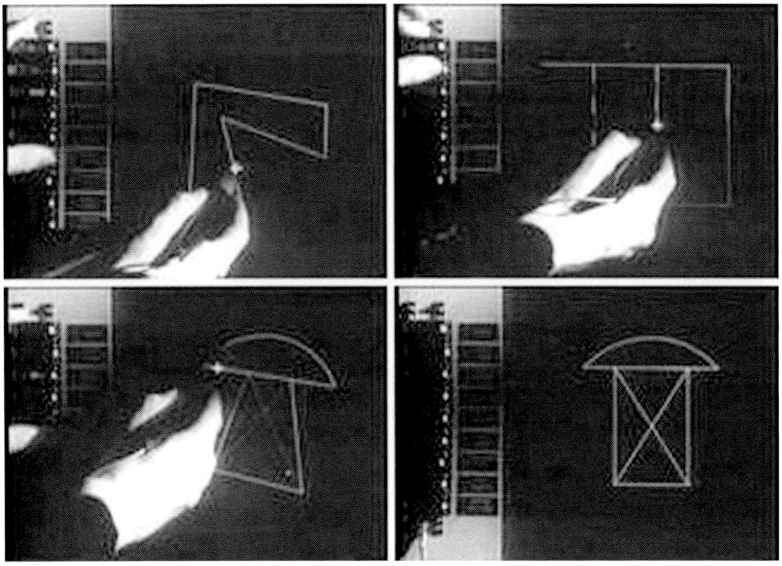

130. Ivan Sutherland's "Sketchpad" program, 1960s. The world's first CAAD program. (Wikimedia Commons)

came to look at design is "a process that ensured that their creations met the design goals, abided by its [sic] constraints, and reduced the likelihood of errors." Further, he contends that design as a cognitive process consists merely of four iterative stages: 1) problem analysis, 2) solutions synthesis, 3) evaluation, and 4) communication.[25] As a scientist and engineer, he is critical of architects' reliance on habitual practices rather than empirical ones. He insists that computer technology will fundamentally alter the way architects do their work. As we have seen, such historically false views are the legacy of over a century of assembly-based thinking among engineers and scientists, not a natural consequence of the progress in architecture as a profession. It is important to understand the history of the development of computer-aided design in order to unpack such attitudes toward both pedagogy and professional practice in the current era.

Computer assistance in the design of machines and circuits was available to scientists in the mid-1950s, when researchers at MIT began to consider ways of improving the generation of graphic information on digital platforms. Algorithms written for early computer languages could "draw" lines and manipulate objects in coordinate systems based upon Euclidian geometry. Before long the virtual space in computers would become as sophisticated as the most advanced 3-D drawings and physical models produced by leading architectural firms.[26]

The invention of CAD is often credited to a French engineer, Pierre Bezier, who created UNISURF, a program to aid in the design of automobile components for the Renault Company, from 1966–68. However, most scholars point

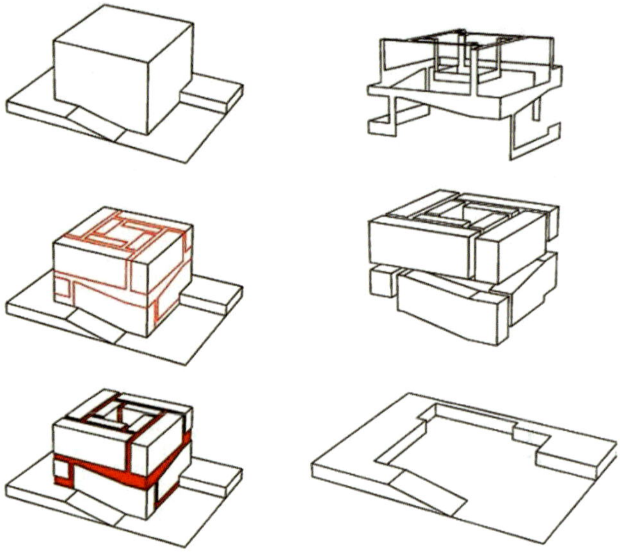

131. Early CAD wireframe drawings were crude and almost useless as design tools. It took over two decades for computer graphics to supplant hand drawing in most architectural offices around the world. (Author collection)

to Ivan Sutherland's SKETCHPAD thesis (1963) at MIT as the point at which science took notice of the graphic capabilities of the computer.[27] Sutherland defined the basic capacity, parameters, and necessary components of drawing software in a concise document that had wide influence. His work laid out many of the mathematical and technical problems that needed solving before a truly sophisticated computer drawing system could be brought to market. [130]

Sutherland was a student at America's leading university for science and engineering in the post-war period. MIT became the acknowledged center of digital design research with the inception of Nicholas Negroponte's famous Media Lab in the late 1960s.[28] Between Negroponte and Sutherland, the Massachusetts Institute of Technology did more to advance computer visualization than any other university or research center. Following its lead, by the early 1980s most architecture schools had a computer lab with access to mainframe machines.[29] Later, William Mitchell came to MIT to pursue his own groundbreaking research, and published extensively on the virtues of digital design. Wonderful things came from all of this work.[30]

Outside of the academy, some computer scientists bet on their ideas and founded companies to sell advanced software. The most successful of these were the founders of Autodesk under John Walker. His company released the most successful mini-computer design software in January of 1982.[31] Though Apple was then introducing the Macintosh desktop system, digital drafting programs (DataCAD, MiniCAD, and ArchiCAD were early competitors) relied on MS-DOS operating systems for more than a decade, eventually also supporting UNIX open source software. [131] Paradoxically, publishing, photography, and graphic design professionals gravitated to Macintosh-based software. The eventual winner in the race to satisfy these "art" professionals was Adobe Systems, Inc. Their flagship program, Photoshop, offered a wide array of image manipulation tools that were quickly adopted by architecture firms for their presentation drawings, often overlaid on vector files generated by drafting software.

This bifurcation of vector based and raster based image generation and processing has defined the cognitive poles of software designers and engineers for over thirty years. There is, in fact, a continuing debate about how "user friendly" vector based programs can be, given both their original DOS platform and their non-intuitive, machine-based "tools" for drawing. Nevertheless, new software such as SketchUp has improved the accessibility of these programs to the general public, and Google has successfully marketed this product in recent years. Similar software is also available to architects, and is used in many studio courses.

In contrast to the relative speed at which software development progressed, early literature on the benefits of CAD in both creative work and education paints a rather confusing picture of its use and adoption in the design professions. Some early papers touted the efficacy of computing machines in basic problem solving using simple algorithms and matrices.[32] Other critics and proponents saw wide applicability to art, interior design, engineering, and architecture.[33] The CAD industry was of course invested in pushing its products in all applicable markets, and architecture was an early target. [131] Unfortunately, the first commercial CAD packages were written by engineers for mechanical design applications such as aircraft engines and automobile assemblies. Building construction was not part of their purview. Moreover, engineers needed not only graphics capability but also information processing, problem solving, and mathematical modeling in their software.[34]

The learning curve required to train CAD draftsmen in the new imaging and drafting software was initially steep, resulting in only limited market penetration among the largest and most sophisticated A/E (architectural and engineering) companies such as SOM in Chicago. By the early 1980s that office had constructed a virtual 3-D model of the loop in Chicago for use in urban design projects, but used proprietary software in doing so. In addition, some software companies were slow to respond to critiques from architects, resulting in negative opinions about the applicability of CAD to professional practice and education. A 1992 survey by social scientists indicated that the majority of Fortune 500 firms using CAD and CAM systems were not utilizing their expensive software for any meaningful production work, mainly due to reluctance by engineers and management to change old habits. The study also pointed to the inadequacy of the current software in living up to its promised design advancements.[35]

Things changed rapidly during the next decade. In a 2002 research study that surveyed most of the architectural schools in the U.S., Alfredo Andia found that CAD software was rarely used by academic studio critics from the mid-1970s until the mid-1980s. Widespread skepticism about its efficacy was evident in many quarters. As he observed:

> Academicians both in the architectural and computer science fields decried the simplistic nature of commercial CAD. Research and computation professors who were close to the design methods community considered commercial CAD a frivolous instrument, one that ignored the informational potential of software design. In parallel, a great number of established studio professors, concerned that students would not acquire traditional drafting and design skills, banned commercial CAD from their design studios.[36]

According to Andia it was not until the early 1990s, when Columbia University brought CAD into its studio curriculum, that "paperless architecture" became a reality for architecture students. As a faculty member at Columbia from 1985–1989 I saw little evidence of computer-based learning, but that changed with the inception of Bernard Tschumi's reign as dean of the GSAPP. There was clearly a shift in the attitude of faculty after younger, computer literate, professors began teaching in studios throughout the U.S., Canada, and Western Europe. By the mid-1990s most schools had hired at least one computer graphics specialist to teach and assist with the installation of machines and software in studios. Many schools eventually replaced drafting desks with smaller, paperless, computer stations. Eventually students would supply their own laptops and plug into university local area networks.[37]

In the 1990s architectural firms were hiring recent graduates based upon their literacy with CAD software as well as their design portfolios. From a purely practical perspective there was no longer a choice for the prospective architect: computers were now part of the design and production toolbox. Learning to use various kinds of visualization and modeling software was a must, and a wave of new technologies caused a profound transformation in the daily practice of architects throughout the world. Moreover, younger and more tech savvy designers were gaining attention for their innovative use of these software tools. Andia suggests that these architects found "academic rationalizations or languages used to describe architectural concepts [to be] obsolesced by digital media," since concepts and design ideas could be tested immediately, allowing decisions to be made "on the spot."[38] Moreover, these designers found a new paradigm for conceptualization in the virtual space of the machine, one that freed them from history, materiality, and Euclidian geometry. Their heroes—Frank Gehry, Zaha Hadid, Greg Lynn, and Coop Himmelblau—were proving that bold new forms could be invented within the virtual worlds made possible by computer graphics.

Modeling and visualization were not the only areas to be transformed by the power of digital media. Information available on the Internet allowed material choices and specifications to be linked to construction documents, and communication software facilitated collaboration over long distances. With advances made possible by the gaming industry, virtual reality simulations of designed spaces helped architects and designers locate their users in prospective environments without building anything physical. Indeed, computers had redefined much of the practice of architecture by the end of the twentieth century.[39] A number of research labs and universities experimented with programs that imbedded evaluative or "intelligent" subroutines within CAD packages, providing assistance to designers. Engineers seemed to appreciate these features more than architects, who wanted more control over the generation of their "solutions."[40]

Following the new millennium things progressed even more swiftly. Normalization of digital design methods in the studio curriculum of most architecture schools proceeded apace, and the grumblings of older faculty seemed less and less relevant to this brave new world. There were plenty of critical assessments of the effects of CAD on design education in journals, however.

Bryan Lawson, dean of the faculty of architecture at the University of Sheffield in the UK, published one such paper in 2002 that raised many of the essential questions about the pros and cons of digital modeling and drafting, especially for training young architects. In contrast to Andria, he found serious problems with the use of these platforms in design education and practice. We might begin our discussion of the relationship between grounded cognition and digital visualization with his acute observations.

Digital Visualization and Visual Judgment

Does digital design software enhance or impede creativity? According to Lawson, a psychologist as well as an architect, computers offer many advantages for designers, but stimulating real creative work and assisting with decision making are not among them. One of the shortcomings of such software is that it does not allow designers to cognitively assess multiple aspects of a design problem, even when rotating models or generating multiple views in virtual space.[41] Another is that the metaphors and terms used in visualization programs do not readily correspond to the languages employed by architects in their daily work. Lawson cites a simple example of roof components and geometries, in which "valley," "hip," "eave," and "ridge" have powerful associations with construction, while vector, tetrahedron, irregular polygonal solid, or polyhedron do not.

He quotes the eminent Dutch architect, Herman Hertzberger, who is known for his social engagement with the public as well as his cutting edge designs. Hertzberger believes that the computer misleads architects by generating impressive looking 3-D models of possible building forms, most of which would make horrible buildings. Without a memory stoked with examples of great buildings, a student is easily seduced by his/her ability to generate interesting looking forms digitally. No reference to architecture appears in an abstract computer model. No principles of design or aesthetics accompany the images as they magically appear on the screen. Students are easily fooled into thinking that they have come up with creative solutions, but Hertzberger considers this fake creativity.[42]

The interactive loop that occurs when architects make diagrammatic drawings, refine them, and fill in details, is circumvented by computer images, which look real almost immediately. There is no conversation between the drawing and the designer. In addition, the completeness of many 3-D images, in perspective, encourages the brain to read them as solids, without the spatial cavities and voids that are necessary in the creation of real architecture.[43] Like physical models, computer models can appear to be solids representing only the exterior masses of a building, leading designers to shortchange the design of interior spaces and details. Finally, it is clear that digitally literate young people are conditioned to see screens as virtual environments, much as they do in sophisticated role-playing games, leading them to consider a drawing finished when it is only a sketch. As Lawson observes, computer graphics tend to impress us with their "dazzle" and panache:

> First, we live in such a televisual age that any information that looks televisual is automatically considered authoritative. Second, the complete mastery of such computer systems is still sufficiently rare and novel that we tend to admire it in the way we admire an animal taught to do tricks. So it is possible to put forward computer presentations that look attractive and even dazzling, that seem authoritative, while the architecture so represented is really quite awful.[44]

Richard Sennet reminds us that, since the Industrial Revolution, machines have threatened to usurp the creative skills of the artist/craftsman. Computers are the most powerful, and most threatening machines ever invented in this regard. He points to many of the drawbacks that architects have observed in their use of CAD: "overdetermination" of design elements prior to their testing in the real world, poor visualization of materials on screen, lack of engagement with the building site, too much abstraction in wire-frame perspectives, and circumvention of the give and take between the brain and its external mnemonic images.

As he explains it, "Abuses of CAD illustrate how, when the head and the hand are separate, it is the head that suffers."[45]

In addition to misleading designers about the nature and completeness of images, digital media also distort key perceptual judgments of scale, distance, size, and direction, all aspects of the visual brain that are fundamental to a sense of space and location. In 1993 two scientists at the University of Washington conducted an experiment in which architects were asked to "tour" a gallery space in both physical and virtual conditions that were purportedly identical. Great care was taken to ensure that the virtual model built in CAD was dimensionally and geometrically similar to the physical space in their laboratory. Twenty-four subjects were guided through three different viewing conditions: viewing a monitor, wearing a mask/head display without head position tracking and with head position tracking. The results were not conclusive, but subsequent research showed similar perceptual distortion.

Each subject was asked to complete three "tasks" while experiencing a series of actual gallery spaces in the Henry Art Gallery, and then to repeat the tasks in three virtual simulations of the galleries. First, each subject was required to estimate the dimensions of the space; second, each was asked to describe his/her position in the space on a map; and third, each was required to describe the feel of the space. As architects, the first two tasks would be relatively simple and the third more challenging. The scientists found that their subjects consistently underestimated the size of spaces and distances in virtual environments in comparison to the actual gallery rooms. Their explanation was based upon settled research in perception of size constancy, where sizes appear smaller when seen from truncated viewpoints. Moreover, though location data was inconclusive, it was clear that the experiment showed the importance of full sensory engagement, throughout the body, when attempting to assess qualities of space and interior volume.[46]

In a theoretical paper, Brian Lonsway of RPI in Troy, New York contested some of this research by insisting that the virtual "dimensionality" in digital simulations, from 2D through 4D (Minkowskian) representations, was merely a convention imbued in architects through their culture and history with drawings as representational types. Encouraging designers to strip away this baggage and embrace the unknown, he advanced a paradigm for formal experimentation that has indeed been embraced by some architects. He did not, however, contest the findings on the basis of any understanding of cognition or the visual cortex. Like most theoretical architects, his interests were in form-making and semiotics, not in the experience of humans in their environments.[47]

Why is all of this important in the training and practice of design, when it is clear that computers are no better than pencils and paper, simple tools that aid designers in problem solving? Don't powerful tools free the architect to imagine more complex and aesthetically exciting forms and spaces? If we listen to proponents of digital design, there is virtually no limit, and no down side, to the use of computers to expand the capacity of architects. Rivka Oxman, a prominent Israeli social scientist, has suggested that digital design will usher in completely new ways of thinking about buildings because of its powerful imaging and analytical capabilities.[48]

On the one hand, familiarity with cyberspace and digital worlds will undoubtedly enable younger people to adapt to virtual environments with greater facility as time goes by. Books by William Mitchell, Michael Benedikt, and Christine Boyer hailed the potentialities of cyber-cities during the 1990s, while brain scientists were probing the mind on a neuronal level.[49] Familiarity with the new media will break down some of the barriers now seen as hampering the creative use of digital technologies in architecture. Global connectivity on the Internet will make sharing and collaboration a matter of access to broadband.

The advantage of digital representation as a technical tool, a super-pencil, has already been established, and there is nothing to suggest that designers will not fully adopt this media in the near future, though a significant segment of the profession still employs hand drawing in schematic design. There will also be increasing integration of 3-D computer models and physical models, allowing firms to "print" their complex visions at virtually any scale with speed and accuracy.[50] REVIT, a form of database that links product information to CAD drawings, has also revolutionized the speed of construction document production. Productivity is the key to profits, and all industries have benefitted from digital platforms in this regard.

There are, however, profound psychological, social, aesthetic, and philosophical questions that must be answered before the design professions jettison or modify traditional means of studying and evaluating solutions to all kinds of visual design problems. Most can be answered by seriously considering the new research on embodiment and cognition. For instance, how much time do students and intern architects who are trained on the computer use in judging the quality and efficacy of their designs? One study found that students spend less time judging or evaluating their designs when using CAD than when sketching or drawing by hand.[51] The authors employed protocol analysis and used familiar design problems in their experiments. They reviewed much of the literature on traditional versus CAD protocols in conceptual design,[52] noting that students were focused on "problem solving" at the expense of creative exploration. They concluded that new research should zero in on the first phase of design activity: concept formation, analysis, or presentation. The latter presents the most obvious for CAD as a tool. This book points to some critical issues that have not yet been addressed.

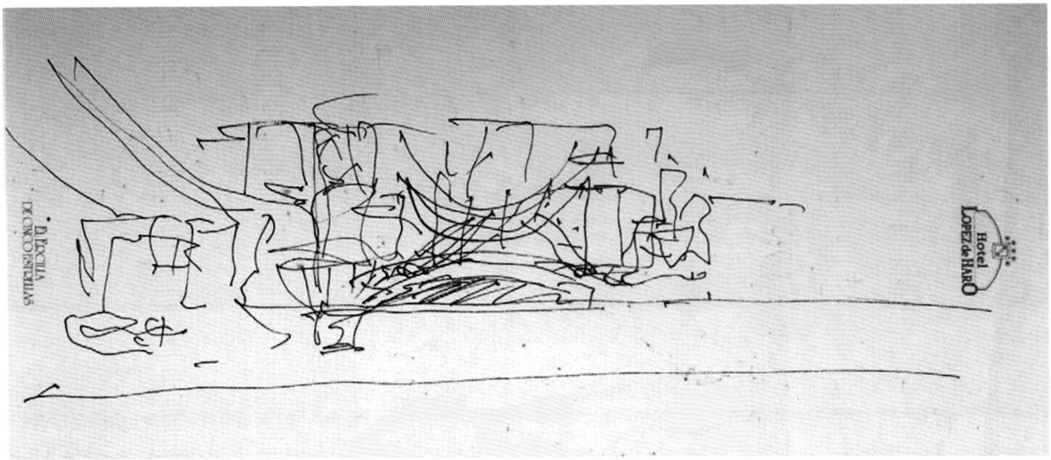

132. Frank Gehry, early "napkin sketch" for Guggenheim museum, Bilbao, Portugal. (Courtesy, Frank. O. Gehry)

In a provocative paper published in the *Journal of Architectural Education*, Eugenia Ellis raised the important question of how designing in digital "hyperspace" might change the entire nature of educating architects. Her thesis was based upon an analysis of the positions of three influential philosophers on the subject of technology: Jose Ortega y Gasset, Jaques Ellul, and Maurice Merleau-Ponty. As she points out, two of these thinkers did not consider the position of the artisan in their formulations of the role of machine production, though Merleau-Ponty placed the body at the center of his phenomenological world. Ortega y Gasset, like many rationalists, believes that invention only occurs when the designer, not the maker, is in control of production. Ellul, like other writers we have considered, finds artisanal society full of invention. Ellis also correctly points out that computers and other electronic media alienate us from the natural world. "Our experience of the world is no longer of nature, but of technology. Indeed, we cannot experience the environment through any digital screen or virtual reality headset. The technology we experience does not reveal its secrets; everything happens as if by magic," just as Ortega y Gasset predicted a century ago when he spoke of "supernature."[53] She goes on to lament the fact that students do not draw by hand, but rather enter the empty void of hyperspace in order to experience their own building designs, which appear on the screen as if in a movie. Haptic, aural, and other sensory cues are absent from their understanding of the architecture they hope to create. Alienation from nature produces the kind of placeless, soul-less buildings that we increasingly find in our technology-crazed world. Ellis concludes that only by learning to draw by hand can an architect directly engage the natural world through her intellect and senses combined.

As Charles Moore and Kent Bloomer pointed out (Chapter 2), the human body is the fundamental reference point for all environmental perception, and

133. Frank Gehry concept elevation sketch for Bilbao Guggenheim. (Courtesy Frank Gehry and Partners)

should therefore be the center of any study of architecture as art, science, or profession. As we have seen in both case studies and research results about the visual brain, any form of external memory symbol, or representational mode, must feed the brain's internal, grounded, cognitive operations. Spatial perception is a complex, embodied process that engages many sensory, motor, and cognitive systems simultaneously. No external representational mode can fully capture these stimuli, let alone their images in the brain. All that designers can ask of any representational mode is that it aid in judging the aesthetic, instrumental, or emotional value of a concept, at whatever stage it happens to manifest itself in the mind.

The close relationship between perceptually constructed images that exist in biological memory, and external symbolic concepts that exist outside the organism, has not received adequate study in either scientific or humanistic research. What we have seen in this historical sampling of architects and their cognitive habits is that more research will yield the kind of understanding that may lead us to find the proper place for both digital and analog forms of representation.[54]

The necessary role for traditional forms of drawing is borne out in the proven efficacy of these modes in the development of visual judgment. No young architect is equipped to assess the aesthetic quality of concepts without a dense network of internal memories, and a means of accessing external symbolic fields. Hand drawing and first hand experience of great buildings, landscapes, and visual art develops this capacity more thoroughly than any other pedagogical vehicle. Every kernel of research in neuroscience, art history, and anthropology, among the pertinent fields of inquiry, supports this conclusion. For this reason alone architects should suspend their infatuation with digital pedagogy and re-evaluate its efficacy as a means of equipping young designers with the tools to engage the challenges of sustainability, energy conservation, and urban development worldwide.

134. Gehry office, models of Guggenheim Bilbao. (Courtesy Frank Gehry and Partners)

Bilbao and PIXAR

Frank Gehry designed his complex masterpiece, the Guggenheim Museum in Bilbao, Spain, (1991–98) using hand sketches and cardboard models. Though Catia aircraft modeling software made the construction and cost estimating of the building easier, Gehry did not bring the design into the computer until it was well defined in non-digital media. Yet the myth among digital architects and computer scientists is that Bilbao proved their claims about the superiority of computer generated images in designing complex built form.

An extraordinary book documenting the entire design process for the museum was published shortly after completion. In it both Frank Gehry and the author, Coosje van Bruggen, relate the story of how an architect trained to sketch his ideas quickly could envision one of the most complex buildings ever designed. Selected sketches, study models, diagrams, and photographs of the studio document the multi-year design and construction process in exacting detail. If this account is accurate, nothing suggests that computer software was employed in the design of the museum.

Frank Gehry began his thinking the way most architects do, with a visit to the site. On two separate trips in May and July 1991, he toured Bilbao and discussed two potential sites with Thomas Krens, director of the Guggenheim in New York. He took photographs, made notes on a tourist map, and questioned planning officials about the urban fabric and history of their city. On the way home from his second visit, he produced a "napkin sketch" on stationery, recording some impressions and visual concepts. [132] While still in the air he instructed his staff to prepare a site model and several small study models of alternative approaches to the design. These cardboard effigies were crude—a pyramid of stepping squares, a group of cubes, some circular pieces crashing together. [133] His associate, Edwin Chan, flew to New York with the models so that Gehry could work with them, in the office of Peter Eisenman, while attending to other projects.[55] [134]

According to the architect this conceptual mode was normative—every project began with such studies. He remembered, at age 10, "building cities"

135. View of completed Guggenheim, Bilbao. (Wikimedia Commons)

with his grandmother using toys and objects from around his family house in Toronto. Like Aalto, he seems to have imprinted these early experiences, associating them with creativity and play. "I'm still doing the same thing with little boxes and trash and pieces of wood," he recalled in 1997. His exograms were study models, giving him the flexibility to look at physical objects for scale and site references, and then think more freely in sketches.[56] During the initial three-way competition, in July and August of 1991, Gehry and his staff worked intensely in the Los Angeles studio, producing more and more detailed models in an interactive sequence. Eventually there would be six main physical models, along with dozens of hand sketches, many illustrated in the van Bruggen book. [134] Following the New York models, the architect worked with a scheme based on "great hall and long gallery" parti using the Puente de la Salve over the Nervion River as a foil. His colored pencil sketches of this scheme provided a rational armature for freer formal ideas that would develop during the coming weeks. Initially inspired by fish and snake metaphors, Gehry abandoned these on July 13, 1991 for the motif that came to define the atrium of the museum: a flower.

His critical sketches, developed in parallel with model studies, show the architect initially playing with an array of rectangular skylight volumes.

Turning to models, the staff began to pile these forms on top of each other, deforming them from rectangles into bent tubes. Polaroid photos were given to Gehry, who would use them to make scribbly sketches. A definitive north elevation sketch, from July 15, 1991, finally caught the mental image Gehry was looking for. From that point, the building had its parti, and Gehry won the competition.[57] [135]

Likewise, sketching out ideas is a crucial step for artists working at Pixar Animation Studios. Every storyboard, character sketch, and setting is first drawn by hand. Before digital tablets were available, these sketches were done in traditional media on paper or board. Today tablets are used along with pencils and paper. The artists use the stylus to produce art just like they did with pencils, pens, and paints.

Thousands of these drawings are made before any computer animation or modeling takes place in their film production process. Production designer Harley Jessup made this clear in a 2006 lecture published in the book *Thinking/Drawing*, edited by Marc Treib. [136] His description of the process of writing a story, inventing characters, designing settings, and animating was little different

136. Harley Jessup, drawing for *Ratatouille*. Hand drawings, whether done on paper or computer tablet are essential to every Pixar film. (Courtesy, Disney-Pixar)

137. Harley Jessup, drawing for Ratatouille. Thousands of hand-drawn sketches were prepared before any animation occurred in this film. (Courtesy, Disney-Pixar)

than the work done by Winsor McCay on Gertie the Dinosaur or Walt Disney on Mickey Mouse earlier in the twentieth century.

Jessup discussed the necessity for detailed, hand-sketched story boards for each sequence in seven Pixar films: "for Brad Bird's film *Ratatouille*, there were 26 sequences encompassing over 1,600 shots." For these alone, 20,000 hand drawings were required. He presented the rough number of drawings done for the studio's feature films to that date: [137]

A Bug's Life: 21,565
Toy Story 2: 28,244
Monsters Inc.: 46,024
Finding Nemo: 43,536
The Incredibles: 21,081
Cars: 47,000
Ratatouille: 72,000

Hundreds of artists work at Pixar, not only in the art and story departments, but also in many of the technical departments, including lighting, layout, and computer animation. Many avail themselves of the many life drawing and painting courses offered by the studio during breaks and lunch hour.

Jessup's powerful argument for hand drawing even extended to the creation of intuitive, on the spot sketches and caricatures. As he concluded: "The act of drawing continues to be the standard medium for communicating visual ideas at Pixar, and although the thousands of drawings we create never appear directly on the screen, they remain the foundation of every feature film we create. The computer is a miraculous tool, but a great story is, in fact, the heart of a Pixar film and to tell that story we always begin with a drawing." These accomplished creative people understand how vital the eye-hand connection is when conceiving any visual art.

Nicholas Carr, one of the most incisive critics of automation and computer deskilling, notes that increasing dependence on machines has resulted in a bias against hand work, suggesting that manual labor required less knowledge than purely cognitive activities. His research suggests otherwise: "All work is knowledge work. The carpenter's mind is no less animated and engaged than the actuary's. The architect's accomplishments depend as much on the body and its senses as the hunter's do. What is true of other animals is also true of us: the mind is not sealed in the skull but extends throughout the body."[58] In crucial respects the computer screen alienates us from our bodies, replacing sensory engagement with a numbing virtual void.

Notes

1 Rudolph Arnheim, *Visual Thinking* (Berkeley, Univ. of California Press: 1969): 1–12. Arnheim, though associated with German gestalt psychology, anticipated much of what neuroscience proved in the 1990s decades earlier. "I shall suggest," he wrote, "that only because perception gathers types of things, that is, concepts, can perceptual material be used for thought; and, inversely, that unless the stuff of the senses remains present the mind has nothing to think with." 1.

2 One book that collected some of this research was Jon Lang, editor, *Designing for Human Behavior: architecture and the behavioral sciences* (Stroudsburg, Dowden, Hutchison & Ross: 1974).

3 A critique of this view can be found in Roger Scuton: *The Aesthetics of Architecture* (London, Methuen: 1979).

4 An interesting compendium of research in this vein was *VIA 2: Structures Implicit and Explicit*, the Journal of the Graduate School of Fine Arts, University of Pennsylvania, in 1971. Included was an early essay on semiotics by Eco, as well as the work of Robert de Ricolais, a structural engineer experimenting with "least material" concepts.

5 Charles Jencks & George Baird, Eds., *Meaning in Architecture* (New York, Braziller: 1969).

6 Karsten Harries, "Theory As Ornament," in A. Vinegar and M.J. Golec, Eds., *Relearning From Las Vegas* (Minneapolis, Univ. of Minnesota Press: 2009): 79–95. The author notes the large influence of both of Venturi's books on a literary and theoretical turn in architecture during the 1970s and 1980s.

7 For a superb analysis of this publication and its aftermath, see Ross Elfline, "The Dematerialization of Architecture: Toward a Taxonomy of Conceptual Practice," *JSAH* 75:2 (June 2016): 201–23.

8 The conference, which occurred on January 17 and 18, can be viewed on YouTube: https://www.youtube.com/watch?v=rqCg1WLUMac. Among those speaking were Charles Jencks, Joseph Rykwert, Peter Eisenman, Colin Rowe, and Dalibor Vesely. In the audience was Bernard Tschumi, a student at the AA at the time.

9 A good overview of this history can be found in Mark Jarzombek, "A Conceptual Introduction to Architecture." *Thresholds*, no. 33, 2008, 16–20.

10 See Branko Mitrovic, *Visuality for Architects* (Charlottesville, Univ. of Virginia Press: 2013): 1–7. Mitrovic's book presents a critique of this theoretical myopia by looking mainly at philosophy and at a few key psychological studies of visual perception. He is correct in assigning much of the blame to language oriented, story-telling design philosophies. As he writes: "The generations of architectural academics and theorists whose intellectual framework was formed on Derrida and Heidegger are unlikely to consider or read contemporary works on the psychology or philosophy of perception."

11 Patrick L. Pinnell, "Ostrakon: Postmodernism, Deconstructivism and the Millennium," *GA Critic* 1992: 8–19, quote on page 9.

12 For a treatment of this trend, see Jordan Kauffman, *Drawing On Architecture: The Object of Lines, 1970–1990* (Cambridge, MIT Press: 2018): 53–274.

13 See Edward Dimendberg, *Diller Scofido + Renfro: Architecture After Images* (Chicago & London, Univ. of Chicago Press: 2013).

14 Juhani Pallasmaa, *The Thinking Hand: Existential and Embodied Wisdom in Architecture* (New York: Wiley: 2009): 22.

15 John Silber, *Architecture of the Absurd: how 'genius' disfigured a practical art* (New York, Quantock Lane Press: 2007).

16 Some of this is highlighted in Harry Francis Mallgrave's excellent book, *Architecture and Embodiment: The Implications of the New Sciences for Humanities and Design* (New York, Routledge: 2013). Mallgrave also discusses Darwinian adaptation and neuroaesthetics.

17 Sarah Robinson, "Introduction: Survival Through Design," in Robinson and Pallasmaa, Eds. *Mind in Architecture: Neuroscience, Embodiment, and the Future of Design* (Cambridge: MIT Press: 2015): 3.

18 A good example of this ignorance is shown in a 2013 paper from a Romanian university. See Dana Pop, "Space Perception and Its Implication in Architectural Design," *Acta Technica Napocensis: Civil Engineering & Architecture* Vol. 56, No. 2, 2013.

19 The new methodology has been called "parametrics." C.f. David Jason Gerber, "Parametric Practice: Models for Design Exploration in Architecture," PhD. Dissertation, Harvard University, May 2007. Some have called the trend in design "parametricism."

20 See, for instance, Garry Stevens, "Reflections of an Apostate CAD Teacher," *Journal of Architectural Education*, Vol. 51, No. 1 (Sep., 1997), 78–80. He says that in 15 years of teaching CAD he does not see value for the profession, given the fact that so many "CAD jockeys" are not trained architects. "To date, I believe, the progress of computing in our field has been driven by the desires of those in the information-technology industries and their academic allies. We owe it to ourselves to recapture our future and use computing to aid the profession, not destroy it."

21 The late William J. Mitchell was the most probing and respected advocate for computer-aided design among architectural educators. While at MIT he developed a broad body of research on logical models of design that benefit from digital processes. See, for instance Mitchell, *The Logic of Architecture: Design, Computation and Cognition* (Cambridge, MIT Press: 1990). As he states, "When a computer-aided design system is used, the data structure and its associated operations establish the design world." 38. Unfortunately, Mitchell died before some of the neuroscience that contradicted his ideas was available. He spends much of his book trying to fit various historical approaches to architecture, such as Palladio's theory of harmonic proportions, the system of the classical orders, the golden section, formal typologies, mathematical theorems, and linguistic theories into a cage defined by the logical operations of machine language. Despite a vast bibliography and a good deal of intellectual gear grinding, he fails.

22 Russell Sturgis, *The Artist's Way of Working*, (New York, Dodd, Mead & Co.: 1905): 5.

23 Will S. Hylton, "BLOW UP: Chuck Close's Self-Imposed Exile," *New York Times Magazine*, July 16, 2016: 33–43, 53, quote on 41.

24 See for instance: Salman, H., Laing, R., & Conniff, A. (2008). "The changing role of CAAD in the architectural design studio." *The Built and Human Environment Review.* 120082539 [Online]. Available from http://www.tbher.org/index.php/tbher/ article/view/3.

25 Yehuda Kalay, *Architecture's New Media: Principles, Methods and Theories of Computer-Aided Design* (Cambridge, MIT Press: 2004): 9–10. Kalay proposes a largely ahistorical reading of the "history" of architectural design, suggesting that this problem/solution oriented method was "practiced for hundreds of years but was only formalized in the 1960s."

26 See Robert Bruegmann, "The Pencil and the Electronic Sketchbook: Architectural Representation and the Computer," in Blau & Kaufman, op. cit., 139–55. The author notes that by 1989 CAD had not become a design tool in architectural offices. Its use was limited to writing specifications and generating working drawings.

27 Ivan Edward Sutherland, "Sketchpad, A Man-machine Graphical Communications System," PhD. Thesis, unpublished, Massachusetts Inst. of Technology, 1963.

28 On Negroponte and his program, URBAN5, see Bruegmann, 146–47. Also N. Negroponte, *The Architecture Machine* (Cambridge, MIT Press: 1970).

29 https://en.wikipedia.org/wiki/Computer-aided_design

30 William J. Mitchell, *Computer Aided Architectural Design* (New York, Petrocelli/Charter: 1977).

31 https://en.wikipedia.org/wiki/Autodesk

32 Larry S. Martin, Robert N. S. Chiang and Donald E. Bergeson, "The Computer as a Tool of Architectural Education" *Journal of Architectural Education* (1947–1974), Vol. 28, No. 1/2, Part 2., 1974 AIA/ACSA, Teacher's Seminar Pedagogical Catalog (1974), 71–72; and T. Willoughby, W. Paterson and G. Drummond "Computer-Aided Architectural Planning" *Operational Research Quarterly* (1970–1977), Vol. 21, No. 1 (Mar., 1970), 91–98

33 See Muriel Cooper, "Computers and Design" *Design Quarterly*, No. 142, (1989), 14–31.

34 For a comprehensive history and analysis, see Kalay, *Architecture's New Media*, 61–81. He observes that "second generation" CAD systems were often worse than those sold to early users in that they offered only drafting capability. Later, visualization and evaluation were added by major manufacturers such as AutoDesk and Microstation.

35 See Jeffrey K. Liker, Mitchell Fleischer & David Arnsdorf, "Fulfilling the Promises of CAD," *Sloan Management Review* 33:3 (Spring 1992): 74–86.

36 Alfredo Andia, "Reconstructing the Effects of Computers on Practice and Education During the Past Three Decades," *Journal of Architectural Education* (56:2) A-Ware \ Digital Technology and Architecture (Nov., 2002), 7–13.

37 Andia's study indicated that there was at least one faculty member in such a position for each 200 students in the majority of U.S. programs by that time. Ratios of student per computer station reached 10 to 1 as well. "Restructuring," 8.

38 Ibid., 8.

39 Yehuda Kalay, *Architecture's New Media: Principles, Theories and Methods of Computer-Aided Design* (MIT, 2004).

40 See Kalay, 71–74. The lack of success in "third generation" software can be attributed to "the reluctance of the architectural community to relinquish any part of the design process to the computer." 74.

41 Garry Stevens, a retired architecture professor, has produced a no holds barred critique of CAD in education on his website, "Dr. Garry's Terrace" for those who want an assessment by an experienced teacher and scientist. He is the author of the popular book *CADD Made Easy*. http://www.archsoc.com/kcas/whatevercad.html

42 Bryan Lawson, "CAD and Creativity: Does the Computer Really Help?" *Leonardo* 35:2 (2002): 327–31.

43 One critical study of student design protocols when using CAAD is Huda S. Salman and Richard Laing, "The impact of computer aided design programs on conceptual design in an educational context," *Design Studies* 35 (2014): 412–39. The authors note that students divert attention from critically evaluating their solutions to a problem because CAAD programs require the input of detail at early stages.

44 Lawson, 329.

45 Sennett, *The Craftsman*, 44.

46 Daniel Henry and Tom Furness, "Spatial Perception in Virtual Environments: Evaluation an Architectural Application," issued via IEEE, Univ. of Washington Human Interface Technology Laboratory, 1993.

47 Brian Lonsway, "The Mistaken Dimensionality of CAD," *Journal of Architectural Education* 56:2 (Nov. 2002): 23–25.

48 Rivka Oxman, "Theory and design in the first digital age," *Design Studies* 27 (2006): 229-265. The article relies entirely on new architectural theory and on the studies of "digerati" invested in the CAAD industry.

49 Michael Benedikt, *Cyberspace: first steps* (Cambridge, MIT Press: 1991); William J. Mitchell, *City of Bits* (Cambridge, MIT Press:1995); M. Christine Boyer *Cybercities: visual perception in the age of electronic communication* (New York: Princeton Univ. Press: 1996).

50 See Rabee M. Reffat, "DIGITAL ARCHITECTURE AND REFORMING THE BUILT ENVIRONMENT." *Journal of Architectural and Planning Research*, vol. 25, no. 2, 2008, 118–29.

51 See Huda Salman, Richard Laing and Anna Conniff, "The impact of computer aided architectural design programs on conceptual design in an educational context," *Design Studies* 35 (2014): 412–39. The authors insist, however, that the computer is still a good tool for use in conceptual design, ignoring the implications of their study.

52 Several studies point to problems with CAD: Bilda, Z., & Demirkan, H. (2003). "An insight on designers' sketching activities in traditional versus digital media." *Design Studies*, 24(1), 27–50; Hanna, R., & Barber, T. (2001). "An inquiry into computers in design: attitudes before—attitudes after. "*Design Studies*, 22(3), 255–81. Schon, D. A., & Wiggins, G. (1992). "Kinds of seeing and their functions in designing." *Design Studies*, 13(2), 135–56.Tang, H. (2001). *Exploring the roles of sketches and knowledge in the design process*. PhD thesis. The University of Sydney.

53 Eugenia Ellis, "Ceci Tuera Cela: Education of the Architect in Hyperspace," *Journal of Architectural Education* 51:1 (Sept. 1997): 37–45.

54 In September, 2016 the University of Notre Dame held a conference on "The Art of Architecture: Hand Drawing and Design." Many of the papers looked at both traditional and digital techniques, but most were critical of the latter.

55 Coosje van Bruggen, *Frank O. Gehry: Guggenheim Museum Bilbao* (New York, Guggenheim Museum: 1997): 36–37.

56 Ibid., 72.

57 Ibid., 79–93.

58 Harley Jessup, "Graphite and Pixels: Drawing at Pixar," in Treib, *Thinking/Drawing*, 170–181. Quote on 181.

59 Nicholas Carr, *The Glass Cage: How Our Computers Are Changing Us* (New York, W.W. Norton: 2014): 148.

[11]

DESIGN WITH EMBODIMENT

The story of design thinking that has been outlined so far suggests that humans began to think differently about fabricating buildings once modern capitalism and industrial processes demanded more efficient means of problem solving and assembly. During the nineteenth century not only manufacturing but also art production became commodified to an extent unknown in previous eras. I have suggested that there were broadly negative consequences resulting from this turn away from both craft based and drawing-based means of cognition, though modern building production could not have advanced without engineering, assembly lines and systems management. No current theorist would suggest that design be stripped of these methods and means of producing the infrastructure required to sustain modern cities and the built environment. However, as designers became increasingly alienated from the humans they served, the environment suffered, leading to the largely sterile and mono-cultural landscape that exists throughout the developed world, and increasingly confronts those in developing nations as well. In the final two chapters of this book, I present what I believe to be an alternate, hybrid means of understanding the nature of architectural design, hoping that the design professions may reconnect with traditions that have sustained *Homo sapiens* for thousands of years.

Embodied Design in the Historical Record

The foregoing narrative has established an arc of cognitive development in designers from prehistory until the present day that hints at material that may be studied in more detail as both science and archaeology expand. From prehistory until the Renaissance, builders used a combination of direct experience with materials, technics, and forms and limited external representations to create both large and small structures. Master builders such as the Romans, South Asians, and Chinese created monumental stone, wood, and brick buildings by following conventional patterns and perfecting technological and logistical methods, but without extensive depiction of all the elements in their buildings. Large labor pools allowed foremen and designers to work at the site, directing crews and following the progress of construction on a daily basis. There was little distance between the designer and the craftsman/builder. In this way design was embodied; those conceiving new buildings already had direct, physical knowledge of all the elements under their control. When information needed to be conveyed, it was relatively simple matter to record it in written or spoken language, or use physical models of the pieces under construction.

We saw evidence in the workshops of the middle ages a continuation of practices perfected by Greek and Roman builders over a period of almost a thousand years. From a cognitive point of view, schemas for creating buildings and other structures were relatively straightforward and linear. Once the steps were assimilated into the brains/bodies of those creating the structures, through patterns passed down by word of mouth or in simple texts, design and construction could proceed. Though sometimes complex, heuristic principles and procedures could be passed on and utilized by cultural groups with similar regional and ethnic traditions. Both primitive hunter-gatherer societies and more advanced agricultural societies evinced similar cognitive schemas. As Claude Levi-Strauss observed, some societies relied on "bricoleurs" or jacks-of-all-trades in order to solve problems and create new artifacts. Like shamans in healing, these special individuals were the master builders in primitive societies.[1] In more advanced societies groups of artisans carried this cultural knowledge from generation to generation, as we saw in Bavaria in Chapter 6.

The crucial break between master builder and architect occurred during the Renaissance, when humans began to imagine themselves as individuals with the agency to change their circumstances and examine their lives. A single mind with creative ideas could invent new forms and objects, as Leonardo did in his famous equestrian statue for the Sforza family. External representations aided the memory of designers in these endeavors, and Leonardo was one of the artists who changed the nature of conceptualization with his magnificent drawings. [138] Models also became more sophisticated as tools for studying new forms and structures. While those making these mnemonic aids were still involved in building the building or fabricating the artifact, their understanding remained embodied. Such knowledge remained with designers throughout the Renaissance and Baroque periods, spurring the creation of some of the most

138. Leonard da Vinci, drawing of rock formations, red and black chalk on paper. The artist demonstrates his intuitive, embodied knowledge about the physical world through drawing. No text or verbal description can convey the information shown here. (Courtesy, Royal Collections, Windsor Castle)

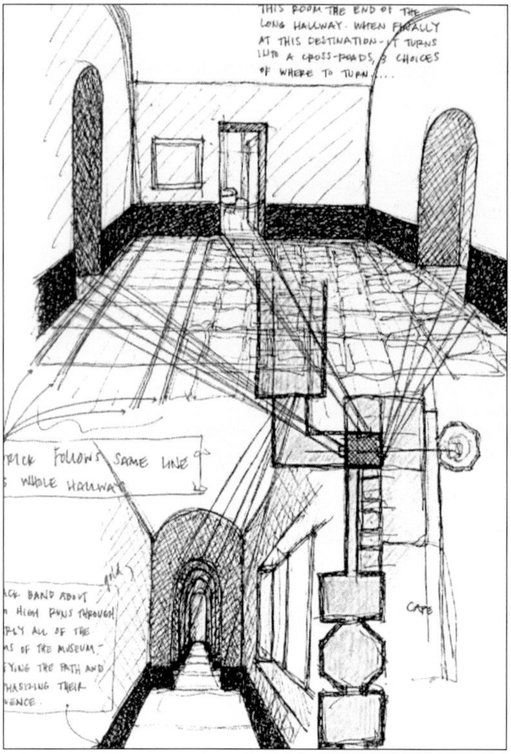

139. Allison Dutoit, professor: student sketch of an interior in Copenhagen. The drawing allows its author to form a "mental" map of the room, place, and location in the city. (Author collection)

extraordinary buildings ever built. We saw that the theater was an important medium in the generation of new forms and ideas for buildings.

Changes in western epistemology in the seventeenth century, the scientific revolution that we now accept as part of our worldview, began to erode the artist's embodied knowledge of the things in his purview. The gradual erosion of artisanal knowledge radically changed the way artifacts were created by distancing the designer from the building or object, requiring a communicative medium (drawing, model, specification) as a mediator between conceptualizer and constructor. Moreover, by the early twentieth century the status of the craftsman or artisan had greatly diminished, whereas that of the engineer and scientist had expanded. A technocratic ethos was now dominant in western societies, and would soon also pervade eastern ones. The artisan's direct experience with artifacts, their forms and materials, had no part in the engineer's practice. Even scientists relied on special lab technicians and machinists to construct their experimental equipment. Rational thought required no haptic connection to the physical world.

Though a few prescient architects such as Alvar Aalto, Louis I. Kahn and Carlo Scarpa maintained direct connections with craftsmen, and thus with embodied knowledge, by the end of the twentieth century the profession had lost the crucial element that connected design to making. Brain science, meanwhile, was examining the visual connection between the nervous system and the body, suggesting that rational cognition was a fallacy buried in Cartesian dualism. Pulling architecture in the opposite direction, CAD systems and digital reality software further distanced designers from the things they wished to see realized in physical form. Design drawing became divorced from *seeing* the intended result. The distinction between architecture and representation was now a blur, "divided" as Dalibor Veseley has contended. How, then, was a human-centered environment to be envisioned?

Sensory Engagement with the Natural and Built Environments

For the past century many architectural schools have stressed a form of learning steeped in theoretical literature and published photographic monographs trumpeting the work of individual architects. The architect was portrayed as a creative inventor striving for genius, just as artists sought to distinguish their work from run-of-the-mill producers of aesthetic objects. During the rise of the Modern Movement in Europe, many architects embraced a form of didactic, antagonistic dissemination of ideas modeled on the avant-garde art of the period. Publications such as *L'Esprit Nouveau*, the CIAM Charters, little Magazines, student journals, and various progressive design periodicals became the dominant platforms for advancing not only new design ideas but also the careers of international leaders in architecture. While the avant-garde had a clear role in changing the artist's position in society for its benefit, this form of media propaganda proved salubrious.[2]

Things changed with the rise of the information society and the World Wide Web. After the millennium, *Wired Magazine* proclaimed the triumph of Big Data. In the information age, dialectic movements ceased to push intellectual and scientific ideas forward but instead seemed to mire them in fruitless quarrels among aging gurus. Even critical theory and deconstructionism turned out to be movements spun by intellectuals with suspect credentials and mysterious pasts. Moreover, new brain science rendered a number of tenaciously held philosophical ideas obsolete, changing the entire landscape of cognitive studies and pushing many areas of inquiry to the periphery.

Today the architectural media has become a vehicle for empty self aggrandizement and ideological posturing on the part of ever more isolated groups of designers with a particular axe to grind against whatever "establishment" they choose to vilify. In the world of the Starchitect as media superhero, there is no straw man to beat with theoretical scythes. Provocateurs like Rem Koolhaas search in vain for enemies while becoming cynical tools of the global financial juggernaut. Much of what is written about architecture is redundant, uncritical, and of marginal help to practicing professionals. Current theory literature is so negative that one critic has described it as an attack on architecture itself.[3] The continuing education system, meanwhile, produces material mainly aimed at hocking miracle building technologies and materials. More words in longer, increasingly opaque, texts will not help designers fight global warming, income inequality, and housing shortages throughout the planet.

If the human sciences are to provide direction for artists and designers struggling to come to terms with information overload and instant messaging fatigue, it will be at the level of direct physical experience rather than abstract conceptualization. At the beginning of my career I taught courses in architectural theory, and touted the benefits of a literate professional elite, trained in the bastions of America's top institutions of higher learning. There is still a place for such an elite, and all architects should be conversant with the best history has had to offer, but I have become convinced that our society has come to point where theory must give way to pragmatic thinking. The physical environment is deteriorating at a rate that was inconceivable during my youth, and all the human sciences and professions must come to terms with the crisis that fossil fuel dependence has brought upon the earth.

Architects must find the most direct, thoughtful, and effective means of engaging with the environment—both human made and human influenced—now best seen as a biosphere. Now that neuroscience has led us to a new understanding of our intellectual capacities, we are poised to embrace the challenges we face with new tools. Embodied cognition offers human investigators a direct means of touching the biosphere more intimately as we act to bring it into balance. There is, as Aldo Leopold has said, a "right relationship" between humans and nature.[4] If our epistemology changes to reject the alienating legacy of post-Enlightenment positivism based upon materialistic views of the world, we have an opportunity to change course, veering toward pragmatic solutions to our crises.

Neuroscience has also given artists a boost by stressing the fundamental benefit of creative thinking for learning and intellectual development. Moreover, active engagement through sensory-motor apprehension of the natural world brings all humans closer to an understanding of the issues facing our species, just as Alexander von Humboldt predicted over two centuries ago. For instance, teaching students to draw with pencils, pens, and paints can directly benefit their abilities to understand urban environments and spaces. Rather than touring a historic city with a camera and a cell phone, students are better off carrying sketchbooks or easels with drawing paper. Allison Dutoit, a Berkeley trained architect and educator, has proved this by leading her students on sketching tours of Copenhagen, assigning them tasks based on cognitive mapping of the city. [139] Her "Mind Map" exercise brings students to a fuller, more embodied understanding of the places they inhabit. "Drawing here attempts to capture the qualities of the "haptic city" that allows us as designers to engage in the life of the city. The city comprises streets and buildings and monuments and parks and bus routes and bollards. Yet our personal city features small elements experienced in our daily routine: the signs in the window of the crazy person's house; the button we push for crossing a street; the clock on the corner building, the street tree that serves as timekeeper for the seasons."[5] Her exercises track well with the "pattern language" precepts of Christopher Alexander, one of Berkeley's most influential architect-educators.[6]

No text, no matter how brilliant and poetic, can impart this kind of perceptual/cognitive grasp of a particular place on the earth. Yes, texts are capable of enriching a sensory experience, adding layers to it, and humans will not give up using language to describe their multi-modal sensory images of places, but they do not flood the imagination so directly as these viseo-symbolic maps, created with pencils on paper. The fundamental connection, as we have seen, is between external symbolic memory in a drawing or other physical art practice, and internal or biological memory—the record of feelings, smells, and semantic fragments. As Walter Benjamin surmised in his wonderful literary memoirs of cities and places, mere words and sentences could not record his full experiences as well as say, films or photographs, despite their limitations. That is one of the reasons he argued so persuasively for forms of art that seemed to engage memory more directly than language.[7] Just as hikers, walking at paces that slow time and force over-stimulated adults to use all their senses to appreciate the scenery, find more meaning and enjoyment in their deliberate journeys, so students who take up pencils when visiting a great building like Fallingwater do not forget the impact of the building on their future work as architects. This has been proven many times over, beginning perhaps as early as the fifteenth century, when astonished Italian architects began to draw Roman ruins, measuring details, and comparing them to others in their notebooks. As Norman Crowe noted in the 1980s, external memory fields that he called "visual notes" are not merely quaint nods to a non-digital stone age.[8] They are necessary in all kinds of design processes.

Reviving Artisanal Knowledge

Artisans comprehend their work in terms of both design and production, and are never forced to choose between the two. Modern designers have generally viewed artisans as quaint practitioners of ancient skills that have little bearing on their lives and work. We saw in Chapter 9 that mimetic practices waned during the Romantic Era, and that today architects and engineers look at design as a process of assembling parts that may or may not be understood as a whole artifact prior to fabrication. As Larry Shiner has explained the negative consequences of losing workshops:

> If the image and status of the artist achieved new heights in the nineteenth century, that of the artisan or craftsperson sank even lower as many were forced to give up their independent workshops. Although the Industrial Revolution used to be talked of as if steam-powered machines swept away the old workshop system overnight, historians now describe a more gradual process, varying from region to region, industry to industry. Nevertheless, the disappearance of tens of thousands of small workshops was inexorable, even if it took most of the nineteenth century to complete.[9]

Yet the concept of embodiment reflects positively on the experience of artisans in connecting thinking with making, just as Aristotle praised the crafts of stone carving in equal measure to those of creating lyric or epic poetry. In addition, craftsmen and women work in collectives that provide an extra measure of social reinforcement for their work. Shiner notes that even in shoemaking workshops during the nineteenth century there was enough leisure time for intellectual discussions among colleagues, who respected each other as equals despite a master, journeyman, apprentice system. In today's Internet marketplace, small workshops are again becoming viable through sales sites such as Etsy and eBay. There is no question that a revival of artisanal knowledge can occur. The question is: at what scale?

Material culture scholars have probed the methods and artifacts of folk and vernacular artisans, finding theoretical threads to explain persistent practices and type forms. There has been little research related directly to brain science that might offer less complex, more human explanations for say, Navaho rug weaving, Mayan textile dying, and Iroquois canoe building. Yet no collector or anthropologist would denigrate the inventive genius of the artisans who made such wonderful objects of use and aesthetic delight. The best products of artisanal knowledge command high prices at auctions and antiques dealers throughout the world, and will continue to grow in value.

The prominent sociologist, Richard Sennet, has argued that non-artists study methods of craft design and making in order to find a sense of community and worth in themselves. Society, he finds, has clung pathologically to a technocratic ethos, impoverishing every form of production, even the computer industry. As he writes, "The ancient ideal of craftsmanship, celebrated in the [Homeric] hymn to Hephaestus, joined skill and community. Traces of that ancient idea are still evident today among [for example] Linux programmers. They seem an unusual, marginal group because of the troubled ways in which craftsmanship is now organized." Those troubles, examined in his book, *The Craftsman*, are a) attempts by corporations to motivate workers to work well in competitive environments; b) skill development that separates the head and the hand, such as when CAD software is used for design; and c) conflicting standards of quality in modern production, one based upon "correctness" and the other on "practical experience." He concludes his study with a plea for new pragmatism in philosophy, and more reverence for body-based knowledge, somewhat similar to the one advocated in this book.[10]

There is no country or culture on earth more dependent on ancient crafts and artisans than Japan, where even individuals with dying traditional skills are given special titles and protection by the government, through "intangible heritage" programs. The tea ceremony embodies many of these unique traditions, from ceramics and carpentry to fine furniture making. The Japanese aesthetic principles of "wabi sabi" and humility are alive in every teahouse, shoji screen, and ceramic cup. All claims to original or novel art are put aside in favor of culturally sanctioned criteria for beauty and appropriate simplicity.

More important, master artisans such as sword and knife makers, and the revered craftsmen who make blades for "kanna" (wooden planes) are viewed not as producers of tools but as artists in their own right. [140] The art of metalsmithing has not been relegated to a position of low status, as it has in most Western cultures. Invention occurs in the most rare and subtle manipulation of the material in hand—precious white and blue steel, soft iron, and hardwoods such as Japanese white oak. All of these masters are, in critical respects, mimicking the practices passed onto them by their teachers, all master artisans carrying centuries of knowledge in their bodies and brains. Trevor H.J. Marchand has documented the patterns of mimetic communication in master woodworkers in Britain, who pursue similar methods in their craft. His research indicates that teachers of such artisanal practices convey more via non-verbal, embodied demonstration to their students than in speech or writing.[11] I benefit from this when I watch popular woodworking videos and learn new skills such as dovetailing and cutting mortises with a chisel.

At the level of building production, our technology is generally far removed from artisanal activity, though many special materials and decorative art objects may be produced by hand. However, it is not far fetched to suggest that architects study the methods of hand craftsmanship and design in order to reconnect with not only the natural world but also with their most creative cognitive faculties.[12] One field that has used embodied practice to enhance creativity in performances is music.

Mimesis in Music

Architects are very like musicians in their patterns or habits of learning. I have spent my life performing as a serious amateur musician in a cappella singing groups, church ensembles, opera companies, and musical theater organizations. My voice teachers are among the most influential artists I have ever met, as

140. Japanese smoothing plane, or *kanna*. All the elements are hand made by traditional craftsmen, using embodied knowledge that is centuries old. These planes can produce slivers of wood measuring only a few microns in depth, and are far more precise than machine made planes.

they must use mimetic practices to teach their art. As we noted in the history of design practice, many if not most of the great innovators in the field were trained by masters in drawing or design, much as a renowned musician learns via great masters who give their time to teaching.[13] These learning experiences, we should remember, were originally passed along in vernacular folkways, such as those embodied in Delta blues and Cuban son-montuno music. [141]

Both traditions come from the people of West Africa who came to the New World as slaves, bringing call-and-response chants and drum rhythms with them. Both employ a structure based upon the simplest harmonic and formal elements—indeed, many tunes are virtually copies of older songs, with identical keys, chord progressions, riffs, and harmonies. Yet the mark of inventive artists like Arsenio Rodriquez, Muddy Waters, and B.B. King is unmistakable in this music. And so the human brain is drawn to these repetitive, familiar tunes as to no other music on earth. If there is a secret to playing call-and-response music, and developing new ideas in old material, it cannot be distilled through intellectual or rational means. The blues musician's personal "mojo" does not come from the cortex, but rather from the primordial brain-body connection. Just as artisans find ideas through hand work, the blues musician finds his innermost expression through his or her voice and chosen instrument, most often a guitar. Repetition, constant strumming, and humming drive the music forward, eventually allowing the new art to emerge. These musicians develop their styles mainly through collective performances among peers, as demonstrated in many of the historic recordings of Smithsonian Folkways and other roots music publishers. It would be hard to find a more primal example of embodied knowledge than the drums of Son-Montuno and the guitar of the Delta. Chano Pozo and Robert Johnson found their genius just this way, and gave their music to the wider culture through recordings.

141. Chano Pozo playing the conga drum. Pozo brought Cuban *son montuno* rhythms to jazz while working with Dizzie Gillespie. All West-African inspired music, including the blues and hip-hop, emerges from the brain-body's mimetic/kinetic faculties, deep below the cortex, and thus immune to "rational" analysis. Most "roots" or folk music is made this way.

A recent book by Arnie Cox presents a similar case for treating music as an embodied practice,[14] arguing for a "mimetic hypothesis" of performance and composition. Citing Merlin Donald's work as well as the pioneering research on language and metaphors of George Lakoff and Mark Johnson,[15] Cox develops a persuasive argument that, through imaginative play and early mimetic behavior in children, music functions as a vehicle for connecting humans with each other. When we hear performances of live music and dance, we use "mimetic motor imagery" to mirror the actions and sounds that we perceive coming from the performers. In this way the brain engages affective, motor, and critical faculties by internal forms of imitation, while also forming empathic bonds between the artist and the listener, just as a painter might with the beholder. Various forms of metaphorical representation, through pitch, rhythm, narrative, and tempo allow musicians to represent space and time abstractly, while also engaging senses other than hearing. His theories invite comparison with architecture and other visual arts. G. Gabrielle Starr has produced a study of the "sister arts" of poetry, painting, and music that makes similar points about neuroaesthetic judgments of beauty. Her emphasis on the importance of mental imagery in art jibes with concepts that we have discussed in this book.[16]

All music instructors spend far less time talking to students than teachers of other arts (the possible exception being dance). This is because the mimetic practice of playing or singing must be shown rather than described to students. When studying voice it is even more essential that the student listen and watch the teacher, as the internal muscles used to produce sound cannot readily be seen when the sound is produced, as it is in most instrumental music. Guitar students and pianists know that watching and imitating the hands of their teachers is critical to learning their instruments. Some teachers use eurythmic methods to get the entire body involved in feeling the pulse of a musical piece. These teaching methods are essentially embodied, and cannot be otherwise.

The contemporary countercultural push for "Biophilia" in the environment uses similar means of imparting natural rhythms and forms into the fabric of everyday life. [142] Though not currently embraced by many architects, this new form of organic design has found sympathetic adopters among corporate leaders, especially in Silicon Valley. Consultants such as Terrapin Bright Green, a New York based practice, show many companies how to use Biophilic Design in their management, manufacturing, and design processes. Christopher Alexander's disciples have also consulted for decades with institutions wanting a more natural and biologically based form of planning, though their projects are relatively small scale. Once these movements connect with resiliency groups, such as Europe's Transition Movement, their influence will undoubtedly spread rapidly.

142. Biophilic design: grass stairs and brick feel organic, and afford actions like climbing and sitting. Alvar Aalto, Saynatsalo Town Hall, Finland. (Author photo 2016)

Practicing Architecture as Making

When architects were among those who fabricated elements used in building, and touched building materials on a daily basis, they were grounded in the craft of making what they conceived. Modern building methods, technologies, and professional specialization long ago dissociated the designer from the tradesman as well as the construction site foreman. Architects such as Alvar Aalto, Carlo Scarpa, and Louis I. Kahn fought to remain connected to the building trades and to individual craftsmen, and their example is inspiring. Even so, it is not practical to assume that practicing architects in developed countries will be able to maintain such close relationships with tradespeople.

And yet, many professionals in the developed world complain of their isolation from handwork, creative thinking, and the satisfaction of seeing the fruit of their labors in physical form. They long to be architects.[17] Paradoxically, though architects have the opportunity to experience the realization of their creative ideas, most remain in the same isolated cubicles as their colleagues in business, law, and engineering. Though they may work in a "studio" among peers, they are far away from the true experience of studio artisans, who maintain their discipline by direct contact with hand-eye (sensory-motor) disciplines. Significantly, a number of Japanese architects refuse to jettison the cultural roots of their craft, and maintain various forms of traditional handwork in their offices. Renzo Piano describes his office as a "Building Workshop," and encourages his employees to participate in the creation of models and prototypes for elements that may (or may not) be used in future projects. Doubtless today, with one of the world's busiest offices, this practice has given way to more pragmatic approaches to design.[18]

There are, however, ways of re-connecting with sensory-motor skills in the day-to-day practice of design. As Harley Jessup has suggested, designers in large offices may attend optional, or better, required classes in drawing, sculpture, woodworking, and other craft disciplines. Instead of going off-site for continuing education, large companies could encourage employees to attend workshops provided in-house, with only nominal expense to the host enterprise. As suggested (but often ignored) by Intern Development Programs, both novice and experienced architects could participate in site inspections and visits to projects under construction. Even group travel excursions, perks not often granted to lower level employees, could be arranged when partners travel to check on material suppliers or specialist fabricators. All such educational/professional experiences produce better designers, and as such the costs are offset by gains in productivity and morale. [143, 144]

It may also prove effective to establish departments within larger offices that deal exclusively with trades and building material suppliers. Though today the Internet provides a gigantic catalog of products that are purportedly available at the click of a mouse button, neither clients nor architects are adequately apprised of the quality of these durable goods. With such a vast range of choices, it is often difficult for the professional to argue for a well-established supplier over a start-up company when a client sees prices and extravagant claims on a website, believing them to be accurate with little justification. Professionals who maintain regular contact with the best trades and crafts can serve as impartial judges for their peers.

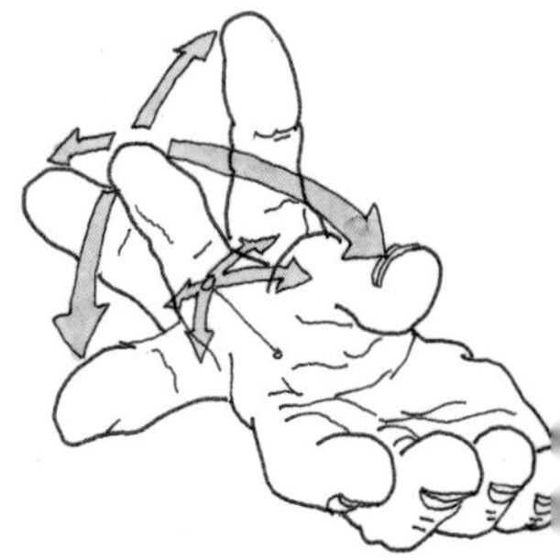

143. The human hand is an intelligent agent, connecting brain, body, and environment (Author, after Frank R. Wilson)

A model for such an atelier would be one that students encounter when they study traditional trades under Les Compagnons du Devoir, a French organization founded by Pierre de Coubertin (who also revived the modern Olympic Games). In these studios young apprentices and journeymen work alongside master artisans in blacksmithing, timber framing, stone cutting, and even baking. After intensive schooling during their middle and high school years, young people are placed in jobs that require them to produce goods that are sold or constructed under professional standards. Les Métailliers Champenois are involved in such training in both France and the U.S. In a similar architectural studio, young craft/designers would always be in touch with a master/head craftsman who could guide their endeavors on a daily basis.[19]

If special studios in large offices were tasked with such things as evaluating new materials such as structural glass, or devising new forms of sun shading with reference to proven models from the past, architects might work side by side with engineers, technical specialists, and machinists who have significant exposure to what is currently in the marketplace. Rather than specifying glass from a sample and a description in an Internet catalog, an architect might have recourse to direct information from his or her own office in the form of reports or even prototypes. Firms such as Kieran/Timberlake in Philadelphia purport to involve their experienced staff in such endeavors, but their work has so far yielded few compelling results.[20] Research that is conducted in a vacuum seldom produces breakthroughs.

The form of embodied creative thinking advocated here is not widely used in any architectural office today. One of the reasons for this stems from the conceptual bias that we noted in Chapter 10. When themes like dematerialization are pursued with compulsive abstraction and disregard for all previous experiments (c.f., Bruno Taut's "crystal" metaphor or the first all glass office tower by Mies van der Rohe), architects venture into the nether reaches of form-making. The restrictions that bound earlier architects—load capacities for steel beams and columns, and tensile strength of glass—are ignored in favor of free association. Whatever may be achieved through a computer model or plastic physical effigy is deemed worthy of converting into a building. Designers float freely in a creative space that is as vacuous as the virtual space of the computer screen.

144. Michelangelo, study, Lazarus raised from the dead. Red chalk on buff paper. Because the positions of the feet require difficult foreshortening, the artist draws them in detail at the bottom of the page (Courtesy, British Museum, London).

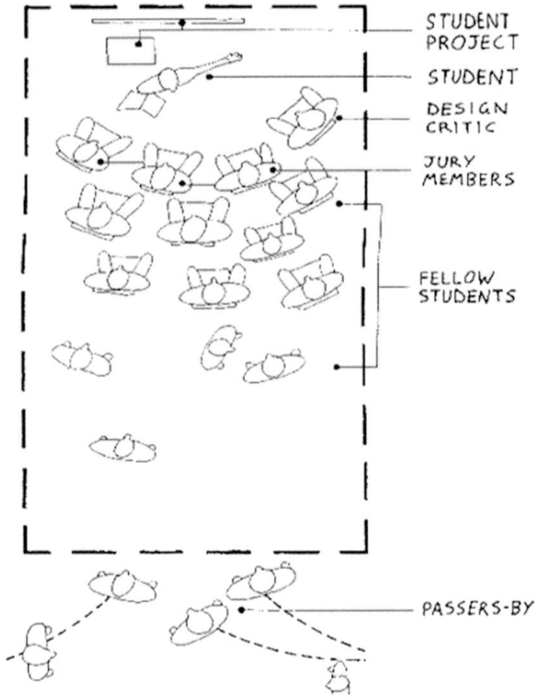

145. Top view showing a typical design jury at an architectural school. Though music schools also have "juries," the evaluators do not face the students or confront them with criticism following their performances. The hierarchical system still used in virtually all design schools has outworn its usefulness and should be abandoned. (Author collection)

Digital printing and other instant forms of physical modeling will not authenticate this condition of disembodied play in creative spaces for making, as the trendy catch words now describe many design endeavors in places like Brooklyn, Milan, or Silicon Valley. Now that pure form of virtually any kind is achievable in both virtual space and plastic simulacra, any architect is free to invent without regard for static limitations or even economic constraints. This is perhaps why so many inventive contemporary buildings by Starchitects cannot be built within established budgets. Frank Gehry, Santiago Calatrava, Norman Foster, Zaha Hadid, and Rafael Vinoly have all been sued by clients for ignoring budgets and other practical parameters that most architects respect out of necessity. The "blobitecture" of Greg Lynn and his followers is but one extreme example of such foolish flights of invention.When John Summerson penned his brilliant essay on "The Case for a Theory of Modern Architecture," he found only one salient characteristic of all mid-twentieth century avant-garde designs: the "authority of the program." Today, even that flimsy conceit has fallen away, and we are left with the pursuit of novel forms and concepts that make no pretense of satisfying the requirements of clients, society, and users.[21] The foolhardy quest for originality in design has pushed architects to the periphery, leaving technologists and engineer-designers in charge of a market sector that traditionally went to architects. As we have seen, their view of design as a linear assembly line process runs against all previous models of design thinking, and has produced nothing but soul-less, machine-like environments for over a century. More significantly, architects have lost credibility as designers by adopting the posture of their colleagues in engineering—today the term "architect" is more often used to describe a software designer than one involved with buildings.

Re-connecting with both hand skills and the materials and methods of building can strengthen the art and science of architecture. Formal invention within strict parameters has been the central task of architects for thousands of years, and abiding by the rules of design that exist in the human brain will continue to generate great ideas for buildings of all kinds. It remains to be seen whether the leaders of the profession today, or those half a century hence, will have the foresight to embrace neuroscience and jettison a century of misguided games with forms and concepts that have in no way enriched the art of architecture.

There is little question that our new understanding of how creative thinking works in the brain will change many human endeavors in the near future. If architecture is to keep pace with these changes, practicing architects and architectural educators will have to step up their adaptive learning. [145] One of the most important changes must come through the re-engagement with mimetic areas of the brain, leaving verbal/linguistic faculties in the background. It matters how architects frame their professional and creative endeavors—entrenched patterns of cognition must be jettisoned before a new paradigm can emerge.

Notes

1 Claude Levi Strauss, *La Pensée Sauvage* (Paris, Pion-Pocket, 1990, c. 1962).

2 See, for instance, Renato Poggioli, *Theory of the Avant Garde* (Cambridge, Harvard Univ. Press, 1969).

3 See Louis Martin, "Against Architecture," *Log*, No. 16 (Spring/Summer 2009), 153–67.

4 See Aldo Leopold, *A Sand County Almanac & other writing about ecology & conservation* (New York, Library of America: 2013).

5 Allison Dutoit, "Looking as Inquiry: Drawing the Implied Urban Realm" in Treib, *Thinking/Drawing*, 156.

6 Christopher Alexander, Sara Ishikawa, Murray Silverstein, et al., *A Pattern Language: Towns, Buildings, Construction* (New York, Oxford Univ. Press: 1977). This revolutionary book is the second in a trilogy that included *The Timeless Way of Building* (v. 1) and *The Oregon Experiment* (v. 3).

7 See, for instance, Benjamin, "Berlin Chronicle," in Demetz, Ed., *Reflections*, 3–60.

8 Norman Crowe and Paul Laseau, *Visual Notes for Architects and Designers*, 2nd Edition (New York, John Wiley & Sons: 2012).

9 Larry Shiner, *The Invention of Art* (Chicago, Univ. of Chicago Press: 2001): 206.

10 Richard Sennett, *The Craftsman* (New Haven, Yale Univ. Press: 2008): 51–52. His position does not explicitly take up the neuroscientific views that we have examined, but in many ways confirms them on a social level.

11 Trevor H.J. Marchand, "Embodied cognition and communication: studies with British fine woodworkers," *Journal of the Royal Archaeological Institute* (Vol. 16, no. 1, May 2010: 100–120.) This article also contains major literature on other forms of embodied learning, such as in sports.

12 Studio faculty routinely use craft practices in model making and other 3-D processes. For instance, see Luis Eduardo Boza, "(Un) Intended Discoveries Crafting the Design Process," *Journal of Architectural Education* 60:2 (Nov. 2006): 4–7.

13 Cross, I. (2001). Music, Cognition, Culture And Evolution. *Annals of the New York Academy of Sciences, 930*, 28–42; and Cross, I. (2008). Musicality and the human capacity for culture. *Musicae Scientiae, Special issue 2008*, 147–67.

14 Arnie Cox, *Music and Embodied Cognition: Listening, Moving, Feeling, and Thinking* (Bloomington, Univ. of Indiana Press: 2016).

15 George Lakoff and Mark Johnson, *The Metaphors We Live By* (Chicago, Univ. of Chicago Press: 1980).

16 G. Gabrielle Starr, *Feeling Beauty: The Neuroscience of Aesthetic Experience* (Cambridge, MIT Press: 2013).

17 A recent book documents the trend toward "making" among younger Americans. Matthew B. Crawford, *Shop Class as Soulcraft: An Inquiry into the Value of Work* (New York, Penguin: 2010).

18 For the most comprehensive treatment to date, see *PIANO: Renzo Piano Building Workshop, Complete Works, 1966–2014* (Berlin, Taschen: 2014).

19 One school in the U.S., The College of the Building Arts, in Charleston, uses these principles. However there is no comparable system in any country but France. See Bernard de Castéra, *Le Compagnonnage* (Paris, Presses Universitaires de France: 1988), for history of the movement and its principles.

20 The best presentation of this work is in Stephen Kieran and James Timberlake, *Prefabricating Architecture: how manufacturing technologies are poised to transform building construction* (New York, McGraw Hill: 2004).

21 John Summerson, "The Case for a Theory of Modern Architecture," in *The Unromantic Castle* (London, Thames & Hudson: 1990): 257–266. This famous essay, by an expert on classical architecture, debunked many myths associated with the Modern Movement.

[12]

TWELVE STEPS

In the previous chapter we discussed the phenomenal nature of architectural design in terms of the concepts introduced in this book: cognitive habits, embodiment, mimetic practices in the arts, and modes of conception. I argued that architectural practice, and education, must re-engage with these historically grounded ways of working if they are to serve humans and the environment in the coming decades. I did not prescribe any means by which this might take place. So, as a coda, let me offer the following twelve suggestions, and conclude with a plea for research into the neuroaesthetics of architecture.

1. Go back to the hand-drawn sketch as the fundamental medium and tool for creating architecture. Hand drawing is a must not only for practicing architects (many of whom have never stopped using them), but for the most technologically oriented students of design. The daily or weekly practice of drawing strengthens neural networks and engages cognitive faculties at many levels, just as playing scales and exercises keeps musicians sharp. Students should sketch from the very beginning of their education, and be required to use sketches to document and develop their visual ideas. [146]

2. Make every effort to see new places, and visit outstanding buildings, landscapes, and urban ensembles as often as possible. Rather than only using a camera or a cell phone, record experiences in a sketchbook, via images with words attached. Travel abroad as often as practicable, to extend the breadth

of direct experience with the spaces and places that have inspired predecessors through the centuries. Architecture programs can learn valuable lessons from Notre Dame, which requires all of its third year students to study in Rome. [147]

3. Make architectural history a requirement in all design programs, and avoid the pitfall of presenting only "modern architecture" in such courses. Beginning such courses in the eighteenth or nineteenth century is as inexcusable today as failing to present non-western buildings, cultures, and artifacts. It is also useful to teach young architects the proportional and grammatical systems associated with classical architecture, Chinese traditional architecture, and other non-Western systems, which may soon prove to be linked to schemata in the brain.

146. Winsor McCay, sketches done while waiting to go on stage for a vaudeville drawing exhibition. His wife has had a new baby, so he sketches images of his son's first days. A few words add detail to his memories. (Author collection)

147. William Walcott, watercolor sketch of the basilica of St. Peter's, Rome. Great artists become adept at portraying complex subjects with quick, bold strokes and dramatic contrasts in luminance. It is probable that they have attuned their performance to sensory, motor and cognitive faculties that are only now being studied experimentally. (Author collection)

4. Require students to engage with building users as soon as possible in their studio experiences. Many schools now offer practicums that allow students to build small projects for clients needing pro-bono services. The Yale Building Project is the most famous of these, but most schools now bring studios to the public and engage in solving real problems for those in need. Drawings are limited in their capacity to inform students about space and materiality. Only by seeing buildings constructed can they grasp the fundamental tectonic qualities of their art.

5. Maintain regular contact with tradespeople, artisans, and makers of building components and materials. The modern continuing education system in the U.S. provides many opportunities for such contact, but longer workshops with master artisans are always more meaningful than distance education or online courses. At the university level, maintain workshops in the architecture program or in allied engineering and material science departments, and make certain that all students are exposed to their projects and processes.

6. Balance linguistic and theoretical dialogue with purely visual and haptic means of presenting architectural ideas. The field has drifted too far into modes of thinking that pull the curtain over the visual brain, literally preventing designers from engaging with the essential qualities of environments, spaces, and type forms. Though all architects must understand the power of physical forms to convey symbolic meaning, focusing mainly on semantic knowledge has led us away from the core concerns of our discipline.

7. Integrate analog and digital tools in the design studio, as PIXAR does in its film production. No designer can afford to throw away powerful computers and software when they improve accuracy and productivity in so many areas. However, there is no reason why much basic design cannot be done with traditional drawings and models, nor is it necessary to ignore low tech and artisanal processes when considering how to make buildings more beautiful and durable.[1]

8. Avoid all forms of virtual representations of designed environments until the presentation stage of design. Research suggests that these "worlds" distort and misconstrue the nature of the spaces and forms they are meant to convey. As software engineers develop more accurate means of representing buildings and natural environments, adopt these means cautiously. Tablets have not yet matched the flexibility and expressive capacity of paints, inks, pencils, and charcoal on paper.

9. Employ digital drafting platforms as adjuncts to hand drawing, using these tools the same way architects used hardline drawings during the nineteenth and twentieth centuries—as means of accurately conveying construction plans, sections, elevations, and details. Avoid smart software, such as REVIT, that layers and attaches proprietary products to the design process before creative decisions can be objectively rendered.

10. In the academy, employ research professors in areas of bona fide applicability to the task of building. Demand evidence based, peer reviewed research products. Remove such non-professional faculty from the studio and employ only practicing architects at all levels of design instruction.

11. Teach basic drawing with constant reference to the most recent research in cognitive science and visual perception. Engage with direct experience as well as with mnemonic models and schemata, and use references from well-known visual artists and art works. [144]

12. Emphasize the collaborative nature of design as a discipline, and foster collaboration in the studio curriculum rather than emphasizing individual innovation as a criterion for architecture. Encourage students to work in teams, and provide opportunities to do so, despite the difficulty of assessing individual contributions during the grading process. [145]

These simple measures, even if implemented first on a small scale in design communities or individual schools, will provide positive proof of the benefits of a design practice based upon embodied cognition rather than purportedly rational or conceptual thinking. The results will be immediately evident in more beautiful, commodious, and healthy environments.

Design, Aesthetics, and the Brain

When an architect has an image in his or her head, he/she draws it to bring into focus, then modifies it, perhaps models it, renders it, and presents it before the process of making begins. Our pedagogical system has always in some way mimicked the process that professionals use to realize their designs. Today, with computers and 3-D printing, there is every reason to believe that studio education will closely track the activities of practicing architects. However, there are reasons to re-think some of the common assumptions about basic design principles and methods of evaluating student work now that neuroscience has provided new aesthetic schemas that appear to be universal.

The root of all aesthetic judgment resides in the brain. If recent discoveries are an indication of new ways of understanding how our sensory-motor system feeds the cortex, then most previous aesthetic theories will need to be recast in neuroscientific terms before they can be properly evaluated. Learning to see critically, and to recognize what may be innate principles of order and pleasing proportions should not be spoon fed by reading texts on architectural history, theory, or aesthetics. Since we experience sensory and cognitive ideas concurrently, it would make sense that our pedagogy reflects the way our brain works—teach architects how to *see* by drawing examples of the environment that are clear examples of lasting beauty and quality. The Brain Science Institute at Johns Hopkins Medical School has established the first center for the study of Neuroaesthetics, and efforts are being made to bring architects into the programs there.[2] There will be challenges, because most neuroscientists cannot access the vast literature and experience that architects have brought to their crafts through the ages.

Semir Zeki's laboratory has gone so far as to create a test for "beauty" using fMRI scans, though its sampling can hardly be called definitive since it used a small number of subjects and very limited stimuli. It isn't so far fetched to hope that new perceptual research may improve upon the laws put forth by Gestalt psychologists during the formative years of the Modern Movement, proving extraordinarily influential in the development of contemporary design education. Researchers have observed that Gestalt principles are still used in the majority of basic design courses throughout the world, as there has been no persuasive alternative since the 1940s.[3] Jon Lang has argued that what passes for an aesthetic philosophy in most schools of architecture is essentially an uncritical reading of

Gestalt perceptual experiments dating back to the 1930s, and their reification in the Bauhaus at Dessau under Walter Gropius.[4]

148. Author, in situ sketch of staircase, ricetto, Laurentian Library, Florence, 1977. I still remember the feeling of the space and its powerful forms as if I were touching them. (Author collection)

Donald Hoffman, among the scientists studying visual systems, has been culling aspects of current research to devise a set of laws that are quite similar to some Gestalt principles.[5] It may be too soon to extrapolate aesthetic principles from these basic experimental findings, but we might expect some prescriptive guidelines to emerge in the very near future. It would be a loss to architects if they were not considered when devising courses in basic design. There would be no necessity for throwing away Gestalt principles that conformed to the most current research—figure ground discernment, for example. At the same time, some discarded aesthetic laws, such as those of symmetry, arrangement, and proportional ratios, might be reintroduced into the canon as they are reified by perceptual experiments.

"Analysis of precedents" has fallen out of favor in many schools of architecture, though it was a required part of studio education during the last quarter of the twentieth century. If such a course were enhanced with sketching buildings and places in situ, students would immerse themselves in an aesthetic without

149. Author, in situ sketch of aedicular "window," ricetto, Laurentian Library, Florence, 1977. Though my drawing skills were still developing I managed to convey information about both form and light in a quick sketch. (Author collection)

any didactic prescriptions about whether one building was better than another. Notre Dame, which sends its juniors to Rome for an entire year, produces some of the most sought after young architects in the U.S., despite its tainted reputation for teaching "classical" architecture. It is likely that these young designers gain more by sketching some of the most beautiful buildings ever created than they do in any design studio.

We have seen that Filippo Juvarra, Alvar Aalto, and Louis Kahn each had a formative experience while sketching historic buildings in their environments. These experiences could not guarantee a successful career or convince clients to build their designs, but there is no question that these architects looked back on their early drawing and travel experiences as essential to their future practice. More important, they had standards by which to measure their progress as designers. Each also had a means of connecting their memories of sketching with their education, and melding the experiences after beginning their professional practice.

We have also noted in Chapter 7 that students exposed to the discipline of drawing from the cast before beginning al fresco sketching were better equipped to use the full range of their perceptual and cognitive faculties, if we connect past pedagogy with current neuroscience. Learning to draw is a fundamental skill that makes conception of visual designs possible—as Carlo Scarpa wisely observed. And, although drawing lines in virtual space on a computer screen is similar to making them on paper, the computer removes a crucial visual and tactile form of feedback from the process.[6]

I find it quite difficult to judge the scale, proportion, and quality of my designs when using AutoCAD to sketch directly, and I am not alone in this perception. Using a pencil or pen to make lines on a porous surface provides direct, tactile confirmation of what we understand in our heads. Further, since we really do not see conceptual images in any phenomenal theater of the mind, there is no point in pretending that a student can translate his or her design vision directly to a black virtual space in a machine. Will future computer visualization programs provide a better interface that will simulate paper drawing? Perhaps, but today there is no digital equivalent, even using a stylus and a tablet.

Analogies to composing music on a MIDI interface through a synthesizer may provide a useful reference for those who believe in a paperless future. The representation of a musical composition generated by sampling through a keyboard provides a reasonable "external memory field" for many composers working today—James Horner has produced numerous film scores using this method. Not every composer is comfortable with the technology, but younger artists in non-classical music have adopted it without much fanfare. Just as CADD generates finished looking designs quickly from similar inputs, fully orchestrated scores can be generated with far less tedious note-scribbling than Beethoven or Mozart required. That said, most music schools still require students to gain proficiency on a traditional instrument before moving on to electronic composition. Forming aesthetic judgments through embodied practice works the same way in music as it does in visual art.

So, it is too soon for architects to give up this most immediate, creative, and pleasurable form of representation perfected by Leonardo and his contemporaries almost 600 years ago, but familiar to cave dwellers thousands of years before that. External representation of the environment still arrives through our five senses, whether generated on a synthesizer or a violin, in a specific space with acoustic properties that are unique to it. Likewise, standing in the Pantheon with a sketchbook in hand, no architectural student will fail to find his or her senses flooded with aesthetic wonders. [148] Challenging himself/herself to record the experience by making a drawing, or writing a poem, or tracing a floor tile, locks those wonders in memory, to be retrieved whenever that representation is visited in the future, or he or she picks up a pencil to create his/her own Pantheon. He/she will draw in order to *see*, to conceive, and to make the vision tangible. And with each successive line he or she will refine the aesthetic, connecting neurons, and discovering a unique capacity to create. [149]

150. Carlo Scarpa working on the design of the Banca Popolare di Verona with his associate Arrigo Rudi and his engineer, Renato Scarazzai, c. 1978. (Courtesy, Marmomac Stoneworks, and George Dodds).

Notes

1 A studio at Catholic University in Washington, D.C. experimented with CNC machines and hand sketching, to good effect. See Luis Eduardo Boza, "[Un]Intended Discoveries Crafting the Design Process," *Journal of Architectural Education* 60: 2, 1:1 (Nov. 2006): 4–7.

2 The lab's website is at www.brainscienceinstitute.org. As one research project on pattern recognition states: "This work is motivated by the hypothesis that 3D aesthetic preferences are driven at least in part by neural mechanisms for 3D visual perception in the brain. Artists and architects must exploit these neural mechanisms to produce profound visual experiences. The specific hypothesis is that highly abstract, minimalist sculptures capture essential forms that resonate with neural mechanisms for 3D shape vision." The researchers are: Ed Connor, Xiaoqin Wang, Steven Hsiao, Steven Yantis, and Charles Limb.

3 Roy R. Behrens, "Art, Design and Gestalt Theory," *Leonardo* 31:4 (1998): 299–303, traces the influence of Gestalt psychology from its origins to the present day, noting its presence in most basic design courses.

4 Jon Lang, "Formal Aesthetics and Visual Perception: Questions Architects Ask," *Visual Arts Research* 10:1 (Spring 1984): 66–71. Lang's research tracks the persistence of Gestalt/Bauhaus aesthetics in current pedagogy. As he notes, "there has been important research on perception and on aesthetics during the past four decades; new hypotheses have been generated; new theories have been developed. But architects have been reluctant to deal with the implications of these observations." 68.

5 See Hoffman, *Visual Intelligence*, op. cit.

6 A particularly acute demonstration is found in a recently published paper. J. Park, Y. Jin, S. Ahn, S. Lee, "The Impact of Design Representation on Visual Perception: Comparing eye tracking data of architectural scenes between photography and line drawing," *Archives of Design Research* 32:1 (2019):5-29. Line drawing significantly improves memory, attention and acuity over photography.

APPENDIX

Anatomy of the Human Brain

Appendix 1. Exterior left side view of the human brain, showing major areas: the Frontal, Temporal, Parietal and Occipital lobes. (Author)

Appendix 2. Medial or cutaway section of the human brain. Areas in color are involved in memory storage, emotional regulation, spatial mapping, and communication with sensory cortices, among other critical functions. Both the Midbrain and the Cortex are shown here. (Author)

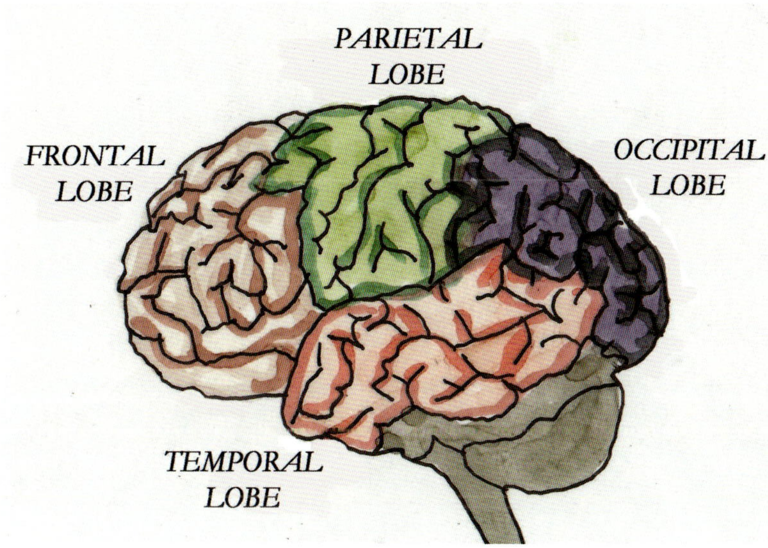

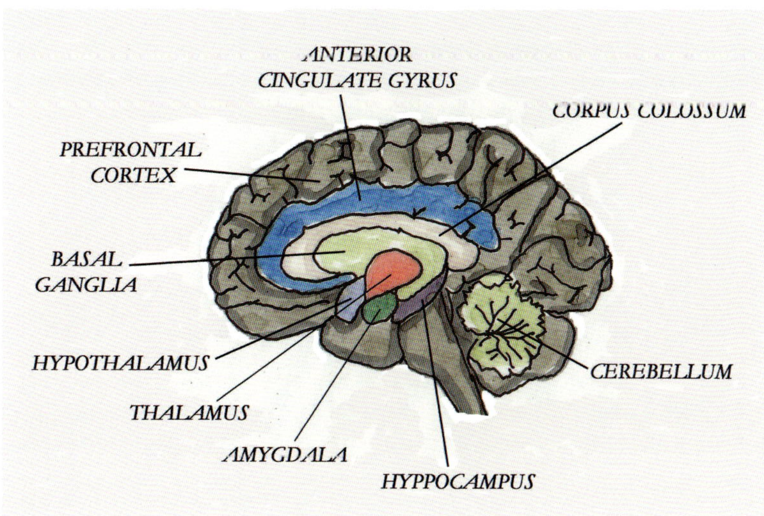

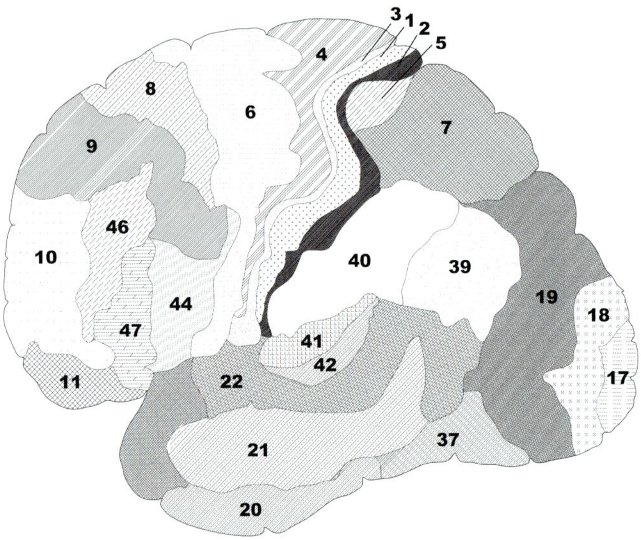

Appendix 3. In the early twentieth century the German brain scientist Korbinian Brodmann identified functional areas of the cortex using the *cytoarchitectonic* method. His 52 areas have been useful to later scientists. This left side view identifies those visible from that vantage point, including areas 1, 2, and 3. These form the primary somatosensory cortex, while area 4 is the motor cortex. Area 17 is the primary visual cortex; areas 41 and 42 form the auditory cortex. (Author)

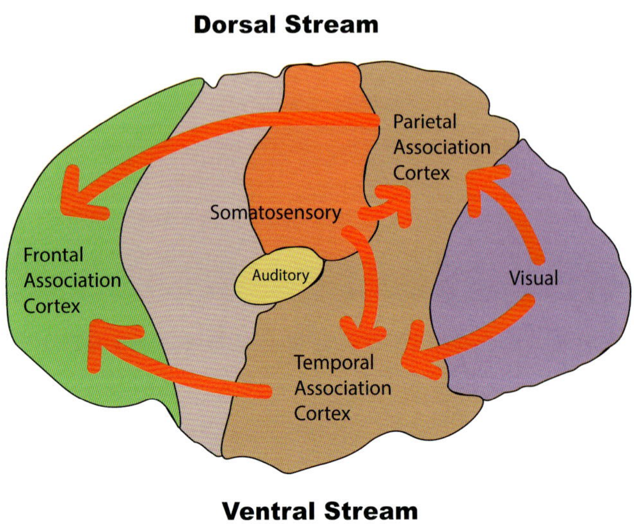

Appendix 4. The major *associational* areas of the cortex are located near the sensory cortices that they serve. When signals move from the sensory areas to nearby motor cortices they follow paths along the Dorsal or Ventral streams. Both memories and sensory-motor activities proceed along these pathways. (Author)

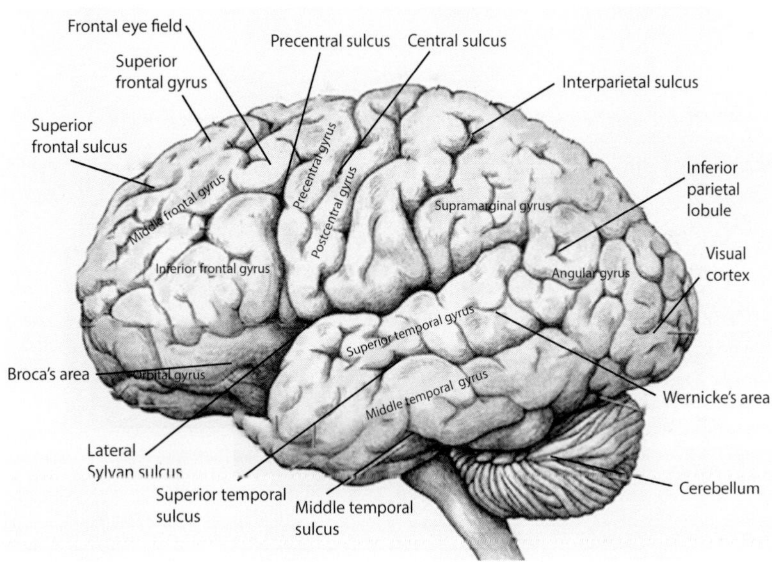

Appendix 5. The cerebral cortex, or outermost layer of the brain, is formed by folds and morphological high areas called *sulci* and *gyri*. A major sulcus divides the brain roughly in half (the Lateral Sylvan Sulcus), while smaller folds such as the three along the parietal lobe (the precentral sulcus, central sulcus, and interparietal sulcus), divide the cortex into major functional areas.

BIBLIOGRAPHY

Cognitive Science and Social Science

Alexiou, Katerina, T. Zamenopoulos and J. H. Johnson, "Exploring the neurological basis of design cognition using brain imaging: some preliminary results," *Design Studies* 30 (2009): 623–647.

Amsel, Ben D., Thomas P. Urbach and Marta Kutas, "Empirically Grounding Grounded Cognition: The Case of Color," *NeuroImage* 99 (2014): 149–156.

Antonis K, Daradoumis T, Papadakis S, Simos C. "Evaluationof the effectiveness of a web based learning design for adult computer science courses." IEEE *Transactions on Education* 54 no. 3 (2011).

Arbib, Michael A. *How the Brain Got Language: The Mirrior System Hypothesis.* New York, Oxford Univ. Press, 2012.

Arbib, M. A. *In Search of the Person: Philosophical Explorations in Cognitive Science.* Amherst, MA: University of Massachusetts Press (1985).

Arbib, M. A. Marr in Perspective. *Journal of Social and Biological Structures,* 10(4), 413-424 (1987).

Arbib, M. A. Neurons, Schemas, Persons and Society—Revisited. In G. Auletta, I. Colagè, & M. Jeannerod (Eds.), *Brains Top Down: Is Top-Down Causation Challenging Neuroscience?* (pp. 57-87). Singapore: World Scientific. (2013b).

Arbib, M. A. (Ed.) *Language, Music and the Brain, a Mysterious Relationship.* Strüngmann Forum Reports, vol. 10. Cambridge, MA: The MIT Press. (2013c).

Arbib, M. A., Gasser, B., & Barrès, V. Language is handy but is it embodied? *Neuropsychologia*, 55, 57-70. (2014).

Arnheim, Rudolph. *Visual Thinking.* University of California Press. Berkley, California, 1969.

Barrett L. F. and Moshe Bar, "See It with Feeling: Affective Predictions during Object Perception," *Philosophical Transactions: Biological Sciences,* Vol. 364, No. 1521, Predictions in the Brain: Using Our Past to Prepare for the Future (May 12, 2009), 1,325–334.

Biederman, Irv "Recognition by Components: A Theory of Human Image Understanding," *Psychological Review* 94:2 (1987): 115–47.

Bilda, Z., & Demirkan, H. "An insight on designers' sketching activities in traditional versus digital media." *Design Studies,* 24(1), 27–50. (2003).

Bourdieu, Pierre *Distinction: A Social Critique of the Judgment of Taste* Cambridge, Harvard Univ. Press: 1984.

Briscoe, Robert. "Depiction, Pictorial Experience, and Vision Science." *Philosophical Topics*, vol. 44, no. 2, 2016, 43–82.

Cerra, Pablo Pando, Jesús M. Suárez González, Bernardo Busto Parra, Diana Rodríguez Ortiz, and Pedro I. Álvarez Peñín. "Can Interactive Web-based CAD Tools Improve the Learning of Engineering Drawing? A Case Study" *Journal of Science Education and Technology* 23 no. 3 (2014).

Chatterjee, Anjan. *The Aesthetic Brain: How we evolved to desire beauty and enjoy art* New York, Oxford Univ. Press: 2014.

Clark, Andy. *Being There: Putting Brain, Body and World Together Again* Cambridge: MIT Press: 2001.

Csikszentmihalyi, Mihaly. *Creativity: the psychology of discovery and invention* New York: Harper Perennial: 1996.

Cohen, Gillian. "Visual Imagery in Thought." *New Literary History* 7:3 (Spring, 1976): 513-523.

Coluccia, Emanuele, Andrea Bosco, Maria A. Brandimonte, "The role of visuo-spatial working memory in map learning: new findings from a map drawing paradigm" *Psychological Research* 71 (2007): 359–372.

Cross, Nigel. *Design Thinking: Understanding How Designers Think and Work* London, Bloomsbury Academic: 2011.

Csikszentmihalyi, Mihaly. *Flow: the psychology of optimal experience* New York: Harper Perennial, 1991.

Damasio, Antonio. *Looking for Spinoza: Joy, Sorrow, and the Feeling Brain.* Harcourt, NewYork, 2003.

De Waal, Frans. *Mama's Last Hug: Animal Emotions and What They Tell Us About Ourselves.* New York, W.W. Norton, 2019.

Dennet, Daniel. *Consciousness Explained.* Boston, Little, Brown and Company, 1991.

Donald, Merlin. "Imitation and Mimesis," in Susan Hurley and Nick Chater, Eds., *Perspectives on Imitation: From Neuroscience to Social Science*, Vol. 2, Cambridge MIT Press: 2005: 283-300.

Donald, Merlin. *A Mind So Rare: the evolution of human consciousness.* New York, W.W. Norton, 2002.

Donald, Merlin. *The Making of the Modern Mind.* Cambridge, Harvard Univ. Press, 1991.

Eberhard, John Paul. *Brain Landscape: the coexistence of neuroscience and architecture.* New York, Oxford Univ. Press: 2008.

Farah, Martha J. 'Psychological evidence for a shared representational medium for mental images and percepts' *Journal of Experimental Psychology: General* 114 (1985): 91–103.

Farah, Martha J. *The Cognitive Neuroscience of Vision* Malden, MA, Blackwell, 2000.

Farrell, Michael P. *Collaborative Circles: Friendship Dynamics and Creative Work.* Chicago, Univ. of Chicago Press, 2001.

Fernandez, Myrna A., Jeffrey D. Wammes, Melissa E. Meade, "The Surprisingly Powerful Influence of Drawing on Memory" *Current Directions in Psychological Science* (Volume: 27 issue: 5): 302-308.

Fink, Andreas and Mathias Benedek, "The Creative Brain: brain correlates underlying the generation of original ideas," in Vartanian et. al, Eds., *Neuroscience of Creativity*: 207-231.

Fish, Jonathan and Stephen Scrivener. "Amplifying the Mind's Eye: Sketching and Visual Cognition," *Leonardo* 23:1 (1990): 117-126.

Friston, Karl J. "Modalities, Modes and Models in Functional Neuroimaging," *SCIENCE* New Series Vol. 326, No. 5951 (Oct. 16, 2009): 399-403.

Frith C. and K. Friston, "Studying brain function with neuroimaging," in Michael D. Rugg, Ed., *Cognitive Neuroscience* (London, Taylor & Francis: 1997): 169-192.

Freud, Sigmund. *Civilization and Its Discontents.* London, Hogarth, 1963.

Gabora, Lian and Apara Ranjan, "How Insight Emerges in a Distributed, Content-addressable Memory," in Vartanian, et. al., *Neuroscience of Creativity*: 19-43.

Gallese, V. "Being like me": self-other identity, mirror neurons and empathy. *In Perspectives on imitation: from cognitive neuroscience to social science,* vol. 1 (eds S. Hurley & N. Chater), pp.101–118. Cambridge, MA: MIT Press. 2006

Gallese, V. A neuroscientific grasp of concepts: from control to representation. *Phil. Trans. R. Soc.* B 358, 1231–1240. 2003.

Gallese, V. Intentional attunement: a neurophysiological perspective on social cognition and its disruption in autism. *Cogn. Brain Res.* 1079, 15–24. 2005

Gallese, V. The manifold nature of interpersonal relations: the quest for a common mechanism. *Phil. Trans. R. Soc.* B 358, 517–528. 2003

Gallese, V. The roots of empathy: the shared manifold hypothesis and the neural basis of intersubjectivity. *Psychopathology* 36, 171–180; 2003.

Gallese, V. Embodied simulation: from neurons to phenomenal experience. *Phenomenol. Cogn. Sci.* 4, 23–48. 2005.

Gallese, V. The "shared manifold" hypothesis: from mirror neurons to empathy. J. *Conscious. Stud.* 8, 33–50. 2001.

Gazzaniga, Michael *Nature's Mind: The Impact of Darwinian Selection on Thinking, Emotions, Sexuality, Language and Intelligence* (New York, Basic Books: 1992).

Gepshtein, Sergei and Michael Kubovy. "The emergence of visual objects in space-time." *PNAS* 97:14, July 5, 2000: 8186-8191.

Gero, J. and McNeil, T. "An approach to the analysis of design protocols," *Design Studies* 19:1 (1998): 21-61.

Gibson, James J. "The Ecological Approach to the Visual Perception of Pictures," *Leonardo,* Vol. 11, No. 3 (Summer, 1978), pp. 227-235.

Gibson, James J. *The Ecological Approach to Visual Perception.* Taylor & Francis Group, New York, 1986.

Gibson, James J. *The Senses Considered as Perceptual Systems.* Ithaca, Cornell Univ. Press, 1966.

Goel, Vinod "Creative Brains: Designing in the real world," *Frontiers in Human Neuroscience* (Volume 8, April 2014, Article 241): 1-14.

Goel, Vinod and Peter Pirolli, "The Structure of Design Problem Spaces," *Cognitive Science* 16 (1992): 395-429.

Goel, Vinod. *Sketches of Thought.* Cambridge, MIT Press: 1995.

Goker, Mehmet H. "Effects of experience during design problem solving," *Design Studies* 18 (1997): 405-426.

Goldschmidt, Gabriela. "The Backtalk of Self-Generated Sketches," *Design Issues* 19:1 (Winter, 2003): 72-88.

Graziano, Michael. *The Spaces Between Us: A Story of Neuroscience, Evolution and Nature.* New York, Oxford Univ. Press: 2018.

Groh, Jennifer. *Making Space: how the brain knows where things are.* Cambridge, Harvard Univ. Press: 2014.

Groleau, Carole, Christiane Demers, Mireille Lalancette and Marcos Barros. "From Hand Drawing to Computer Visuals: Confronting Situated and Institutionalized Practices in an Architecture Firm" *Organization Science* 23 no. 3 (2012).

Gur, Moshe. "Space reconstruction by primary visual cortex activity: a parallel, non-computational mechanism of object representation." *Cell Press.* Haifa, Israel, 2015.

Hanna, R., & Barber, T. "An inquiry into computers in design: attitudes before--attitudes after. "*Design Studies*, 22(3), 255-281. (2001).

Henderson, John M. and Andrew Hollingsworth, "High-Level Scene Perception," *Annual Review of Psychology* 50 (1999): 243-271.

Hubel David, *Eye, Brain and Vision,* Scientific American Library Vol. 22. New York, W. H. Freeman: 1988.

Huette S., Kello C.T, Rhodes T., Spivey M.J. Drawing from Memory: Hand-Eye Coordination at Multiple Scales. *PLoS ONE* 8(3): e58464. (2013)

Jain, Anshul. "Interactions between auditory and visual motion mechanisms and their role of attention: psychophysics and quantitative models." dissertation. New Brunswick: New Jersey, 2008.

Jonides John and Edward E. Smith, "The architecture of working memory," in Michael D. Rugg, Ed., *Cognitive Neuroscience* (Hove, UK, Taylor & Francis: 1997).

Joost, X. M. & Groh, J. "Multisensory guidance of orienting behavior." *Hearing Research.* 258 (2009) 106–112.

Jorgenson, Lyric, A. et al. "The BRAIN Initiative: developing technology to catalyse neuroscience discovery." *Phil. Trans. R Soc. B* 370. (2014).

Jung, Rex E. and Richard J. Haier, "Creativity and Intelligence: brain networks that link and differentiate the expression of genius," in Vartanian et. al., *Neuroscience of Creativity*, 232-254.

Kandel, Eric. *Reductionism in Art and Brain Science: bridging the two cultures.* New York: Columbia Univ. Press, 2016.

Kandel, Eric, et al, Editors. *Principles of Neural Science, 5th Edition.* New York, McGraw Hill, 2013.

Kubovy, Michael. *The Psychology of Perspective and Renaissance Art.* Cambridge, Cambridge Univ. Press, 1986.

Kurzban, Robert *Why Everyone (Else) Is a Hypocrite: Evolution and the Modular Mind* Princeton, Princeton University Press: 2010.

Lakoff, George and Mark Johnson. *Metaphors We Live By.* Chicago, Univ. of Chicago Press, 2003.

Lang, Jon, Ed. *Designing for Human Behavior: architecture and the behavioral sciences.* Stroudburg, Dowden, Hutchison & Ross, 1974.

Lawson, Brian. "Does the Computer Really Help?" *Leonardo* 35:3 (2002).

Leopold, Aldo. *A Sand County Almanac & other writing about ecology & conservation.* New York, Library of America: 2013.

Levi Strauss, Claude. *La Pensée Sauvage.* Paris, Pion-Pocket, 1990.

Levin, Helen. "A Response to William Mitchell on 'The Death of Drawing.'" *Leonardo* 35:1 (2002).

Livingstone, Margaret. *Vision and Art: the biology of seeing,* revised and expanded edition. New York, Abrams: 2014.

Mallgrave, Harry Francis. *The Architect's Brain. Wiley*-Blackwell, New York, 2007.

Marcus, Gary. *Guitar Zero.* Penguin, New York and London, 2012.

Marr, David, *Vision* (San Francisco, W.H. Freeman: 1982.

McCullough, Malcolm. *Digital Ground: Architecture, Pervasive Computing, and Environmental Knowing.* Cambridge, MIT Press, 2005.

Mitchell, William J. and Malcolm McCullough. *Digital Design Media.* New York, Van Nostrand Reinhold, 1995.

Mitchell, William J. *The Logic of Architecture: design, computation and cognition.* Cambridge, MIT Press, 1990.

Mitrovic, Branko. *Visuality for Architects.* University of Virginia Press. Charlottesville, Virginia, 2013.

Moser, Edvard I. et al., "Place Cells, Grid Cells, and the Brain's Spatial Representation System," *Annual Review of Neuroscience* 31 (2008): 69-89.

Negroponte, Nicholas. *The Architecture Machine.* Cambridge, MIT Press, 1970.

Nikolić, M., Škec, S., Martinec, T., Horvat, N. "The Role of Sketching Activities and Outcomes in Conceptual Design Phase," in *Proceedings of the 22nd International Conference on Engineering Design* (ICED19), Delft, The Netherlands, 5-8 August 2019.

O'Keefe, John "A Review of the Hippocampal Place Cells," *Progress in Neurobiology* 13, no. 4, (2009): 419-439.

Park, J., Jin, Y., Ahn, S., & Lee, S. "The Impact of Design Representation on Visual Perception: Comparing Eye-Tracking Data of Architectural Scenes Between Photography and Line Drawing." *Archives of Design Research,* 32 (1), 2019, 5-29.

Petroski, Henry *Invention by Design: how engineers get from thought to things* Cambridge, Harvard Univ. Press: 199).

Petroski, Henry *The Evolution of Useful Things* New York, Vintage: 1994.

Petroski, Henry. "Engineering: Drawing Bridges." *American Scientist,* vol. 87, no. 4, 1999, pp. 302–306.

Picon, Antoine. "Architecture and the Virtual: towards a new materiality." *PRAXIS: Journal of Writing Building,* no. 6, 2004, pp. 114–121.

Pierce, Charles Sanders *Collected Papers of Charles Sanders Pierce, Vols. 1-6,* 1931-1935. Cambridge, Harvard Univ. Press, 1958.

Pinker, Stephen. *How the Mind Works.* Norton, New York, 1999.

Renfrew, Colin. *Prehistory: The Making of the Human Mind.* London: Folio Society: 2013.

Rizzolatti, G. & Craighero, L. The mirror neuron system. *Annu. Rev. Neurosci.* 27, 169–192. 2004.

Rizzolatti, G., Fadiga, L., Fogassi, L. & Gallese, V. The space around us. Science 277, 190–191. 1997

Rizzolatti, G., Fadiga, L., Gallese, V. & Fogassi, L. Premotor cortex and the recognition of motor actions. *Cogn. Brain Res.* 3, 131–141. 1996

Rizzolatti, G., Fogassi, L. & Gallese, V. Neurophysiological mechanisms underlying the understanding and imitation of action. *Nat. Neurosci.* Rev. 2 (2001): 661–670.

Rugg, Michael D., Ed. *Cognitive Neuroscience.* Cambridge, MIT Press, 1977.

Sachs, Hanns. *The Creative Unconscious: studies in the psychoanalysis of art.* Cambridge, Sci-Art Publishers, 1951.

Salman, Huda S. and Richard Laing, "The impact of computer aided design programs on conceptual design in an educational context," *Design Studies* 35 (2014): 412-439.

Schon, D. A., & Wiggins, G. (1992). "Kinds of seeing and their functions in designing." *Design Studies*, 13(2), 135-156.

Shapiro, Lawrence A. *Embodied Cognition.* New York, Routledge, 2009.

Shuwairi, Sarah M., Rebecca Bainbridge, and Gregory L. Murphy, "Concept formation and categorization of complex, asymmetric and impossible figures," *Attention Perception Psychophysics* 76 (2014): 1789-1802.

Sussman Ann and Justin Hollander. *Cognitive architecture: Designing for how we respond to the built environment.* New York, Routledge, 2014.

Thompson, Evan *Mind in Life: Biology, Phenomenology and the Sciences of Mind.* Cambridge, Belkapp Harvard: 2007.

Tibbets, Paul E. "Where Does Cognition Occur? In One's Head or in One's Embodied/Extended Environment," *Quarterly Review of Biology*, 89:4 (December 2014): 359-368.

Tsow, David and Linda Beamer. "Verbalization and Visualization: A Need in Architecture Education." *Journal of Architectural Education* 40:2 (1987).

Tversky, Barbara. *Mind In Motion: how action shapes thought.* New York, Basic Books: 2019.

Vartanian, O., A. Bristol and J.C. Kaufman, Eds., *Neuroscience of Creativity.* Cambridge, MIT Press, 2013.

Verhoef, Bram-Ernst, Kaitlin S. Bohon, and Bevil R. Conway. "Functional Architecture for Disparity in Macaque Inferior Temporal Cortex and its Relationship to the Architecture for Faces, Color, Senses, and the Visual Field." *The Journal of Neuroscience* 35:17 (2015).

Visser, Willemien. "Two functions of analogical reasoning in design: a cognitive-psychology approach," *Design Studies* 17 (1996): 417-434.

Watkins, Trevor. "Building houses, framing concepts, constructing worlds," *Paléorient*, 30:1 (2004): 5-23.

Wilder, John. "The Influence of Complexity on the Detection of Contours." dissertation. Rutgers University, New Brunswick: New Jersey, 2013.

Wilford, John Noble. *The Mapmakers.* London: Folio Society, 2016.

Wilson, Frank R. *The Hand: how its use shapes the brain, language and human culture.* New York, Vintage Books, 1999.

Wright, Jessica. "Neural Mechanisms of Position Perception." dissertation. Rutgers, New Brunswick: New Jersey, 2014.

Wright, Robert *The Moral Animal, Why We Are the Way We Are: The New Science of Evolutionary Psychology.* New York: Pantheon, 1994.

Zeki, Semir. *A Vision of the Brain.* Blackwell, London, 1993.

Zeki, Semir. *Inner Vision: An Exploration of Art and the Brain.* Oxford, Oxford University Press, 1999.

Zeki, Semir. *Splendors and Miseries of the Brain.* London, Wiley-Blackwell, 2009.

Zeki, Semir "Neural Concept Formation in Art: Dante, Michelangelo, Wagner," *Journal of Consciousness Studies 9:3* (2002): 53-76.

Architecture, Art History, Humanities

Ackerman, James S. "Architectural Practice in the Italian Renaissance." *Journal of the Society of Architectural Historians.* 13, 1954.

Ackerman, James S. *Distance Points: Essays in Theory and Renaissance Art and Architecture.* Cambridge, MIT Press, 1991.

Ackerman, James S. *Origins, Imitation, Conventions.* MIT Press, Cambridge, 2002.

Ackerman, James S. *Origins, Invention, Revision: Studying the history of art and architecture.* New Haven, Yale Univ. Press, 2016.

Ackerman, James S. *Palladio.* Harmondsworth and Baltimore, Penguin, 1966.

Ackerman, James S. *The Architecture of Michelangelo.* With a catalogue of Michelangelo's works by James S. Ackerman and John Newman. Harmondsworth, England: PenguinBooks, 1970.

Ackerman, James S."Ars Sine Scientia Nihil Est: Gothic Theory of Architecture at the Cathedral of Milan," *The Art Bulletin* 31 (1949): 84-111.

Adams, Nicholas. *Baldassare Peruzzi: Architect to the Republic of Sienna.* Ph.D. dissertation. New York: New York University, 1977.

Agostinho, Francisco Santos. "Architecture as Drawing, Perception and Cognition: Background for an exercise of computer modeling applied to the Church of Sta. Maria de Belem-Lisboa" Faculdade de Arquitectura da Universidade Tecnica de Lisboa – Portugal.

Akin, Omer *A Psychology of Architectural Design* London, Pioneer: 1986.

Akin, Omer and Eleanor F. Weinel, Eds. *Representation and Architecture.* Silver Spring, MD: Information Dynamics, 1982.

Alexander, Christopher, Sara Ishikawa, Murray Silverstein, et. al., *A Pattern Language: Towns, Buildings, Construction.* New York, Oxford Univ. Press, 1977.

Alberti, Leon Battista *On the Art of Building in Ten Books,* tans. Joseph Rykwert, Neil Leach and Robert Tavernor Cambridge, MIT Press, 1991.

Alberti, Leon Battista. *On Painting.* Translated with introduction and noted by John R. Spencer. New Haven, Yale University Press, 1956.

Ambroziak, Brian M. *Michael Graves: Images of a Grand Tour* Princeton, Princeton Architectural Press: 2005.

Ames-Lewis, Francis. "Drapery 'Pattern' – Drawings in Ghirlandaio's Workshop and Ghirlandaio's Early Apprenticeship," *Art Bulletin* 63 (March 1981).

Ames-Lewis, Francis. *Drawing in Renaissance Italy.* New Haven, Yale Univ Press, 1981.

Andray, M. "Les Architectes des Cathedrals," *Les Cahiers techniques de l'art,*" 1 (1947).

Anzivino, C.L. *Jacopo Barozzi il Vignola e gli architetti Italiani del Cinquecento.* Repertorio bibliografico. Bologna: Tamari, 1974.

Argan G.C. "Il 'libro extraordinario' di Sebastiano Serlio," *Festschrift fur Walter Friedlaender* (1933).

Argan, Giulio Carlo. *"Filippo Juvarra."* *Messagero,* 13, December 1963.

Argan, Giulio Carlo. "The Architecture of Brunelleschi and the Origins of Perspective Theory in the Fifteenth Century," *Journal of the Warburg and Courtauld Institutes* (1946).

Arnheim, R. "Brunelleschi's Peepshow," *Zeitschrift fur Konstgeschichte* 41 (1978).

Ashby, Thomas. "Sixteenth Century Drawings of Roman Buildings Attributed to Andreas Corner." *Papers of the British School at Rome* 2 (1904).

Asimow, Morris. *Introduction to Design.* Prentice Hall, Englewood Cliffs, NJ, 1962.

Aubert, M. "Building Yards and Master Builders in the Middle Ages," *Liturgical Arts,* 19 (1951).

Bagrow, Leo. *History of Cartography,* revised and enlarged by R. A. Shelton, Reprint of 1985 edition, New Brunswick, Transaction Publishers: 2010.

Baird, George and Charles Jencks, Eds., *Meaning in Architecture.* New York, Braziller: 1969.

Baker, Paul R. *Richard Morris Hunt.* MIT Press: Cambridge, 1985.

Balardi, Paulo. *Why Architects Still Draw* Cambridge: MIT Press, 2015.

Baldwin, Neil *Edison: Intenting the Century* New York, Hyperion: 1995.

Bambach, Carmen, Ed. *Leonardo da Vinci: Master Draftsman.* Metropolitan Museum of Art and Yale Univ. Press, New York, 2003.

Bambach, Carmen. *Michelangelo: Divine Draftsman and Designer.* New York, Metropolitan Museum of Art, 2017.

Banham, Reyner *Theory and Design in the First Machine Age.* New York, Praeger,1967.

Barbieri, F. and L. Patti. *Tutta l'architettura di Michelangelo.* Milan: 1962.

Barocchi, Paola. *Michelangelo e la suo scuola.* Vols. 1-2, *I disegni della Casa Buonarroti e degli Uffizi.* Florence: 1962. Vol. 3, *I disegni dell' Archivio Buonarroti.* Florence: 1964.

Baroni, C. "Elementi stilistici fiorentini negli studi Vinciani." *Atti del 10 Congresso nazionale di storia dell'architettura.* Florence: 1936.

Baxandall, Michael. *Painting and Experience in Fifteenth Century Italy.* 2nd Edition. Oxford Univ. Press, Oxford, 1988.

Beaujouan, Guy. "Reflexions sur les rapports entre theorie et practique au moyen age," in *The Cultural Context of Medieval Learning,* ed. by J.E. Murdoch and E.D. Sylla. Dordrecht: D. Reidel, 1975.

Behrens, Roy R. "Art, Design and Gestalt Theory," *Leonardo* 31:4 (1998): 299-303.

Beltrame, R. "Gli esperimenti prospettici del Brunelleschi," *Rendiconi della sedute dell'Accademia Nazionale dei Lincei,* Ser. 28. (1973, 1974).

Beltrami, Luca. "Michelangelo e la facciata di San Lorenzo in Firenze." *Rassegna d'Arte.* 1 (1901).

Benedikt, Michael. *Cyberspace: first steps.* Cambridge, MIT Press: 1991.

Bentivoglio, Enzo. "Disegni nel 'libro' di Giuliano da Sangallo collegabelli a progetti per il S. Giovanni dei Fiorentini a Roma," *Mitteilungen des Kunsthistorisches Institus in Florenz* 19 (1975).

Benvolo, Leonardo. *The Architecture of the Renaissance.* 2 vols. London: Routledge and Kegan Paul, 1978.

Bernheimer, Richard, "Gothic Survival and Revival in Bologna." *Art Bulletin.* 36 (1954).

Bertol, Daniela. "'Architecture of Images': An Investigation of Architectural Representation and the Visual Perception of Three-Dimensional Space." *Leonardo* 29:2 (1996).

Bertram Grosvenor Goodhue: architect and master of many arts New York, Press of the American Institute of Architects: 1925.

Betts, Richard J. "On the Chronology of Francesco di Giorgio's Treatises: New Evidence from an Unpublished Manuscript," *Journal of the Society of Architectural Historians* 36 (1977).

Betts, Richard J. "The Architectural Theories of Francesco di Giorgio." 2 vols. Ph.D. Dissertation. Princeton, N.J.: Princeton University, 1971.

Bilda, Zafer, John S. Gero, and Terry Purcell, "To sketch or not to sketch? That is the question," *Design Studies* 27 (2006): 587-613.

Bland, David. *A History of Book Illustration: The Illuminated Manuscript and Printed Book.* 2d rev. ed. Berkeley: University of California Press, 1969.

Blanning, Tim *The Romantic Revolution: A History* New York, Modern Library: 2012.

Blau, Eve & Edward Kaufman. *Architecture and Its Image: Four Centuries of Architectural Representation.* Cambridge, MIT Press, 1989.

Bloomer, Kent and Charles W. Moore, *Body, Memory and Architecture.* New Haven, Yale Univ. Press, 1977.

Blunt, Anthony. *Philibert de l'Orme.* London, Zwemmer, 1958.

Bork, Robert O. *The geometry of creation: architectural drawing and the dynamics of Gothic design.* Burlington, VT, Ashgate: 2011.

Boscarino, Salvatore. "Filippo Juvarra tra Scienza e Technica." *Restauro,* 7 (1978).

Boscarino, Salvatore. *Juvarra architetto.* Rome, 1973.

Bork, Robert O. *The Geometry of Creation: architectural drawing and the dynamics of Gothic design.* Farnham, UK, Ashgate: 2011.

Bottari, Stefano. "Il Quattro Libri dell'Architettura," *Centro di Studi di Architettura Andrea Palladio Bolletino,* 5 (1963).

Boza, Luis Eduardo "(Un)Intended Discoveries Crafting the Design Process," *Journal of Architectural Education* 60:2 (Nov. 2006): 4-7.

Brailov, Aleksandr Yurievich. *Engineering Graphics: Theoretical Foundations of Engineering Geometry for Design.* Springer, 2016.

Branner, Robert. "A Note on Gothic Architects and Scholars," *Burlington Magazine* (1957).

Branner, Robert. "Drawings from a Thirteenth-Century Architect's Shop: The Reims Palimpsest," *Journal of the Society of Architectural Historians,* 27 (1958).

Branner, Robert. "Frabrica opus and the Dating of Medieval Monuments," *Gesta,* 15 (1976).

Branner, Robert. "Gothic Architecture," *Journal of the Society of Architectural Historians,* 32 (1973).

Branner, Robert. "Three Problems from the Villard de Honnecourt Manuscript," *Art Bulletin,* 39 (1957).

Branner, Robert. "Villard de Honnecourt, Reims, and the Origin of Gothic Architectural Drawing," *Gazette des Beaux Arts,* 61 (1963).

Branner, Robert. *St. Louis and the Court Style in Gothic Architecture.* London, 1965.

Brinton, S. *Francesco di Giorgio Martini of Siena.* 2 vols. London: Besan, 1934-35.

Brothers, Cammy. *Michelangelo. Drawing, and the Invention of Architecture.* New Haven, Yale Univ. Press, 2008.

Brown, Horatio F. *The Venetian Printing Press.* London, 1891.

Bucher, Francois, "The Dresden Sketchbook of Vault Projection," *Proceedings of the XXII Congress of Art History* (Budapest, 1972).

Bucher, Francois. "Design in Gothic Architecture: A Preliminary Assessment," *Journal of the Society of Architectural Historians,* 25 (1968).

Bucher, Francois. "Medieval Architecture Design Methods, 800-1500," *Gesta,* Vol. 11, No. 2 (1972).

Bucher, Francois. "Micro-Architecture as the "Idea" of Gothic Theory and Style," *Gesta,* 15 (1976).

Bucher, Francois. "Rediscovered Tracing by Villard de Honnecourt," *Art Bulletin,* 59 (September 1977).

Burger, Fritz. "Uber zwei Architekturzeichnungen Michelangelos in der Casa Buonarroti in Florenz." *Repertorium fur Kunstwissenschaft.* 31, 1908..

Burns, Howard. "A Peruzzi Drawing in Ferrara." *Mitteilungen des Kunsthistorian Institutes in Florenz.* 12, December 1966,

Burns, Howard. "Progetti di Francesco di Giorgio per i conventi di San Bernardino e Santa Chiara di Urbino," *in Congresso internazionale di studi bramanteschi,* Milan, Urbino, Roma (Rome: De Luca, 1974).

Burns, Howard. "Quattrocento Architecture and the Antique: Some Problems," *Classical Influences on European Culture, A.D. 500-1500 A.D.,* edited by R. R. Bolgar. Cambridge: University Press, 1971.

Caraffa, Costanza. "II disegni di Filippo Juvarra." *Disegno di architettura,* 4:8 (1993).

Carboneri, Nino, ed. *Antologia artistica del Monregalese.* Istituto Bancario San Paolo di Torino. Turin, 1971.

Carboneri, Nino. *Andrea Pozzo architetto (1642-1709).* Trento, 1961.

Carboneri, Nino. *La Reale Chiesa di Superga di Filippo Juvarra.* Turin, 1979.

Carlhian, Jean Paul. *Americans in Paris: foundations of America's architectural gilded age: architecture students at the Ecole des Beaux Arts, 1846-1946.* Rizzoli, New York, 2014.

Carpo, Mario. *Architecture in the Age of Printing: orality, writing, typography and printed images in the history of architectural theory.* Cambridge, MIT Press, 2001.

Carpo, Mario. "Drawing with Numbers: Geometry and Numeracy in Early Modern Architectural Design," *Journal of the Society of Architectural Historians,* 62:4 (Dec., 2003): 448-469.

Carr, Nicholas *The Shallows: What the Internet is doing to our brains.* New York, W.W. Norton: 2011.

Charrington, Harry and Vezio Nava, Eds. *Alvar Aalto: The Mark of the Hand.* Helsinki, Rakennustieto Publishing: 2011.

Chauvet, Jean-Marie et al., *Chauvet Cave: the discovery of the world's oldest paintings* London, Thames & Hudson: 2001.

Ching, Frank. *Architectural Graphics.* Van Nostrand Reinhold, New York, 1985.

Clarke, Matthew F. "Jean Labatut and Éducation à Pied D'œuvre: The Princeton Architectural Laboratory." *The Princeton University Library Chronicle,* 74:2 (2013):178–209.

Clute, Eugene. *Drafting Room Practice*. Pencil Points Press, New York. 1928.

Coffin, David R. "Some Architectual Drawings of Giovan Battista Aleotti." *Journal of the Society of Architectural Historians*. 1962.

Cohen, Jeffrey A. et al. *Graced Places: The Architecture of Wilson Eyre,* University of Pennsylvania, electronic exhibition catalogue (unpublished), 1994.

Collins, Peter. *Architectural Judgment*. Faber & Faber, London, 1971.

Conant, Kenneth. "The After-Life of Vitruvius in the Middle Ages," *Journal of the Society of Architectural Historians*, 37 (1968).

Connally, Earnest Allen. *Printed Books on Architecture 1485-1805*. Urbana, Illinois: University of Illinois Library, 1960.

Coolidge, John Phillips. "Vignola and the Little Domes of St. Peter's." *Marsyas*. 2, 1942, pp. 63-123

Coolidge, John Phillips. "Vignola's Character and Achievement." *Journal of the Society of Architectural Historians*. 9, December 1950.

Cooper, James G. "Michelangelo's Laurentian Library: Drawings and Design Process." *Architectural History*, vol. 54, (2011): 49-90.

Cooper, John M. *Plato: The Complete Works*. Hackett Publishing Company, Indianapolis andCambridge, 1997.

Coulton, J.J. *Ancient Greek Architects at Work: Problems of Structure and Design*. New York, Cornell Univ. Press, Ithaca, 1977.

Cox, Arnie. *Music and Embodied Cognition: Listening, Moving, Feeling, and Thinking* Bloomington, Univ. of Indiana Press: 2016.

Crawford, Matthew B. *Shop Class as Soulcraft: An Inquiry into the Value of Work*. New York, Penguin: 2010.

Cret, Paul P. "The Ecole des Beaux-Arts and Architectural Education." *Journal of the Society of Architectural Historians* 1:2 (1941).

Crowe, Norman A. and Steven W. Hurtt, "Visual Notes and the Acquisition of Architectural Knowledge." *Journal of Architectural Education* 39:3 (1986).

Crowe, Norman and Paul Laseau. *Visual Notes for Architects and Designers*, 2nd Edition, New York, John Wiley & Sons: 2012.

dal Co, Francesco et. al., Eds. *Carlo Scarpa: Opera complete*. Milan, Electa Editrice, 1984.

Dal Poggetto, Paolo. *I disegni murali di Michelangelo e della sua scuola nella Sagrestia Nuova di San Lorenzo*. Firenze: Centre Di, 1978.

Damasio, Antonio. *Descartes' Error: emotion, reason and the human brain* New York, Putnam: 1994.

de Botton, Alain. *The Architecture of Happiness*. London, Penguin, 2006.

DeLong David & David Brownlee, Eds. *Out of The Ordinary*. New Haven, Yale Univ. Press, 2001.

Dennett, Daniel *From Bacteria to Bach and Back: The Evolution of Minds*. New York, Norton, 2017.

Dernie, David. *Architectural Drawing*. Lawrence King, London, 2010.

Disegni di Michelangelo: 103 disegni in facsimile. Foreward by M. Salmi, introd. by Charles de Tolnay, comments by P. Barocchi. Milan: 1964.

Dimendberg, Edward. *Diller Scofido + Renfro: Architecture After Images*. Chicago & London, Univ. of Chicago Press: 2013.

Dorst, Kees. "The core of 'design thinking' and its application." *Design Studies* 32 (2011): 521-532.

Drexler, Arthur, Ed. *The Architecture of the Ecole des Beaux Arts*. MIT Press, Cambridge, 1977.

Duhigg, Charles. *The Power of Habit*. Random House, New York, 2014.

Dussler, Luipold. *Die Zeichnungen des Michelangelo*. Berlin: 1959.

Dutton, Denis *The Art Instinct: Beauty, pleasure and human evolution* London, Bloomsbury: 2009.

Edgerton, Samuel Y., Jr. "Brunelleschi's First Perspective Picture," *Arte Lombarda*, 18, 1973.

Edgerton, Samuel Y., Jr. *The Renaissance Rediscovery of Linear Perspective*. New York: Basic Books, 1975.

Edminster, C. Franklin, *Architectural drawing*, 3d ed., enl. Brooklyn, The Author, 1905.

Egnazio Danti Due Regole della Prospettiva Practica, 1583.

Elderfield, John. "Drawing as Suspended Narrative." *Leonardo*, vol. 4, no. 1, (1971): 13–22.

Elliot, T.S. *On Poetry and Poets*. Farrar Strauss & Giroux, New York, 2009.

Ellis, Eugenia. "Ceci Tuera Cela: Education of the Architect in Hyperspace," *Journal of Architectural Education* 51:1 (1997): 37-45.

Elwood, Franklin George. *Architectural drawing plates* (Peoria, Ill., Manual arts Press, 1919).

Essenwein, A. "Hans Schmuttermayer's Fialenbuchlein," *Anzieger fur Kunde der deutschen Vorzeit,* n.f. 28 (1881).

Ettlinger, Leopold D. "The Emergence of the Italian Architect during the Fifteenth Century," in Kostof, *The Architect: Chapters in the History of the Profession* (New York, Oxford Univ. Press: 1977): 96-123.

Fabriczy, Cornel von. "Chronologischer prospect der Lebensdaten und Werkes Giuliano's daSangallo." *Beiheft zum Jahrbuch der Kgl. Preus*. Kunstsammlungen. 1902.

Fabriczy, Cornel von. *Die Handzeichnungen des Giuliano da Sangallo*. Stuttgart: 1902.

Fasolo, Vincenzo. "Disegni architettonici di Michelangelo." *Architecture e arti decorative*. 6, 1926-7.

Fergusson, Frances D. "Leonardo da Vinci and the Tiburio of Milan Cathedral," *Architectura,* 7. (1977), 175-192.

Ferri, P.N. and E. Jacobsen, "Disegni conosciuti di Michelangelo." *Miscellanea d'arte*. 1, 1903.

Ferri, P.N. and E. Jacobsen. "Nuovi disegni conosciuti di Michelangelo." *Rivsita d'Arte*. 2, 1904.

Filarete, Antonio Averlino, known as. *Filarete's Treatise on Architecture* ed. and trans. by J.R. Spencer, 2 vols. New Haven: Yale University Press, 1965.

Fiore, Francesco Paolo. *Citta e macchine del '400 nei disegni di Francesco di Giorgio Martini*. Firenze: L.S. Olschki, 1978.

Fitchen, John. *Building Construction Before Mechanization*. Cambridge, MIT Press: 1986.

Flieg, Karl, "The Architect in his Workshop," *Alvar Aalto* 1963-70 vol. 2. Zurich, 1971.

Forssman, E. "Palladio e Daniele Barbaro." *Centro Internazionale di Studi di Architettura Andrea Palladio Bolletino*, 1966.

Francesca, Piero della. *De perspectiva pingendi*. Ed. by G. Nicco Fasola. Florence, 1942.

Frank, Marie "The Theory of Pure Design and American Architectural Education in the Early Twentieth Century," *JSAH*, v. 67, n. 2 (June 2008): 249-273.

Frankl, Paul "The Secrets of the Mediaeval Masons," *Art Bulletin* 27:1 (1945): 46-65.

Frankl, Paul T. *Principles of Architectural History*. Cambridge, MIT Press, 1969.

Frascari, Marco; Jonathan Hale & Bradley Starkey, Eds. *From Models to Drawings: imagination and representation in architecture*. Abingdon, UK, Routledge: 2007.

Frasier, Iain and Rod Henmi, *Envisioning Architecture: an analysis of drawing*. Van Nostrand Reinhold, New York, 1993.

Frey, Karl. Die *Handzeichnungen Michelangelos Buonarroti. hrsg, und mit kritischer apparate versehen von Karl Frey*. Berlin: 1909-1911.

Frommel, C.L. "Baldassare Peruzzi als Maler under Zeichner." *Romisches Jahrbuch fur Kunstgeschichte* 1967-8.

Frommel, C.L. *Die Farnesina und Peruzzis architektonisches Fruhwerk*. Berlin: 1961.

Gadol, Joan. Leon Battista Alberti, *Universal Man of the Early Renaissance*. Chicago: University of Chicago Press, 1973.

Gebhard, David and Deborah Nevins. *200 Years of American Architecture Drawing*. Whitney Library of Design, New York, 1977.

Giordani, Paolo. "Ricerche intorno alla villa di papa Giulio." *L'Arte*. 10, 1907.

Gloton, Jean Jaques. "Vignole et Palladio." *Centro Internazionale di Studi di Architettura Andrea Palladio Bolletino*. 8, 1966, pp. 82-100.

Goldhagen, Sarah Williams. *Welcome to Your World: How the built environment shapes our lives*. New York, Harper Collins: 2017.

Gonzales, Daniel Benjamin. "Visual Perception in Architecture." PhD dissertation, University of Maryland School of Architecture 2011.

Gourdon, Marie Therese. "Les livres d'architecture francais au XVI siècle," *Portique*, 8 (1951).

Gournay, Isabelle, and Marie-Laure Crosnier Leconte. "American Architecture Students in Belle Epoque Paris: Scholastic Strategies and Achievements at the Ecole des Beaux-Arts," *The Journal of the Gilded Age and Progressive Era* 12:2 (April 2013),154-198.

Grafton, Anthony, Glenn W. Most, & Salvatore Settis. *The Classical Tradition*. Harvard University Press, Cambridge, 2010.

Grayson, Cecil. "L.B. Alberti's 'construzione legittima,'" *Italian Studies*, 19, 1964, 22ff.

Greenberg, Allan. "Design Paradigms in the Eighteenth and Twentieth Centuries." *Via 3: Ornament*, University of Pennsylvania, 1977, 62-81.

Griseri, Andreina. "Juvarra e la pitturad' Arcadia." *Studi piemontesi*, 12,1983, 332-384.

Griseri, Andreina. *La Palazzina di Stupinigi*. Istituto Geografico de Agostini. Novara,68-77.

Griseri, Andreina. *Le metamorfosi del Barocco*. Turin, 1967.

Gritella, Gian Franco. "La cupola della chiesa juvarriana di Soperga a Torino." *Analisi*, 327-339.

Gritella, Gian Franco. "La Palazzina di Cacciadi Stupinigi." *Studi piemontese*, 13, 1984.

Gritella, Gian Franco. *Juvarra: architettura*. 2 vols. Modena, 1992.

Gross, D., Ellen Yi-Luen Do, "A theoretical framework of critiquing in architectural studios," *Design Studies* 34 (2013): 302-325.

Grossman, Elizabeth Greenwell. *The Civic Architecture of Paul Cret*. Cambridge Univ. Press, Cambridge, 1996.

Guerry, Liliane. "Un architecte 'moderne,' Philibert Delorme," *Jardin des Arts*, No. 25 (1956).

Gunter Dittmar, et al. "Architecture and Depiction." *Design Quarterly*, no. 113/114, 1980, pp. 4–7.

Guo, Quingua. "Yingzao Fashi: Twelfth Century Chinese Building Manual" *Architectural History* 41, 1998.

Gutheim, Frederick. *Alvar Aalto*. George Braziller, New York, 1960.

Graf, Ulrich and Barner, Martin. *Darstellende Geometrie*. Heidelberg: Quelle & Meyer, 1961.

Hahnloser, Hans R. "Entwurfe eines Architekten um 1250 aus Reims," *XXIII Congres international d'histoire de l'art*. Stockholm, 1953.

Hale, J.R. "Andrea Palladio, Polybius and Julius Caesar," *Journal of the Warburg and Courtauld Institutes*, 40 (1977).

Hamberg, P.G. "The Villa Lorenzo il Magnifico at Poggioa Caiano and the Origin of Palladianism."Figura. ns. 1 *Idea and Form*, 1959.

Harbeson, John. *The Study of Architectural Design*. New York, The Pencil Points Press, 1926.

Hartt, Frederick. *Michelangelo Drawings*. New York: Harry N. Abrams, 1975.

Hart, Vaughan, Ed. *Paper Palaces: the rise of the Renaissance architectural treatise*. New Haven, Yale Univ. Press, 1998.

Harvey, John. "The Tracing Floor of York Minister," *Friends of York Minister*, 40th Report, 1968.

Harvey, John. *The Medieval Architect*. London, 1972.

Hendrix, John S. "Humanism and Disegno: Neoplatonism at the Accademia di San Lucca in Rome," (2007) *School of Architecture, Art and Historic Preservation Publications*, Paper 1.

Henry, Daniel and Tom Furness. "Spatial Perception in Virtual Environments: Evaluating an Architectural Application." Seattle, University of Washington 1993.

Herbert, Daniel M., *Architectural Study Drawings*. New York, Van Nostrand Reinhold: 1993.

Herbert, Robert L. *The Art Criticism of John Ruskin*, Anchor Books, New York, 1964.

Hersey, George L. *Alfonso II and the Artistic Renewal of Naples, 1485-1495*. New Haven: Yale University Press, 1969.

Hersey, George L. *Phythagorean Palaces: Magic and Architecture in the Italian Renaissance*. Ithaca: Cornell University Press, 1976.

Hersey, George L. *Art and Geometry in the Age of the Baroque*. Chicago, Univ. of Chicago Press: 2000.

Hersey, George L. *The Lost Meaning of Classical Architecture: speculations on ornament from Vitruvius to Venturi*. Cambridge, MIT Press, 1988.

Hewitt, Mark Alan. "Architects, Drawings and Modes of Conception," Marc Treib, Ed. *Drawing/Thinking: Confronting an Electronic Age*, Routledge, New York, 2008, 28-35.

Hewitt, Mark Alan. "Representational Forms and Modes of Conception: An Approach to the History of Architectural Drawing." *Journal of Architectural Education* 39:2 (Winter 1985): 2-9.

Hewitt, Mark Alan. *The Architect and the American Country House 1890-1940*. New Haven and London, Yale University Press, 1990.

Hewitt, Mark Alan. *Gustav Stickley's Craftsman Farms: The Quest for an Arts and Crafts Utopia*. Syracuse, Syracuse Univ. Press, 2001.

Hewitt, Mark Alan. "Juvarra's Drawings for Central Plan Buildings." *Dichotomy*, The University of Detroit School of Architecture, 1983: 17-29.

Hewitt, Mark Alan. "Sketches as Cognitive Traces: Alvar Aalto at Imatra." *New Design Ideas* 3:1 (2019): 1-16.

Heyman, J. "On the Rubber Vaults of the Middle Ages and Other Matters," *Gazette des Beaux Arts*, Ser. 6, Vol. 71 (1968).

Howard, Deborah. "Sebastiano Serlio's Venetian Copyrights," *Burlington Magazine*, 115 (August 1973).

Hughes, Robert. *The Shock of the New*. New York, Knopf, 1991.

Huppert, Ann C. "Envisioning St. Peter's: Perspectival Drawings and the Process of Design," *Journal of the Society of Architectural Historians* 68:2 (June 2009): 158-177.

Huppert, Ann C. *Becoming and Architect in Renaissance Italy: Art, Science and the Career of Baldassare Peruzzi*. New Haven, Yale Univ. Press: 2015.

Hutchison, David W. *Edison at Work: the Thomas A. Edison Laboratory at West Orange, New Jersey* (New York, Hastings House: 1969).

Isaacson, Walter *Leonardo da Vinci* New York, Simon & Schuster: 2017.

Ivins, W.M., Jr. "Illustrated Books of the Early Sixteenth Century," *Arts and Decoration*, 9 (1918).

Ivins, W.M., Jr. "Renaissance Books on Architecture," *Metropolitan Museum of Art Bulletin*, 1 (December 1942).

Jacobsen, Emil. "Die Handzeichnungen der Uffizien in ihren Beziehungen zu Gemalden, Skulpturen, und Gebauden in Florenz." *Repertorium fur Kunstwissenschaft*. 27,1904.

Jarzombek, Mark "A Conceptual Introduction to Architecture." *Thresholds*, no. 33, 2008, pp. 16–20.

Jennings, Jan. "Drawing on the Vernacular Interior." *Winterthur Portfolio*, vol. 27, no. 4, 1992, pp. 255–279.

Jessup, Harley. "Graphite and Pixels: Drawing at Pixar," in Treib, *Drawing/Thinking*, 170-181.

Johnston, George Barnett. *Drafting Culture: A Social History of Architectural Graphic Standards*. MIT Press, Cambridge, 2008.

Johnson, Paul. *The Birth of the Modern: World Society 1815-1830*. New York, Harper Collins: 1991.

Jones, Barry, Andrea Sereni and Massimo Ricci, "Building Brunelleschi's Dome: A Practical Methodology Verified by Experiment," *Journal of the Society of Architectural Historians* 69:1 (March 2010): 39-61.

Jung, Carl. "Psychology and Literature," in *The Spirit in Man, Art and Literature* (Princeton, Bollingen Series, 1966).

Kahnemann, Daniel. *Thinking Fast and Slow*. New York, Penguin Books, 2011.

Kalay, Yehuda. *Architecture's New Media: Principles, Theories and Methods of Computer-Aided Design*. Cambridge, MIT Press, 2004.

Kauffman, Jordan. *Drawing on Architecture: The Object of Lines, 1970-1990*. Cambridge, MIT Press: 2018.

Kavakli, Manolya "Sketching as mental imagery processing," *Design Studies* 22 (2001): 347-364.

Keil, H. "Die architektonischen Wandzeichnungen in der apsis der Sagrestia Nouva, SanLorenzo, Florenz." *Pantheon*. 35, April 1977.

Kemp, Martin. "Leonardo and the Visual Pyramid," *Journal of the Warburg and Courtauld Institutes*, 40, 1977, 128-49.

Kemp, Martin. "Science, Nonscience, and Nonsense: The Interpretation of Brunelleschi's Perspective," *Art History*, 1, 1978.

Kemp, Martin. *The Science of Art: Optical Themes in Western Art from Brunelleschi to Seurat.* New Haven, Yale Univ. Press: 1990.

Kemp, Martin Ed., *Leonardo on Painting* New Haven, Yale Univ. Pres: 1987.

Kemp, Martin. "Geometrical Perspective from Brunelleschi to Desargues. A Pictorial Means to an Intellectual End?" *Proceedings of the British Academy,* 70,1984 (1985), pp. 89-132.

Kemp, Martin. "Perspective and Meaning: Illusion, Allusion and Collusion." *Philosophy and the Visual Arts.* M. Harrison, ed. Dordrecht, 1987, pp. 255-268.

Kemp, Martin. "Science, Non-science and Nonsense. The Interpretation of Brunelleschi's Perspective." *Art History,* 1,1978, pp. 134-161.

Kieran, Stephen and James Timberlake. *Prefabricating Architecture: how manufacturing technologies are poised to transform building construction.* New York, McGraw Hill, 2004.

King, Ross. *Brunelleschi's Dome: How a Renaissance Genius Reinvented Architecture.* London, Bloomsbury Books: 2000.

Kitao, T. L. "Prejudice in Perspective: A Study of Vignola's Perspective Treatise," *Art Bulletin*, 44 (September 1962).

Krautheimer, Richard, in collaboration with Trude Krautheimer-Hess. *Lorenzo Ghiberti.* Second printing with corrections and new preface. Princeton, NJ: Princeton University Press, 1970.

Kruft, Hanno-Walter. "Concerning the Date of the Codex Escurialensis." *Burlington Magazine*, 122, 1970.

Lamberini, Daniela. "Machines in Perspective: Technical Drawings in Unpublished Treatises and Notebooks of the Italian Renaissance" *Studies in the History of Art*, Vol. 59, Symposium Papers XXXVI: The Treatise on Perspective: Published and Unpublished (2003), pp. 212-233.

Lang, S. "Leonardo's Architectural Designs and the Sforza Mausoleum." *Journal of the Warburg and Courtauld Institutes.* 31, 1968.

Lang, Jon. "Formal Aesthetics and Visual Perception: Questions Architects Ask," *Visual Arts Research* 10:1 (Spring 1984): 66-71.

Lanzara, Paolo. "La vigna di Papa Giulio III." *Capitolium.* 31, 1956, pp. 303-315.

Lawson, Bryan. *How Designers Think* (London, Architectural Press): 1980.

Lazzaro-Bruno, Claudia. "The Villa Lante at Bagnaia: An Allegory of Art and Nature." *Art Bulletin.* 59, December 1977, pp. 553-60.

Lemaire, Jean, et al. *Drawing: History of an Art.* New York, 1979.

Leski, Kyna. *The Storm of Creativity* Cambridge, MIT Press: 2015.

Lethaby, W.R. *Architecture, mysticism and myth.* London, Percival: 1892.

Lever, Jill, and Margaret Richardson. *The Architect as Artist.* New York, 1984.

Lewis, Douglas. "A New Book of Drawings by Francesco Muttoni," *Arte veneta*, 30 (1976).

Lewis, Francis Ames. *Drawing in Renaissance Italy*. Yale Univ. Press, New Haven, 1981.

Light, Meg "I Ragionamenti - Visualizing St. Peter's," *JSAH* 44:2 (May 1985): 111-128.

Liu, Chen "Between Perception and Expression: The Codex Coner and the Genre of Architectural Sketchbooks," Unpublished Ph.D. Dissertation, Princeton University, 2011.

Loach, Judi "Architecture, Science and Purity." *Being Modern: The Cultural Impact of Science in the Early Twentieth Century*, edited by Robert Bud et al., UCL Press, London, 2018, pp. 207–244.

Long, Birch Burdette. "Individual Styles of Rendering." *The Architectural Review*, Boston 12, 1905.

Lonsway, Brian. "The Mistaken Dimensionality of CAD." *Journal of Architectural Education* 56 no. 2 (2002).

Los, Sergio *Carlo Scarpa architetto poeta* Venice, 1967.

Lotz, Wolfgang and Deborah Howard, *Architecture in Italy 1500-1600* New Haven, Yale Univ. Press: 1995.

Lotz, Wolfgang. "Das Raumbild in der italienischen Architektur-zeichunung der Renaissance." *Mitteilungen des Kunsthistorischen Institutes in Florenz.* 7, July 1956

Lotz, Wolfgang. "Observationi intorno ai disegni Pallaiani," *Bolletino del Centro Internationaledi Studi di Architettura Andrea Palladio*, 4 (1962).

Lotz, Wolfgang. "Observations on Palladio's Drawings," in his *Studies in Italian Renaissance Architecture.* Cambridge, 1977.

Lotz, Wolfgang. "The Quattro Libri di Architettura," in Ludwig H. Heydenreich and Wolfganf Lotz. *Architecture in Italy 1400 to 1600*, trans. by Mary Hottinger. Harmondsworth, England: Penguin Books, 1974.

Lotz, Wolfgang. "Zu Michelangelos Kopien nach dem Codex Corner." Stil und Uberlieferung in der Kunst des Abendlandes. *Akten des 21 Internationalen Kongresses fur Kunstgeschichte in Bonn,* 1964. vol.2, Berlin: 1976.

Lowry, Bates. "High Renaissance Architecture." *College Art Journal.* 17, 1957-58.

Luce, Kristina "The Collision of Process and Form: Drawing's Imprint on Peter Eisenman's House VI." *Getty Research Journal*, no. 2, 2010, pp. 125–137.

Lynch, Kevin. *The Image of the City.* Cambridge, MIT Press: 1960.

Maffei, Scipione. "Elogio del Signor Abate D. Filippo Juvarra architetto." *Osservazioni Letterarie*, 3:6, p. 193. Verona, Stamperia di J. Vallarsi, 1738.

Mallgrave, Harry Francis. *Gottfried Semper: Architect of the Nineteenth Century* New Haven, Yale Univ. Press: 1996.

Mallgrave, Harry Francis. *The architecture of embodiment: the implications of the new science and humanities for design. London,* Routledge, 2013.

Mallgrave, Harry Francis. F*rom Object to Experience: the new culture of architectural design.* London, Bloomsbury, 2018.

Marconi, P. "Il problema della forma della citta nei teorici di architettura del Rinascimento." *Palladio.* n.s. 22, January-December 1972.

Marchand, Trevor H.J. "Embodied cognition and communication: studies with British fine woodworkers," *Journal of the Royal Archaeological Institute* 16:1 (May 2010): S100-120.

Mark, Robert. "Gothic Cathedrals and Structural Rationalism," *Transactions.* New York Academy of Sciences, Ser. 2, Vol. 33 (1971), 1.

Mark, Robert. "The Church of St. Ouen, Rouen: A Reexamination of Gothic Structure," *American Scientist*, 61 (1968).

Mark, Robert. "The Structural Analysis of Gothic Cathedrals," *Scientific American* (November 1972).

Mark, Robert. and R. A. Prentke. "Model Analysis of Gothic Structure," *Journal of the Society of Architectural Historians*, 27 (1968).

Mark, Robert, Ed. *Architectural Technology Up To The Scientific Revolution.* Cambridge, MIT Press, 1994.

Martin, C. Leslie. *Architectural Graphics.* Macmillan, New York, 1970.

Martini, Francesco di Giorgio. *Trattati di architettura, ingegneria e arte militare.* Ed. Corrado Maltese. Milan: Edizioni il Polifilo, 1967.

Martini, Francesco di Giorgio. *Trattato di architettura civile e militare.* Ed. C. Promis. 2 vols. Turin: Chiro e Mina, 1841.

Maynard, Patrick. "Perspective's Places." *The Journal of Aesthetics and Art Criticism*, vol. 54, no. 1, 1996, 23–40.

McKean, John Maule. "Sebastiano Serlio" *Architectural Association Quarterly*, Vol. 11, No. 4 (1979).

McKewen, Indra Kagis "Claude Perrault," in Vaughan Hart, Ed., *Paper Palaces: The Rise of the Renaissance Architectural Treatise* (New Haven and London, Yale Univ. Press: 1998): 321-337.

Meeks, Carol L.V. "Books and Buildings, 1449-1949, One Hundred Great Architectural Books Most Influential in Shaping the Architecture of the Western World," *Journal of the Society of Architectural Historians*, 8 (January- June 1949).

Merkel, Jane. *Eero Saarinen.* London, Phaidon, 2005.

Merrill, Elizabeth "The *Professione di Architetto* in Renaissance Italy," *Journal of the Society of Architectural Historians* 70: 1 (March 2017): 13-35.

Michelini-Tocci, Luigi. "Disegni e appunti autografi di Francesco di Giorgio in un codice del Taccola," *Scritti di storia dell'arte in onore di Mario Salmi* (2 vols.; Rome: 1962),II.

Michels, Eileen. "Late Nineteenth Century Published American Perspective Drawing." *Journal of the Society of Architectural Historians* 31:4. 1971.

Middeldorf, Ulrich. "Giuliano da Sangallo and Andrea Sansovino." *Art Bulletin.* 16, June 1934.

Millon, H.A. and C.H. Smyth. "Design by Michelangelo for a City Gate: Further Notes on the Lille Sketch." *Burlington Magazine.* 117, March 1975.

Millon, Henry A. "Filippo Juvarra and Architectural Education in Rome in the Early Eighteenth Century." *Bulletin: The American Academy of Arts and Sciences,* 35 (1982): 27-45.

Millon, Henry A. "Filippo Juvarra and the Accademia di San Luca in Rome in the Early Eighteenth Century." *Projects and Monuments in the Period of the Roman Baroque.* Edited by Hellmut Hager and Susan S. Munshower. Papers in Art History from The Pennsylvania State University, 13-24.

Millon, Henry A. *Filippo Juvarra: Drawings from the Roman Period,* 1704-1714. Part I, Corpus Juvarrianum. Rome, 1984.

Millon, Henry, Ed. *The Triumph of the Baroque: architecture in Europe, 1600-1750.* New York, Rizzoli Internatonal, 1999.

Millon, Henry. "The Architectural Theory of Francesco di Giorgio," *Art Bulletin*, 40 (1958): 257-61.

Mischiati, O. and Mercedes Viale Ferrero. "Disegni e incisioni di Filippo Juvarra per edizioni romane del primo Settecento." *Atti della Accademia delle Scienze di Torino* 110 (1975-1976): 211-256.

Mitchell, William R. "The T-Square Club in Philadelphia, 1883-1938." University of Delaware, M.A. Thesis, 1967.

Moore, F.L. "Contribution to the Study of the Villa Giulia." *Romisches Jahrbuch fur Kunstgeschichte.* 12 (1969):171-94.

Moore, Richard A. "Academic Design Theory in France After the Reorganization of 1863," *Journal of the Society of Architectural Historians* 37: 3 (1977).

Mortet, Victor. "La mesure des colonnes a la fin de l'epoque romaine d'apres un tres ancien formulaire," *Bibliotheque de l'Ecole des Chartres*, 57 (1896).

Mortet, Victor. "La mesure et les proportions des colonnes antiques d'apres quelques compilations et commentaires anterieurs au xii siècle," *Bibliotheque de l'Ecole des Chartres*, 59 (1898).

Munshower, Susan Scott and Helmut Hager, Eds. *Architectural Fantasy and Reality: Drawings from the Accademia Nazionale di San Luca in Rome, Concorsi Clementini 1700-1750.* The Pennsylvania State University. University Park, Pennsylvania, 1982.

Murray, Stephen. *Plotting Gothic*, Chicago, Univ. of Chicago Press, 2014.

Murphy, Keith M., Jonas Ivarsson and Gustav Lymer, "Embodied reasoning in architectural critique," *Design Studies* 33 (2012): 530-556.

Myers, Mary. *Architectural and Ornament Drawings: Juvarra, Vanvitelli, The Bibiena Family, and Other Italian Draughtsmen.* The Metropolitan Museum of Art. New York, 1975.

Nachod, R. "A Recently Discovered Architectural Sketchbook of an Intimate Assistant of Bramante in the Construction of St. Peter's," *Rare Books* 8, 1955, no. 1.

Noe, Alva *Out of Our Heads: Why You Are Not Your Brain and Other Lessons from the Biology of Consciousness* New York, Hill & Wang: 2009.

Noe, Alva. *Action in Perception.* Cambridge, MIT Press, 2004.

Oechslin, Werner. "Il contributo dei Bibiena. Nuove attivita architettoniche." *Bollettino del Centro Internazionale di Studi di Architettura Andrea Palladio* 17 (1975):131-160.

Ochsendorf, John. *Guastavino Vaulting: The Art of Structural Tile.* New York, Princeton Architectural Press, 2013.

Oliver, Richard. *The Making of an Architect: 1881-1981.* New York, Rizzoli, 1981.

Oliver, Richard. *Bertram Grosvenor Goodhue.* Cambridge, MIT Press, 1983.

Onians, John, *European Art: A Neuroarthistory*, New Haven, Yale University Press: 2016.

Onians, John. *Neuroarthistory: from Aristotle and Pliny to Baxandall and Zeki.* New Haven, Yale Univ. Press, 2007.

Onians, John. *Bearers of Meaning: The Classical Orders in Antiquity, the Middle Ages, and the Renaissance.* Princeton, Princeton Univ. Press: 1988

Oxman, Rivka. "Design by re-representation: a model of visual reasoning in design," *Design Studies* 18 (1997): 329-347.

Oxman, Rivka. "Theory and design in the first digital age," *Design Studies* 27 (2006): 229-265.

Pallasmaa, Juhani. *The Thinking Hand: Existential and Embodied Wisdom in Architecture.* John Wiley and Sons, New York, 2009.

Pallasmaa, Juhani, and Sarah Robinson, Eds. *Mind in Architecture: Neuroscience, embodiment and the future of design.* Cambridge, MIT Press, 2015.

Pampaloni, Guido. *Palazzo Strozzi.* Rome: 1963.

Pane, Roberto. "I Quattro Libri," in *Bolletino di Centro Internazionale di Studi di Architettura Andrea Palladio,* 9 (1967).

Panofsky, Erwin *Gothic Architecture and Scholasticism.* New York, New American Library, 1976.

Panofsky, Erwin *Idea: A Concept in Art Theory.* Columbia, Univ. of South Carolina Press, 1968.

Panofsky, Erwin. *Perspective as Symbolic Form.* Institute of Fine Arts, New York, 1924-25. Paperback MIT Press, New York, 1991/97.

Panofsky, Erwin. *Renaissance and Renascences in Western Art.* Stockholm: Almquist & Wiksell, 1960.

Panzera, Anthony *The Leonardo Series: Drawings by Anthony Panzera Based on Leonardo's da Vinci's Work on Human Proportion* Albany, SUNY Press: 2015.

Papini, Roberto. "Francesco di Giorgio," *Architetto.* 3 vols. Florence: 1946.

Park, Helen. *A List of Architectural Books in America before the Revolution,* New ed., rev. and enl., with a foreward by Adolf K. Placzek. Los Angeles: Hennessey and Ingalls, 1973.

Parker, K.T. Catalogue of the Collection of Drawings in the Ashmolean Museum. Vol. II. *Italian Studies.* Oxford: 1956.

Parronchi, Alessandro. "La construzione legittima e ugale alla construzione con punti di distanza," *Rinascimento*, 4 (1964).

Parronechi, Alessandro. "Il mondello del Palazzo Strozzi." *Rinascimento.* 9 (1964).

Partridge, Loren. "Vignola and the Villa Farnese at Caprarola." *Art Bulletin.* 52 (March 1970).

Pause, P. *Gothische Architekturzeuchnungen in Deutschland.* Dissertation. Bonn, 1973.

Payne, Alina. *The Architectural Treatise in the Italian Renaissance.* Cambridge Univ. Press, Cambridge, 1999.

Perez Gomez, Alberto. *Architecture and the Crisis of Modern Science.* Cambridge, MIT Press, 1983.

Perez Gomez, Alberto. *Attunement: Architectural Meaning After the Crisis of Modern Science* Cambridge, MIT Press, 2016.

Perez-Gomez, Alberto and Louise Pelletier. *Architectural Representation and the Perspective Hinge.* Cambridge, MIT Press, 1997.

Perrault, Claude *Ordonnance of the five kinds of columns after the method of the ancients* Chicago, Getty Center and Univ. of Chicago Press, 1993.

Pevsner, Nikolas. "Terms of Architectural Planning in the Middle Ages," *Journal of the Warburg and Courtauld Institutes* 5 (1942): 232-237.

Pevsner, Nikolas. *Academies of Art, Past and Present.* New York, Da Capo Press: 1973.

Pierce, James Smith. "Architectural Drawings and the Intent of the Architect." *Art Journal*, 27:1 (1967): 48–59.

Pinnell, Patrick L. "Ostrakon: Postmodernism, Deconstructivism and the Millennium," *GA Critic* 1992: 8-19.

Poggioli, Renato. *Theory of the Avant Garde.* Cambridge, Harvard Univ. Press, 1969.

Polacco, Luigi. "La posizione de Andrea Palladio di fronte all'antichita," *Bolletino di Centro Internazionale di Studi di Architettura Andrea Palladio,* 7:2 (1960).

Pommer, Richard. *Eighteenth-Century Architecture in Piedmont. The Open Structures of Juvarra, Alfieri, and Vittone.* New York and London, 1967.

Pop, Dana. "Space Perception and Its Implications in Architectural Design," *Act Technica Napocensis:* Civil Engineering and Architecture 56:2 (2013).

Portoghesi, Paolo and Bruno Zevi, eds. *Michelangelo architetto.* Turin: 1964.

Powell, Helen & David Leatherbarrow:, Eds., *Masterpieces of Architectural Drawing.* London, Orbis 1982. Press. New York, New York, 2009.

Puppi, Lionello "Il 'sesto libro' di Sebastiano Serlio," *Arte Veneta*, 1967.

Puppi, Lionello. "Il Trattato del Palladio a la sua fortuna in Italia a ella 'estero," *Bolletino di Centro Internazionale di Studi di Architettura Andrea Palladio*. 12 (1970).

Pye, David *The Nature and & Aesthetics of Design* Bethel, CT, Cambium Press: 1978.

Radford, Antony & Garry Stevens. *CADD Made Easy: a comprehensive guide for architects and designers*. New York, McGraw-Hill, 1987.

Rappoli, Mark & Robert Violette, Eds. *Gehry Draws*. Cambridge, MIT Press, 2004.

Reffat, Rabee M. "Digital architecture and reforming the built environment." *Journal of Architectural and Planning Research*, vol. 25, no. 2, 2008, pp. 118–129.

Reische, Emil. "Die Zeichnungen der Cyriacus im Codex Barberini des Giuliano di San Gallo," *Kaiserlich Deutsches Archaeologisches Institut in Athens. Mitteilungen. Athenische abtheilung*, 14, 1889.

Rohan, Timothy. *The Architecture of Paul Rudolph*. New Haven, Yale Univ. Press, 2014.

Rosenblatt, Nina. "Empathy and Anaesthesia: On the Origins of a French Machine Aesthetic," *Grey Room*, No. 2 (Winter, 2001),78-97.

Rosenau, Helen. *Design and Medieval Architecture*. London, 1934.

Rosenfeld, Myra Nan. "Review of Marco Rosci, Il Trattato di Architettura de Sebastiano Serlio, Vol. 1; Sesto libro delle habitationi di tutti li gradi degli homini," *Art Bulletin*, 52 (September 1970).

Rosenfeld, Myra Nan. "Sebastiano Serlio's Drawings in the Nationalbibliothek in Vienna for his Seventh Book on Architecture," *Art Bulletin*, 56 (1974).

Rowland, Benjamin. *The Art and Architecture of India: Buddhist, Hindu, Jain*. Harmondsworth and Baltimore, Pelican History of Art: 1981.

Rowland, Ingrid and Thomas N. Howe. *Vitruvius: Ten Books on Architecture*. Cambridge Univ. Press, Cambridge.

Rushton, Joseph George. *Italian Renaissance Figurative Sketchbooks, 1450-1520*. Doctoral dissertation. Minneapolis: University of Minnesota, 1976.

Saalman, Howard. "The Baltimore and Urbino Panels," *Burlington Magazine* 110 (1968).

Saint, Andrew *Architect and Engineer: A Study in Sibling Rivalry*. New Haven and London, Yale Univ. Press: 2007.

Salmi, Mario. *Disegni di Francesco di Giorgio nella collezione Chigi Saracini*. Siena: 1947.

Sander, M. *Le livre a figures italiens depuis 1467 jusqu'en 1530*. Milan, 1942-1969.

Sangallo, Giuliano. *Il Taccuino Senese di Giuliano da San Gallo*. 50 facsimili di disegni d'architettura, scultura ed arte applicate; ... Sienna: Stabilimento fotolitografico di Rodolfo Falb, 1902.

Sangallo, Guiliano da. *Il libro di Giuliano's da Sangallo: Codice Vaticano Barberiniano Latino* 4424. edited by Christian Hulsen. 2 vols. Leipzig-Turin, 1910.

Sangallo, Guiliano da. *Il taccuino senese di Giuliano da Sangallo*, edited by Rudolfo Falb. Siena, 1902.

Sangalo, Giuliano da. *Il Libro di Giuliano da Sangallo*, codice vaticano Barberiniano latino 4424 riprodotto in fototipia con introduzione e note di Cristiano Huelsen Lipsis: O. Harrassowitz, 1910.

Sanpoalesi, Paolo. *Brunelleschi*. Firenze, 1962.

Saunier, Charles. "Sur certains livres d'architecture du XVIe siècle," Melanges offerets a Henry Lemmonnier, *Archives de l'art francais, nouv. Per.*, 7 (1913).

Scaglia, Gustina. "Newly Discovered Drawing of Monasteries by Francesco di Giorgio Martini," *Architecture* 2 (1974): 112-124.

Scaglia, Gustina. "Three Renaissance Drawings of Church Façades." *Art Bulletin*. 47 (June 1965).

Scarpa, Carlo. *Banca populare di Verona head offices, Verona, Italy 1973-81*. Tokyo, A.D.A. Edita Tokyo, 1983.

Schildt, Goran. *Alvar Aalto: The Early Years*. Rizzoli International, New York, 1984.

Shiner, Larry. *The Invention of Art*. Chicago, Univ. of Chicago Press: 2001.

Schon, Donald *The Reflective Practitioner: How Professionals Think in Action* New York, Basic Books: 1984.

Schumacher, Patrik. "Parametricism And the Autopoiesis Of Architecture." *Log*, no. 21, 2011, pp. 62–79

Schwartz, Gary. "The Battle of the Medici Chapel Basement." *Art News*. 75 (October 1976).

Scolari, Massimo. *Oblique Drawing: A History of Anti-Perspective*. MIT Press, Cambridge, 2012.

Scully, Vincent, Jr. "Michelangelo's Fortification Drawings, A Study in the Reflex Diagonal." *Perspecta*. 1, 1952.

Scruton, Roger. *The Aesthetics of Architecture*. Princeton, Princeton Univ. Press: 2013.

Sennett, Richard *The Craftsman*. New Haven, Yale Univ. Press, 2008.

Serlio, Sebastiano. *On Domestic Architecture: The Sixteenth Century Manuscript of Book VI inthe Avery Library of Colombia University*. Text by Myra Nan Rosenfeld. Cambridge,Massachusetts: MIT Press, 1978.

Shearman, John. "Raphael, Rome and the Codex Escurialensis," *Master Drawings*, 15 (Summer 1977).

Shelby, Lon R. "Medieval Mason' Tools, II. Compass and Square." *Technology and Culture*, (1965).

Shelby, Lon R. "Medieval Masons' Templates," *Journal of the Society of Architectural Historians*, 30 (1971).

Shelby, Lon R. "The Education of Medieval English Master Masons," *Medieval Studies*, 32 (1972).

Shelby, Lon R. "The Geometrical Knowledge of the Medieval Master Masons," *Speculum*, 47 (July 1972).

Shelby, Lon R. "The Secret of the Medieval Masons," *On Pre-Modern Technology and Science: A Volume of Studies to Lynn White, Jr.*, ed. by Bert S. Hall and Delno C. West. Malibu, California: Undena Publications, 1976, I.

Shelby, Lon. R. *Gothic Design Techniques: The Fifteenth Century Design Booklets of Mathes Roriczer and Hans Schmuttermayer.* Carbondale: Southern Illinois University Press, 1977.

Siebenhuner, H. "Zur Entwicklung der renaissance-prespektive." *Kunstchronik*, 7, 1954.

Silber, John. *Architecture of the Absurd: how 'genius' disfigured a practical art.* New York, Quantock Lane Press, 2007.

SKETCHBOOK: "Menicantonio" Sketchbook, ca. 1513-1520. 81 leaves. New York: Pierpont Morgan Library.

SKETCHBOOK: Codex Coner, dated variously from the second through the fourth decades of the sixteenth century. Over 100 leaves. London: Sir John Soane's Museum.

SKETCHBOOK: Giuliano da Sangallo, late 15th- early 16th century. 52 leaves, pen, wash, and some watercolor on parchment. Siena: Biblioteca Comunale, Cod. S.IV.8.

Smith, Kendra Schank *Architects' Sketches: dialogue and design.* Oxford, Architectural Press: 2008.

Smith, Pamela H. *Body of the Artisan*, Chicago, Univ. of Chicago Press: 2004.

Somol, R. E. "No Place like Home: Domesticating Assemblages." *Assemblage* 13 (1990): 60–71.

Spretnak Charlene. *Relational Reality: New discoveries of interrelatedness that are transforming the modern world* Topsham, Maine, Green Horizons Books: 2011.

Stamps III, Arthur E. "Jungian Epistemological Balance: A Framework for Conceptualizing Architectural Education?" *Journal of Architectal Education* 48:2 (1994)

Starr, G. Gabrielle *Feeling Beauty: The Neuroscience of Aesthetic Experience* Cambridge, MIT Press: 2013.

Stegmann, C. von, and H. von Geymuller. *Die Architektur der Renaissance in Toscana.* 11 vols. Munich: Bruckmann, 1885-1908.

Stein, Susan R. *The Architecture of Richard Morris Hunt.* Univ. of Chicago Press, Chicago, 1986.

Steinetz, Kate T. "A Pageant of Proportion in Illustrated Books of the Fifteenth and Sixteenth Centuries in the Elmer Belt Library of Vinciana," *Centaurus*, 1 (1951).

Stevens, Garry. "Reflections of an Apostate CAD Teacher." *Journal of Architectural Education* 51:1 (1997).

Stevens, Garry. *The Favored Circle: The Social Foundations of Architectural Distinction.* Cambridge, MIT Press, 1998.

Stierli, Martino. *Montage and the Metropolis: architecture, modernity and the representation of space.* New Haven, Yale University Press, 2018.

Sturgis, Russell. *The Artist's Way of Working*, New York, Dodd, Mead & Co., 1905.

Sublette, Ned. *Cuba and Its Music: from the first drums to the mambo.* Chicago, Univ. of Chicago Press, 2004.

Suwa, Masaki and Barbara Tversky, "What do architects and students perceive in their design sketches? A protocol analysis," *Design Studies* 18 (1997): 385-403.

Svensen, Carl L. *Architectural drafting.* New York, D. Van Nostrand, 1929.

Tafuri, Manfredo. "J. Barozzi da Vignola e la crisi del Manierismo a Roma." *Centro Internationale di Studi di Architettura Andrea Palladio. Bollettino.* 9, 1967.

Tang, H. *Exploring the roles of sketches and knowledge in the design process.* PhD thesis. The University of Sydney, 2001.

Thapar, Binda. *Introduction to Indian Architecture.* Singapore, Periplus Editions: 2004.

Thomas, George E. Ed. *Drawing Towards Building: Philadelphia Architectural Graphics, 1732-1986*, Pennsylvania Academy of Fine Arts, Philadelphia, 1986.

Thornton, Peter. *Seventeenth Century Interior Decoration in England, France and Holland.* New Haven: Yale University Press, 1978.

Toca, Mircea. "I disegni di Baldassare Peruzzi per il trattati d'architettura." *Necropoli*: 1971.

Toker, Franklin B. "Gothic Architecture by Remote Control: An Illustrated Design Contract of 1340," *Art Bulletin LXVII: 1* (March 1985): 67-94.

Tolnay, Charles de. "Die Handzeichnungen Michelangelos im Archivio Buonartoti." *Munchner Jahrbuch der bildenden Kunst.* 5, 1928.

Tolnay, Charles de. "Michelange et la façade de St. Lorenzo." *Gazette des Beaux Arts.* 6 ser., 11, 1934.

Tolnay, Charles de. "Michelangelo Studies." *Art Bulletin.* 22 (1940).

Tolnay, Charles de. "Unknown Sketches by Michelangelo." *Burlington Magazine* 98 (October 1956).

Tolnay, Charles de. *Disegni di Michelangelo nelle collezione italiani.* Florence: Centro Di, 1975.

Treib, Marc, Ed. *Drawing/Thinking: Confronting an Electronic Age*, Routledge, New York, 2008.

Troy, Nancy J. *The De Stijl environment.* Cambridge, MIT Press, 1983.

Tunnard, Christopher & Boris Pushkarev, *Man Made America: Chaos or Control?* New Haven, Yale Univ. Press. 1963.

Wulf, Andrea. *The Invention of Nature: Alexander von Humboldt's new world.* New York, Vintage Books: 2016.

van Bruggen, Coosje. *Frank O. Gehry: Guggenheim Museum Bilbao.* New York, Guggenheim Museum, 1997.

Veltman, K. "Renaissance Optics and Perspective: A Study of the Problems of Size and Distance." Ph.D. Thesis, London, 1975.

Venturi, A. Architetti dal XV al XVIII secolo: *Francesco di Giorgio*. Rome, Ferri e Rocchi, 1925.

Venturi, A. Storia dell'arte italiana. VIII. *L'Architettura del quattrocento*. Part I. Milan, Hoepli, 1923.

Verga, C. "La prima prospettiva del Brunelleschi," *Critica d'art*, 42, 1977, No. 154-56, 71-83u, No. 157-59, 119-32.

Vesely, Dalibor. *Architecture in the Age of Divided Representation: The Question of Creativity in the Shadow of Production* Cambridge, MIT Press, 2006.

Viale Ferrero, Mercedes. *Filippo Juvarra scenografo e architetto teatrale*. Turin, 1970.

Vico, Giambattista. *New Science*. Harmondsworth, Penguin Books, 1999.

Vidler, Anthony. "Diagrams of Diagrams: Architectural Abstraction and Modern Representation." *Representations*, no. 72, 2000, pp. 1–20.

Vignola, Giacomo Barozzio. *Alcune opera d'architettura di Iacomo Barotio da Vignola*, raccolte poste in luce da Francesco Villamena. Roma: 1617.

Vignola, Giacomo Barozzio. *L'Architettura di Jacopo Barozzi da Vignola, ridotta a facile metodo per mezzo di observazioni a profitto de student*. Venezia: Presso Giambatista Pasquali, 1748.

Vignola, Giacomo Barozzio. *Le cinque ordini di architettura et agivnta del opera del eccmo. M. Giacomo Barocio da Vignola*. Con un ragionamenta alli architeti di M. Ottaviano Ridolfi… Con la nouva agiunta di Michel Angelo Bona Rotti Fiorentino. Venezia: Nel negotio Remondini, 1648.

Vignola, Giacomo Barozzio. *Le due regole della prospetiva practica di M. Iacomo Barozzi da Vignola*, con i commentarij del r.p.m. Egnatio Danti, dell'ordine de Predicatori, matematico dello studio di Bologna… Roma: Francesco Zannetti, 1583.

Vignola, Jacopo Barozzi. *Regole delli Cinque Ordini d' Architettura* 1562,

Villard de Honnecourt. *Album de Villard de Honnecourt, architecte du XIIIe siècle*. Paris: Berthaud Freres, 1906.

Villard de Honnecourt. *Facsimile of the Sketchbook of Wilars de Honecort*. Published by M.Alfred Darcel from the Ms. of M.J.B.A. Lassus. Trans. ed. and augmented by the Rev. Robert Willis. London: John Henry and James Parker, 1859.

Wallace, William. *Michelangelo: The Artist, the Man, and His Times*. Cambridge, Cambridge Univ. Press: 2010.

Weisstein, Ulrich Ed., *The Essence of Opera*. New York, W.W. Norton: 1964.

White, John. *The Birth and Rebirth of Pictorial Space*, 2d ed. London, Faber and Faber, 1967.

Wick, Peter A. *Sixteenth Century Architectural Books from Italy and France*. Cambridge, Massachusetts: Harvard College Library, Department of Printing and Graphic Arts, 1971.

Wickhoff, F. *Die Italienischen Handzeichnungen der Albertina*. Vienna: 1892.

Wilde, Joannes. *Italian Drawings in the Department of Prints and Drawings in the British Museum: Michelangelo and His Studio*. London: 1953.

Williams, Robert *Art, Theory and Culture in Sixteenth Century Italy* Cambridge, Cambridge Univ. Press: 1997.

Willich, Hans. *Giacomo Barozzi da Vignola*. Strassburg: 1906.

Wilson, Edward O. *Consilience: the unity of all knowledge* New York: Knopf: 1998.

Wingler, Hans Maria *The Bauhaus: Weimar Dessau, Berlin, Chicago*. Cambridge, MIT Press, 1979.

Wittkower, Rudolf. "Michelangelo's Biblioteca Laurenziana." *Art Bulletin*. 16 (1934).

Wittkower, Rudolf. *Idea and Image: Studies in the Italian Renaissance*. London, Thames & Hudson, 1978.

Wittkower, Rudolf. "Brunelleschi and 'Proportion in Perspective'," *Journal of the Warburgand Courtauld Institutes*, 16, 1953.

Zink, Clifford W. *The Roebling Legacy*. Princeton, Princeton Landmark Publications, 2011.

Zorzi, Giangiorgio. "I Disegni delle Opere Palladiane Pubblicate ne 'I Quattro Libri' a il Loro Significo Rispetto alle Opere Essequite," *Bolletino di Centro Internazionale di Studi di Architettura Andrea Palladio*, 3 (1961).

INDEX

"Mark Hewitt's book presents architectural design throughout history as no less than a form of human cognition. Drawing on the most recent discoveries in neuroscience, he documents the role of memory and the importance of hand drawing to the mental process of design."

—Stuart Cohen, FAIA, Professor Emeritus of Architecture, University of Illinois, Chicago Circle

"When Carlo Scarpa (1906–1978) explained, "I draw in order to see," he was restating in the 1960s what artists and architects believed since the sixteenth century—that disegno is a process of "drawing out" that which is unseen, making the not-yet-visible, visible. Hewitt's book is no less a drawing out of the rich history and meaning of the architect's task, and how the mimesis of drawing is different in kind from the mimicry of digital renderings."

—George Dodds, Alvin and Sally Beaman Professor of Architecture, University of Tennessee, author of *Building Desire: On the Barcelona Pavilion*